Communications in Computer and Information Science 3012

Series Editors

Gang Li, *School of Information Technology, Deakin University, Burwood, VIC, Australia*
Joaquim Filipe, *Polytechnic Institute of Setúbal, Setúbal, Portugal*
Zhiwei Xu, *Chinese Academy of Sciences, Beijing, China*

AF213915

Rationale

The CCIS series is devoted to the publication of proceedings of computer science conferences. Its aim is to efficiently disseminate original research results in informatics in printed and electronic form. While the focus is on publication of peer-reviewed full papers presenting mature work, inclusion of reviewed short papers reporting on work in progress is welcome, too. Besides globally relevant meetings with internationally representative program committees guaranteeing a strict peer-reviewing and paper selection process, conferences run by societies or of high regional or national relevance are also considered for publication.

Topics

The topical scope of CCIS spans the entire spectrum of informatics ranging from foundational topics in the theory of computing to information and communications science and technology and a broad variety of interdisciplinary application fields.

Information for Volume Editors and Authors

Publication in CCIS is free of charge. No royalties are paid, however, we offer registered conference participants temporary free access to the online version of the conference proceedings on SpringerLink (http://link.springer.com) by means of an http referrer from the conference website and/or a number of complimentary printed copies, as specified in the official acceptance email of the event.

CCIS proceedings can be published in time for distribution at conferences or as post-proceedings, and delivered in the form of printed books and/or electronically as USBs and/or e-content licenses for accessing proceedings at SpringerLink. Furthermore, CCIS proceedings are included in the CCIS electronic book series hosted in the SpringerLink digital library at http://link.springer.com/bookseries/7899. Conferences publishing in CCIS are allowed to use our online conference service (Meteor) for managing the whole proceedings lifecycle (from submission and reviewing to preparing for publication) free of charge.

Publication process

The language of publication is exclusively English. Authors publishing in CCIS have to sign the Springer CCIS copyright transfer form, however, they are free to use their material published in CCIS for substantially changed, more elaborate subsequent publications elsewhere. For the preparation of the camera-ready papers/files, authors have to strictly adhere to the Springer CCIS Authors' Instructions and are strongly encouraged to use the CCIS LaTeX style files or templates.

Abstracting/Indexing

CCIS is abstracted/indexed in DBLP, Google Scholar, EI-Compendex, Mathematical Reviews, SCImago, Scopus. CCIS volumes are also submitted for the inclusion in ISI Proceedings.

How to start

To start the evaluation of your proposal for inclusion in the CCIS series, please send an e-mail to ccis@springer.com

Karen Renaud · Steffen Wendzel ·
Andreas Heinemann
Editors

Cybersecurity

10th European Interdisciplinary Cybersecurity Conference, EICC 2026
Darmstadt, Germany, June 17–18, 2026
Proceedings

Springer

Editors
Karen Renaud
Strathclyde University
Glasgow, UK

Steffen Wendzel
Ulm University
Ulm, Germany

Andreas Heinemann
Darmstadt University of Applied Sciences
Darmstadt, Germany

ISSN 1865-0929 ISSN 1865-0937 (electronic)
Communications in Computer and Information Science
ISBN 978-3-032-28956-8 ISBN 978-3-032-28957-5 (eBook)
https://doi.org/10.1007/978-3-032-28957-5

© The Editor(s) (if applicable) and The Author(s), under exclusive license
to Springer Nature Switzerland AG 2027

This work is subject to copyright. All rights are solely and exclusively licensed by the Publisher, whether the whole or part of the material is concerned, specifically the rights of translation, reprinting, reuse of illustrations, recitation, broadcasting, reproduction on microfilms or in any other physical way, and transmission or information storage and retrieval, electronic adaptation, computer software, or by similar or dissimilar methodology now known or hereafter developed.
The use of general descriptive names, registered names, trademarks, service marks, etc. in this publication does not imply, even in the absence of a specific statement, that such names are exempt from the relevant protective laws and regulations and therefore free for general use.
The publisher, the authors and the editors are safe to assume that the advice and information in this book are believed to be true and accurate at the date of publication. Neither the publisher nor the authors or the editors give a warranty, expressed or implied, with respect to the material contained herein or for any errors or omissions that may have been made. The publisher remains neutral with regard to jurisdictional claims in published maps and institutional affiliations.

This Springer imprint is published by the registered company Springer Nature Switzerland AG
The registered company address is: Gewerbestrasse 11, 6330 Cham, Switzerland

If disposing of this product, please recycle the paper.

Preface

The European Interdisciplinary Cybersecurity Conference (EICC) aims to establish a premium venue to support the exchange of information and the sharing of research findings on cybersecurity and its many aspects between European academics and practitioners. EICC encourages dialogue between computer scientists and researchers in all other fields related to cybersecurity, such as behavioral sciences, sociology, criminology, investigations, and law. Interdisciplinary contributions are especially welcome.

EICC 2026 was the 10th edition of the conference, which, until 2019, was known as the Central European Cybersecurity Conference – CECC, being rebranded to highlight the interdisciplinarity of the conference and to widen its reach. This 10th edition of EICC was held at Hochschule Darmstadt (h_da), Darmstadt, Germany, on the 17th and 18th of June, 2026.

One special session proposal (CyFRP, see below) was submitted to EICC 2026, and, after a thoughtful review process, was accepted. The evaluation criteria for the selection of the special sessions included the relevance to the EICC, the timeliness and expected interest in the proposed topics, the ability of the organizers to lead a successful workshop, and their balance and synergy.

The EICC 2026 proceedings include the accepted papers from both the main track and special sessions on *Advances in Interdisciplinary Cybersecurity: Insights from Funded Research Projects – CyFRP*.

A total of 50 papers were submitted to EICC 2026 and the special session from 25 different countries. Two submissions were desk-rejected. The main track and the CyFRP session had an independent Program Committee, which was in charge of reviewing and selecting the submitted papers. All the papers submitted to the main track underwent double-blind review. The papers submitted to the CyFRP special session underwent a single-blind review. Each paper received at least three reviews (171 reviews in total). Of the 43 submissions that went through to the review phase, 17 were selected to be presented at EICC 2026 main track and included in these proceedings (15 full papers, 2 short papers), yielding an acceptance rate of 39.5%. This edition of the conference featured two keynotes. One by Ganna Pogrebna (David Trimble Chair in Leadership and Organisational Transformation, Queen's University Belfast, UK). She delivered a talk titled *Cognitive Bias as a Cybersecurity Vector*. The second keynote was delivered by Dominik Herrmann (Privacy and Security in Information Systems Group, University of Bamberg, Germany). His talk was entitled $n = 7$ *– and That's the Point*.

Many people contributed to the success of the EICC 2026 program. We would like to express our gratitude to all who supported this event: the special session organizers for their commitment, the authors who contributed to this volume, the reviewers for their help in assessing the papers, the two keynote speakers and, last but not least, the EICC participants.

We would also like to thank all members of the EICC Steering Committee for their patience and continuous support. Finally, many thanks to the Springer staff who provided professional support through all the phases that led to this volume.

June 2026

Karen Renaud
Steffen Wendzel
Andreas Heinemann

Organization

Conference Chair

Andreas Heinemann	Darmstadt University of Applied Sciences, Germany

Local Organizing Chairs

Sara Hahn	Darmstadt University of Applied Sciences, Germany
Andreas Heinemann	Darmstadt University of Applied Sciences, Germany

Program Committee Chairs

Karen Renaud	Strathclyde University, UK
Steffen Wendzel	University of Ulm, Germany

Treasury

Simone Zimmermann	Competence Center for Applied Security Technology, Darmstadt, Germany

Steering Committee

Igor Bernik	University of Maribor, Slovenia
Bela Genge	University of Medicine, Pharmacy, Sciences and Technology of Târgu Mureş, Romania
Jörg Keller	FernUniversität in Hagen, Germany
Blaž Markelj	University of Maribor, Slovenia
Aleksandra Mileva	University Goce Delcev, North Macedonia
Wojciech Mazurczyk	Warsaw University of Technology, Poland
Simon Vrhovec	University of Maribor, Slovenia

Program Committee

Habtamu Abie	Norwegian Computing Centre, Norway
Marios Anagnostopoulos	Democritus University of Thrace, Greece
Niyati Baliyan	National Institute of Technology Kurukshetra, India
João Paulo Barraca	University of Aveiro, Portugal
Simona Bernardi	Universidad de Zaragoza, Spain
Luca Caviglione	CNR - IMATI, Italy
Michał Choraś	Bydgoszcz University of Science and Technology, Poland
Kovila Coopamootoo	King's College London, UK
Luca Davoli	University of Parma, Italy
Changyu Dong	Newcastle University, UK
Pavlos Efraimidis	Democritus University of Thrace, Greece
Tobias Eggendorfer	TH Ingolstadt, Germany
Sergio Esposito	Università degli Studi di Catania, Italy
Virginia Franqueira	University of Kent, UK
Damjan Fujs	University of Ljubljana, Slovenia
Steven Furnell	University of Nottingham, UK
Dieter Gollmann	Hamburg University of Technology, Germany
Petra Grd	University of Zagreb, Croatia
Nils Gruschka	University of Oslo, Norway
Piroska Haller	Petru Maior University, Romania
Marko Hölbl	University of Maribor, Slovenia
Pedro Inácio	Universidade da Beira Interior, Portugal
Martin Gilje Jaatun	University of Stavanger & SINTEF Digital, Norway
Georgios Karopoulos	European Commission, Joint Research Centre, Italy
Stefan Katzenbeisser	University of Passau, Germany
Wenjuan Li	Education University of Hong Kong, China
Shigang Liu	CSIRO's Data61, Australia
Brad Malin	Vanderbilt University, US
David Megías	Internet Interdisciplinary Institute (IN3) & Universitat Oberta de Catalunya, Spain
Weizhi Meng	Lancaster University, UK
Pal-Stefan Murvay	Politehnica University of Timisoara, Romania
Sebastian Pape	Social Engineering Academy GmbH, Germany
Sergio Pastrana	Universidad Carlos III de Madrid, Spain
Isabel Praça	GECAD/ISEP, Portugal
Tobias Schmidbauer	Nuremberg Institute of Technology, Germany

Bernardo Sequeiros	Universidade da Beira Interior, Portugal
Anastasia Sergeeva	University of Luxembourg, Luxembourg
Juraj Sikra	University Of Strathclyde, UK
Gerardo Simari	Universidad Nacional del Sur & CONICET, Argentina
Kai Simon	Fraunhofer-Gesellschaft, Germany
Michael Sirivianos	Cyprus University of Technology, Cyprus
Florian Skopik	Austrian Institute of Technology, Austria
Hung-Min Sun	National Tsing Hua University, Taiwan
Sotirios Terzis	University of Strathclyde, UK
Igor Tomičić	University of Zagreb, Croatia
Fernando Tricas	Universidad de Zaragoza, Spain
Nicolas Tsapatsoulis	Cyprus University of Technology, Cyprus
Christos Xenakis	University of Piraeus, Greece
Nicola Zannone	Eindhoven University of Technology, Netherlands
Sebastian Zillien	University of Ulm, Germany

Subreviewers

Roland Bolboaca
Víctor Garcia-Font
Hicham Hammouchi
Thodoris Ioannidis
Christos Chrysanthos Nikolaidis
Rui Pinto
Aikaterini Psychogyiou
Tiago Roxo
Jordi Serra-Ruiz
Ioannis Stylianou
Theodoros Tsiolakis
Anastassios Voudouris

Special Session: Advances in Interdisciplinary Cybersecurity: Insights from Funded Research Projects – CyFRP

CyFRP Program Committee Chairs

Damjan Fujs	University of Ljubljana, Slovenia
Simon Vrhovec	University of Ljubljana, Slovenia

CyFRP Organizing Committee

Nina Bernik	University of Ljubljana, Slovenia

CyFRP Program Committee

Luca Caviglione	CNR - IMATI, Italy
Michał Choraś	Bydgoszcz University of Science and Technology, Poland
Anastasija Collen	University of Geneva, Switzerland
Tiago Cruz	University of Coimbra, Portugal
George Drosatos	Athena Research Center, Greece
Pavlos Efraimidis	Democritus University of Thrace, Greece
Petra Grd	University of Zagreb, Croatia
Piroska Haller	Petru Maior University, Romania
Marko Hölbl	University of Maribor, Slovenia
David Megías	Universitat Oberta de Catalunya, Spain
Pal-Stefan Murvay	Politehnica University of Timisoara, Romania
Niels Alexander Nijdam	University of Geneva, Switzerland
Konstantinos Rantos	Department of Computer Science, Greece
Michael Sirivianos	Cyprus University of Technology, Cyprus
Aimilia Tasidou	CESI LINEACT UR7527, France
Christos Xenakis	University of Piraeus, Greece

Subreviewer

Theodoros Tsiolakis

Keynotes

Cognitive Bias as a Cybersecurity Vector

Ganna Pogrebna

Queen's University Belfast, Northern Ireland, UK

Cybersecurity discussions frequently describe attacks that exploit human behaviour under the label of social engineering. While useful, this framing obscures a deeper mechanism: many cyber incidents succeed because attackers exploit predictable cognitive biases in human judgement.

This talk argues that cognitive bias should be understood as a distinct cybersecurity vector embedded within modern digital systems. Biases such as confirmation bias, automation bias, authority bias, and availability bias influence how analysts interpret alerts, how organisations prioritise threats, and how users respond to suspicious signals. These behavioural tendencies shape decision-making across the entire cybersecurity pipeline—from threat detection and incident response to strategic risk assessment. As AI-assisted tools increasingly support cyber operations, these biases may be amplified rather than reduced, particularly when humans over-trust automated recommendations or fail to question machine-generated outputs. Recognising cognitive bias as a structural vulnerability enables a broader interdisciplinary perspective on cybersecurity, integrating insights from behavioural science, decision theory, and human-machine interaction into the design of more resilient cyber defence systems.

$n = 7$ – and that's the Point

Dominik Herrmann

University of Bamberg, Germany

In a German household in 2023, a man proudly demonstrates his robot vacuum's surveillance camera to an ethnographer. His partner asks him three times to turn it off. He bought the device. It sits in their living room. She does not want it. This is not a security incident in any conventional sense – yet no authentication scheme, no threat model, and no behavioural-bias taxonomy captures what is actually happening here.

In this talk, I report from the Bavarian research consortium ForDaySec[1], where computer scientists and ethnographers, among others, learned how to collaboratively study how people actually live with their digital devices. I will argue that *lovingly neglected* home networks, devices arriving as unwanted gifts, and security delegated to sons-in-law are not edge cases but the norm – while our field has been studying lab-based approximations of these situations for decades. I will also speak honestly about the costs: publication friction, method clashes, career risk for junior researchers. And I will end with a small request to a community that prides itself on rigour: dare more qualitative research. $n = 7$ is sometimes exactly the right number.

[1] https://fordaysec.de/en.

Contents

Hardware and IoT Security

Code and Binary Analysis

Monitoring and Forensics

CyFRP – Special Session

Human-Centered Security

Managing Learning and Difficulty Satisfaction in Cybersecurity Training Challenges

Lovisa Nyholm(✉), Teodor Sommestad, Henrik Karlzén, and Hanna Kvist

Swedish Defence Research Agency FOI, Linköping, Sweden
{lovisa.nyholm,teodor.sommestad,henrik.karlzen,hanna.kvist}@foi.se
https://www.foi.se

Abstract. This paper describes Skade, a web-based system used to expose technical challenges in a cyber range to trainees, without the involvement of instructors or cyber range engineers. Skade has been evaluated by 45 prospective trainees (e.g., junior threat hunters), each spending 2–5 h on one of three different threat hunting challenges. The trainees' assessed Skade to be acceptable on the System Usability Scale. Skade is designed to manage difficulty and learning using optional clues, but contrary to our hypotheses, the availability of clues did not affect trainees' satisfaction with difficulty or perceived learning. Meanwhile, many trainees reported feedback concerning the difficulty level and getting stuck on details. It is concluded that managing the difficulty level of technical educational cyber challenges is important, and remains to be solved.

Keywords: cyber threat hunting · cyber defence exercises · training · cyber range

1 Introduction

Cybersecurity competence remains a scarcity in Europe [10] and the US [27]. The European Union Agency for Cybersecurity (ENISA) stated that "While the demand for people with ICT and cybersecurity skills is rapidly increasing, the cybersecurity skills and talent shortage is growing too" [10]. The field requires a wide range of competencies, among which the role of the cybersecurity analyst is particularly in demand. A report by OECD [26], which analysed the job market in Poland, France and Germany, found that "both France and Germany have seen an increase in the demand for cybersecurity analysts, individuals providing insights to support planning, operations, and maintenance of systems security."

Technical competencies like these require both a particular mindset and technical skills. This is especially true for the activity of cyber threat hunting. Threat hunting has been described as a "highly unstructured task that demands deep technical know-how, data analytics savvy, and out of the box thinking" [24].

© The Author(s), under exclusive license to Springer Nature Switzerland AG 2027
K. Renaud et al. (Eds.): EICC 2026, CCIS 3012, pp. 3–21, 2027.
https://doi.org/10.1007/978-3-032-28957-5_1

A recent survey by SANS [32] found that 61% of respondents across the globe mention workforce limitations as a problem for threat hunting, and 40% are increasing their staffing investment for threat hunting.

Proficiency in complex cybersecurity tasks such as threat hunting cannot be developed through theoretical instruction and lectures alone [7,11,18,19]. At the same time, training these skills in operational environments is challenging. When hunting real threats, there is limited opportunity for experimentation, while conducting exercises in production systems may negatively affect business processes. The use of isolated training environments such as cyber ranges is an option, and similar to the separation of development, test and production environments, as recommended by ISO 27001 Annex A 8.31. Nevertheless, designing meaningful technical challenges in cyber range training environments requires considerable effort. First, because threat hunting involves active exploration of systems, training environments must be sufficiently complex and realistic to replicate a meaningful portion of the conditions and options present in real networks. Access to good hands-on labs has been identified as a barrier to learning in the cybersecurity field [2]. Second, because the tasks are complex and varied, adaptation of challenges to trainees' proficiency levels is difficult. Balancing the difficulty level is known to be important [15].

Skade [33] is a system designed to address these challenges. It automates the instantiation of cyber defence challenges in a cyber range, removing manual challenge setup and the need for human instructors. Skade's library of challenges can also be filled with optional clues to trainees. These clues may provide the trainee with threat intelligence, technical tips, or mindset guidance. This allows trainees to dynamically lower the challenge barrier when needed, without breaking the threat hunting workflow. The availability of clues is expected to increase trainee satisfaction with the difficulty and perceived learning, and these factors are expected to be associated with higher usability of Skade.

In this paper, we present Skade and tests of the theories that the system is built upon. More specifically, we empirically test four hypotheses that link satisfaction of the difficulty level, usability, and perceived learning:

H1. Satisfaction with difficulty level is associated with higher usability.
H2. Perceived learning is associated with higher usability.
H3. The availability of investigative clues increases satisfaction with difficulty.
H4. The availability of investigative clues increases perceived learning.

The evaluation target system administrators and novice threat hunters.

Section 2 presents the challenge management system Skade. Section 3 presents related works. Section 4 describes the evaluation method. Section 5 presents the results. Section 6 discusses these results, and Sect. 7 concludes the paper.

2 The Challenge Management System Skade

The sections below introduce the design principles Skade is built upon, Skade's main components, and three threat hunting challenges used in the test.

2.1 Design Principles

Skade was developed based on four ideas, drawn from pedagogical research: ensuring constructive alignment, covering multiple learning dimensions, providing feedback and assessment, and supporting a motivating setting. These ideas were converted to a set of seven features (in the earlier development of Skade [33]).

None of the ideas or features are novel or controversial. For instance, most hands-on training covers multiple learning dimensions (e.g., abstract conceptualization and active experimentation) and it is common to provide feedback and assessments to trainees (e.g., if they were successful).

The focus of this research is the feature *Enabling optional recommendations and hints at all stages*. This feature was derived from the expectation that the availability of clues has a strong link to trainee motivation to learn. This was based on ideas such as the six Cs [35], three of which suggest that educational motivation is influenced by: choice, challenge, and control. Although the framework was initially developed for children, research indicates that it is also relevant and applicable to teaching in higher education [30].

2.2 Main Components

The main components of Skade are a front-end user interface written in Angular, a back-end written in Python, and Json-specifications that detail the challenge and its chosen configurations of the environment and threat actor. Figure 1 shows the main components and integrations of Skade.

Skade is built on top of the cyber range platform Crate [12]. It uses the Crate API to instantiate the cyber range specified using previous snapshots. It also uses the APIs of Sved [14] and Lore [13] to start simulated users and threat agents in the cyber range. Threats can be simulated with dynamic features that allow trainees to train over and over again on the same challenge but with small changes to the challenge. For Sved this is solved by using a dynamic configuration where, for example, filenames and attacked machines can differ. Lore is ML-based and will always perform slightly different actions each run.

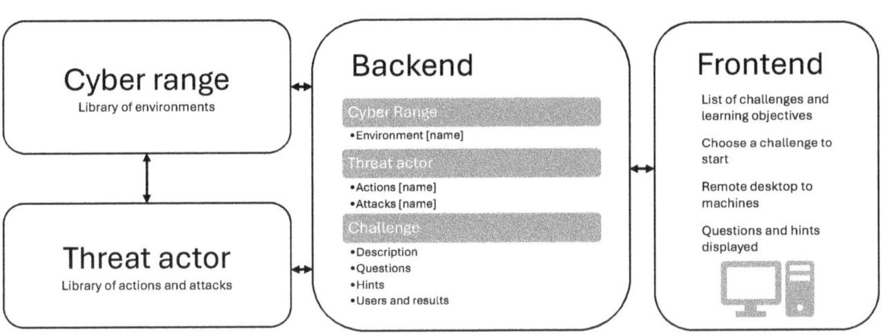

Fig. 1. The main components and integrations of Skade.

An important part of the Json-specifications are the questions used for the challenge. These include response alternatives, details on how to check answers, when the question should be displayed to the trainee, learning objectives, and hints. Most of this is prepared in advance, while parts of are dynamically generated at runtime, e.g., by interacting with the logs of the threat agents to get the correct answer for questions. In the code snippet below, the *segments_ affected* and *correct_ segments* are references to Python-methods extracting data from a Sved log:

```
{
    "name": "Looking for network traffic",
    "text": "From which segments in Curea can you identify the malicious traffic?",
    "alternatives":{
      "response_type": "MULTIPLE_ANSWER",
      "alternatives": {
        "0": "HMI (10.100.1.0/24)",
        "1": "HQCLIENT (198.51.100.0/24)",
        "2": "SCADA (10.100.7.0/24)",
        "3": "SRV (203.0.113.0/24)"
      }
    },
    "answer":{
      "correct_answer": "segments_affected",
      "correction_method": "correct_segments"
    }
 }
```

2.3 Threat Hunting Challenges

In our tests we cover three pairs (or packages) of threat hunting challenges. The packages were separated into two distinct parts because these parts were manipulated differently in the experiment. During the study, each trainee completed one of the three packages. These are briefly described below, with expected time for a trainee in parentheses:

- Package A: The first challenge (2 h) focused on baselining of a small military network with 30 machines using OpenSearch; the second challenge (3 h) had the trainee threat hunt in the network.
- Package B: Both challenges focused on intelligence driven threat hunting in a small company with 40 machines. The first (2 h) focused on host logs; the second (2 h) on the use of a SIEM system.
- Package C: The challenges (1 h + 1 h) were theoretical and asked the trainees questions about typical situations and proper reasoning during threat hunting.

We here give some examples of how Skade represents the challenges. The examples are drawn from the second part of package A, a challenge inspired by the military cyber threat scenario described in [34].

Trainees get access to challenges after a scheduling stage. During the scheduling stage, each trainee is offered multiple challenges to choose from, and each challenge is displayed with a short description and its learning objectives. Our example challenge uses these learning objectives: *uncover new patterns and TTPs*, and *investigate via tools and techniques.*

Each challenge includes background information and network data, presented to the trainee at the start. In our example, the trainee is provided with a network map containing 30 machines and is advised to focus on accessing the SIEM server with full log access. The example challenge has a background text of approximately 500 words, detailing assets to protect, the threat environment, and how to safely access the network SIEM system.

Trainees connect to machines in the cyber range with RDP via Skade, running in their browser. When working on the challenge, trainees are presented with investigative questions that guide them toward identifying baseline and anomalous behaviors. Hints, when available, may be requested at any time. One question and associated hints are shown in Fig. 2. The first hint describes baseline behavior and reminds the trainee of results from the first challenge in the package. The second and third hints guide the trainee toward investigating the relevant machines.

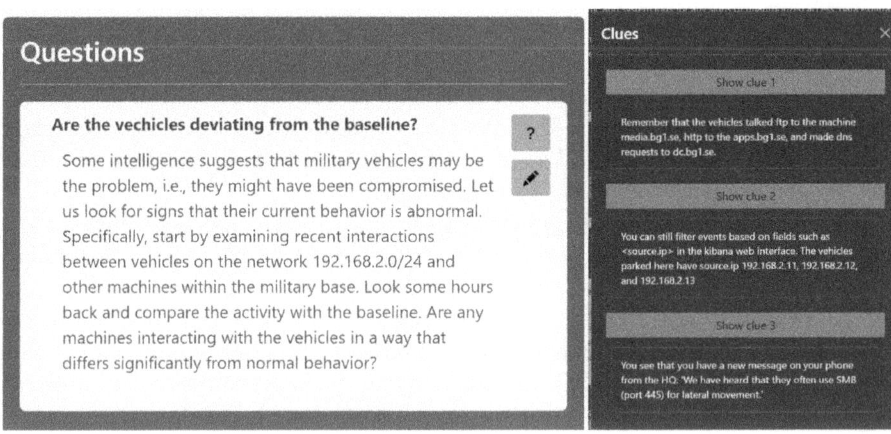

Fig. 2. An example of a question from package A and the hints associated with it.

After completion of the challenge, trainees unlock a reflecting assignment, specifically designed to ask follow-up questions.

3 Related Works

The sections below describe research about how learning depends on training at a satisfactory difficulty level, as well as on how the difficulty level is managed in practice.

3.1 The Importance of the Difficulty Level

The use of simulated cyber environments to support learning is common [6, 11]. The importance of managing the difficulty level in such trainings is widely recognized among cybersecurity researchers. It is reasonable to expect that the suitable difficulty level depends on the proficiency of the trainee. Indeed, Vykopal et al. write that "matching learners proficiency and exercise difficulty is a key success factor of the whole exercise" [37]. Additionally, when Pfaller et al. [28] list challenges associated with cyber exercises the first one enumerated is "adaption for different skill levels".

The idea that the training outcomes are improved if the training is adapted to trainees' proficiency levels is by no means unique to cybersecurity. Alessi [1] has formulated a general hypothesis concerning the relationship between simulation fidelity, trainee proficiency level, and learning. The hypothesis suggests that fidelity induces complexity that would become an obstacle for learning at a certain point. This has been made even sharper in research on training of aviators, where it is said that flight simulators can have too much fidelity to be optimal if the trainee is a non-expert [25].

While this is a widespread idea, we are aware of only one quantitative empirical test in the cybersecurity domain supporting the importance of managing the difficulty level [15]. This study asked trainees whether they found the exercise educational (a kind of perceived learning) and if it was challenging. Scoring either very high or very low on the latter was interpreted as a result of the wrong (unbalanced) difficulty level. The study found a medium correlation ($r = 0.38$) between perceived learning and balanced difficulty.

More generally, perceived usability of learning systems has been suggested to influence learners' ability to acquire knowledge, as poor usability may increase extraneous cognitive load and divert attention away from the learning task itself [23]. Accordingly, the System Usability Scale (SUS) is often used in studies of educational technology across diverse contexts and user groups [36]. In our research, we measure usability of Skade using SUS. In addition, we measure trainees' perceived learning and if they are satisfied with the difficulty level.

3.2 Managing the Difficulty Level

Pfaller et al. [28] suggest that exercises should be flexible and adapted to each team or participant. However, much of the research on cyber security training does not explicitly mention any approach to managing difficulty levels. For instance, Miazi et al. [24] describe a university campus threat hunting competition. Two of three challenges were not solved by any participant, indicating the problem of finding a balanced difficulty level. Similarly, ICSTASY [20] is a "scenario-based, interactive, and immersive cybersecurity training platform", but features for adjusting the difficulty level are not mentioned in its description. Additionally, neither the paper about the platform CRACK [31] nor the documentation about the platform TopoMojo [5] mention how to manage or document challenge skill requirements.

There are some exceptions that introduce difficulty levels, although not by trainee choice. Dobson et al. [9] describe a training and exercise model called R-EACTR that manages complexity with variables such as difficulty level of attacks, obfuscation efforts, deception, and noise. Wei et al. [39] present a university course that uses six threat hunting hands-on assignments on instrumented virtual machines. The assignments were static and categorized as low, medium, or high difficulty level, and designed for different types of students. Beuran et al. [3] propose a lab platform with automated challenge progression that activates certain components in the training based on the progress of the trainee. In other work, Beuran et al. integrate practical hands-on challenges with a learning management system and include the option of hints in systems [4]. Vykopal et al. [38]collect student actions within a virtual environment and responses submitted through a training portal. By analyzing this data, a system determines the most appropriate next task for the student, enabling adaptive learning based on performance. None of these evaluate if hints are appreciated or improve learning.

It is difficult to find details on how difficulty levels are managed in commercial platforms and services for hands-on cybersecurity training. Most offerings are lab-based and provide small, self-contained tasks that run on one or a few virtual machines, or they distribute downloadable artifacts such as logs, disk images, and PCAP files. Examples of such lab-based platforms include CyberDefenders, LetsDefend, and BlueTeamLabs Online. These challenges seem to target different proficiency levels in their design rather than adapting instructions to trainees dynamically.

Platforms such as CyberBit and Cloud Range are similar to Skade in their use of realistic networks instantiated in a cyber range, and they offer training for both individuals and teams at different levels. It is not entirely clear how the training content is adapted to differing proficiency levels. However, the idea of providing clues appears to be present. In particular, CyberBit offers an option to "allow hints on quiz" during exercise planning [8]. Consequently, their approach to difficulty adaptation includes a mechanism similar to the one evaluated in this paper.

4 Evaluation Method

As described in Sect. 2, Skade rests on multiple ideas concerning how cybersecurity analysts should be trained. Figure 3 presents an overview of the study design and the measurements used. The following sections give further details on the study design, how measurements were collected, the sample (trainees) in the tests, and how the data were analyzed.

4.1 Study Design

As illustrated in Fig. 3, a within-sample crossover design was used. In this design, each trainee got two challenges in a package: one challenge with clues available and one challenge without clues. Before each trainee started their training it was randomized (50/50) if clues should be available in the trainee's first or second challenge. Bias that may be introduced by the order in which clues are offered are balanced out, because both sequences have (approximately) as many subjects. When planning the study, we expected effect sizes close to Cohen's d $= 0.5$, and estimated that 34 trainees would be sufficient for the test ($1 - \beta = 0.80$, $\alpha = 0.05$).

According to the ethical review practices in Sweden, a study with this design is exempt from the requirement of review by an ethical review board. Nevertheless, basic ethical requirements have been upheld by acquiring informed consent from the participants, avoiding putting pressure to participate, minimizing risk, and by following data protection procedures.

4.2 Measurements

As illustrated in Fig. 3, four questionnaires (surveys) were administered.

The first one included information about the study, questions concerning informed consent, and demographic questions (e.g., years of experience in cybersecurity).

The second and third questionnaires were identical and asked questions related to the previous challenge. It had two scales. Three items adapted from the After-Scenario Questionnaire [21] were used to measure trainee satisfaction with the difficulty level, e.g., "Overall, I am satisfied with the ease of completing the tasks in this challenge". The reliability (Cronbach's α) was good in both questionnaires (0.837 and 0.892). Four items were used to measure perceived learning. This was inspired by Bandura's self-efficacy scale but re-phrased in terms of a change in ability, to measure the situation after taking each challenge. After each challenge, we asked trainees to provide their agreement on the following statements: a) "I am more ready to do threat hunting than before", b) "I have changed the way I think about how to investigate threats", c) "I feel better prepared to proactively search for threats", and d) "My threat hunting skills have improved". This way of measuring learning effects is common, see for example [29]. Our variant scale had high reliability (0.966 and 0.906).

The final questionnaire used ten items adapted from the System Usability Scale [22] to measure system usability. An example item was "I thought the system was easy to use." The questionnaire also asked questions concerning potential technical problems, and asked for open ended feedback on the overall experience. All survey questions from the second, third and final questionnaire are presented in Table 4 and Table 5 in the appendix.

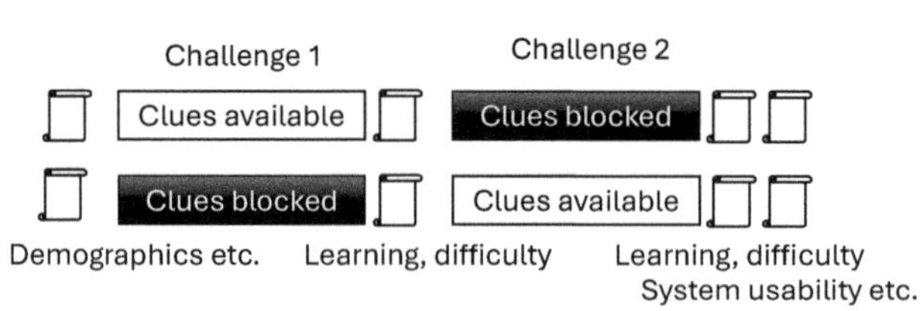

Fig. 3. The challenges in the two study sequences and the four questionnaires administered.

In addition to the four questionnaires, a set of measures were collected from the training platform: various completion timers, how clues were used, and if questions were answered correctly.

4.3 Sample

The target users of the Skade application are trainees who are potential cybersecurity analysts. The challenges included in the present test were designed for training security professionals to become threat hunters (e.g., tier one in a SOC) or for training computer network generalists (e.g., network administrators) in the basics of threat hunting. The sampling frame was set accordingly.

The Swedish Civil Contingencies Agency (MSB) provided support in the recruitment process by inviting persons fitting this sample frame. This was done by distributing a brief description of the training and the study to professional networks that MSB administers. These networks cover cyber-related personnel within organizations considered societally important in Sweden, e.g., specific government agencies and electrical power companies. A similar invite was sent out to a smaller group of threat hunting professionals within CISA, US Department of Homeland Security. The trainees were allowed to select which of the three packages of challenges they wanted to enroll in. All trainees were informed about the study and their personal data rights under the GDPR.

Recruitment was closed after 17 weeks, when 54 persons had chosen to enroll and scheduled to take a package of two challenges. Seven gave up after the first challenge. Two had serious technical issues that made the results incomplete, and were therefore excluded, leaving 45 trainees. The trainees chose their challenge package, with 14 selecting package A (5 h), 12 selecting package B (4 h), and 19 selecting package C (2 h). Participants were not provided with any compensation for their participation.

There was considerable variance among the trainees in their years of work experience in different parts of the field. Most trainees had 0-10 years of work experience in relevant fields. The overall averages and standard deviations (in parentheses) were, per field: cybersecurity ($\mu = 3.98$, $\sigma = 4.42$), incident handling ($\mu = 4.62$, $\sigma = 6.43$), endpoint and host analysis ($\mu = 2.62$, $\sigma = 4.28$), and network event analysis ($\mu = 2.47$, $\sigma = 3.31$). Almost half of trainees had work experience greater than 10 years in system and network administration ($\mu = 11.43$, $\sigma = 9.14$). There were 38 trainees (84%) who knew what threat hunting was before they signed up for the study.

4.4 Analysis Method

The four hypotheses were tested using Jamovi [17]. H1 and H2 were tested with correlational analyses and regression models. Because the crossover design is more or less balanced, H3 and H4 were tested by comparing mean values for the different conditions. This was complemented by a mixed model that controlled for potential effects of the challenge package and whether it was the first or second challenge. In all tests, $p < 0.05$ was considered statistically significant.

Post-hoc analyses included additional tests and simple comparisons of mean values. The post-hoc analyses were also supported by the responses to open ended questions in the questionnaires, e.g., because they indicated problems that trainees had with the challenges.

5 Results

Our evaluation focuses on the four hypotheses presented in Sect. 1: two related to the perceived usability of the system and two related to the utility of optional clues. The sections below describe the tests of these hypotheses.

5.1 Effects on System Usability

Usability was measured with the widely adopted System Usability Scale. This scale can be converted to a value 0–100, for which benchmarks exists. On this scale, Skade received an overall mean of 67.2, the range was 17.5–90.0, and the standard deviation was 16.3. The spread in ratings is illustrated in Fig. 4. H1 says that usability is associated with trainee satisfaction with the difficulty level of the challenges, and H2 says that usability is associated with the perceived learning. Both satisfaction with the difficulty level and the perceived learning were measured twice (after each challenge) and here we use the mean value of each rating. As outlined in Table 1, the Pearson correlations support both hypotheses. The correlation coefficient for system usability and difficulty satisfaction was $r = 0.481$ ($p < 0.001$) and the correlation coefficient between the system usability and perceived learning was $r = 0.334$ ($p = 0.025$).

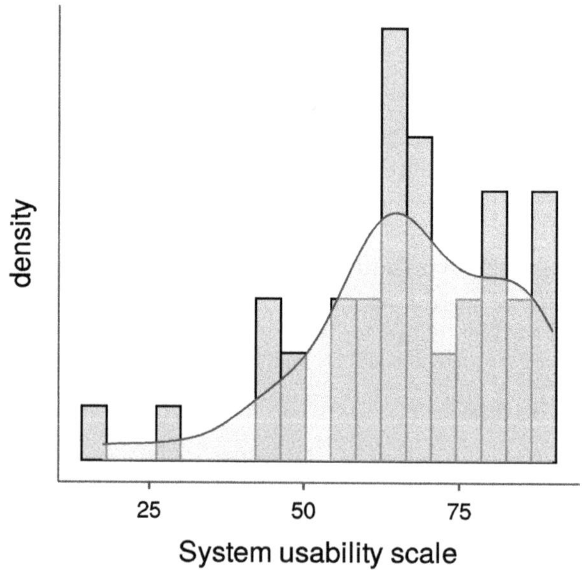

Fig. 4. The trainees' ratings of Skade with the System Usability Scale.

When both difficulty satisfaction and perceived learning are entered into a linear regression model to predict score from the *System Usability Scale*, the

Table 1. Pearson correlations between the difficulty satisfaction, perceived learning, and the System Usability Scale (SUS).

	Perceived learning	System usability
Difficulty satisfaction	0.372	0.481
Perceived learning	—	0.334

coefficient of multiple correlation is moderately high (R = 0.509). However, only *difficulty satisfaction* is a significant predictor ($\beta = 0.414$, p = 0.006), while *perceived learning* ($\beta = 0.180$, p = 0.215) is not. On interpretation is that *perceived learning* influences system usability, but the overlap (collinearity) with *difficulty satisfaction* make the relationship weaker when both variables are used as predictor variables.

5.2 Effects from Clue Availability

The paired values from measures with or without clues were used to test H3 and H4. The overall mean of these values are at the bottom of Table 2, and are almost the same regardless of condition. Accordingly, no significant difference is present in the pairwise comparisons. For *difficulty satisfaction* the mean difference is d = 0.10 (p = 0.506); for *perceived learning* the difference is d = 0.02 (p = 0.915). Thus, there is no clear effect from clue availability on either satisfaction with the difficulty level (H3) or perceived learning (H4). The table also indicates inconsistencies in effects due to when clues were available. For example, trainees performing the first challenge with clues were more satisfied with the difficulty level than trainees performing the first challenge without clues ($\mu = 4.49$ vs. $\mu = 4.03$), but the opposite is true for the second challenge ($\mu = 3.31$ vs. $\mu = 3.35$). Figure 5 shows the results for *difficulty satisfaction* and *perceived learning* for conditions with and without clues.

A mixed model was set up to control for the influence of different variables. The model clustered responses based on trainee (two responses each) and used two predictor variables other than clues: the challenge number (first or second) and the challenge package (A, B, or C). In the model, both control variables are statistically significant predictors of *difficulty satisfaction*. Thus, it matters whether the challenge was the first or second ($p < 0.001$), with higher values for the first challenge. This order effect can be due to many reasons, e.g., that the second challenge of a package was too easy or too difficult, or that fatigue set in. The package also has an effect, with a clear difference between package A and C (p = 0.005). When controlling for these other variables, the presence of clues has no effect (p = 0.231). The same tendencies are found in a mixed model predicting *perceived learning*, but here no variables were statistically significant.

Overall, the fixed effects in the mixed models explain 23% of the variance in difficulty satisfaction and 10% for perceived learning. Considering also random effects, each model explains 70%. Thus, most variance is between trainees, and not between how a specific trainee rated the two challenges.

Table 2. Conditions and their mean values.

Conditions	Difficulty satisfaction (μ)	Perceived learning (μ)
First challenge (clues)	4.49	3.89
First challenge (no clues)	4.03	3.66
Second challenge (clues)	3.31	3.47
Second challenge (no clues)	3.35	3.64
Either challenge (clues)	3.86	3.67
Either challenge (no clues)	3.71	3.65

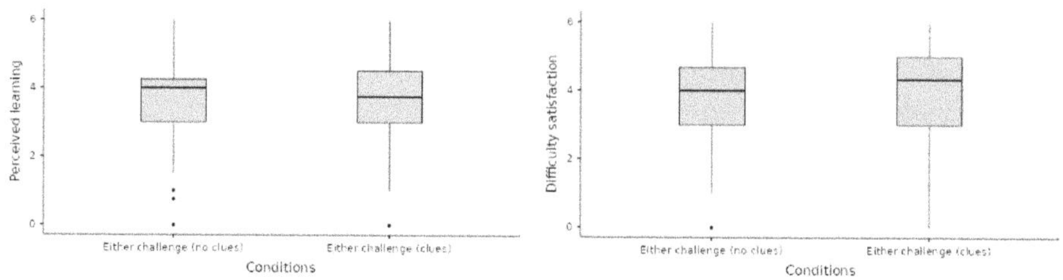

(a) Perceived learning (b) Difficulty satisfaction

Fig. 5. Trainees' responses on perceived learning and difficulty satisfaction with and without clues.

6 Discussion

The sections below discuss the usability of Skade, why the availability of clues did not help in the tests, and the overall prospect of automating hands-on cybersecurity training.

6.1 The Usability of Skade

In the tests we used the widely adopted System Usability Scale. This scale is associated with benchmarks. The mean rating of Skade (67.2) is on a par with public facing websites (67) and business-to-business software (67.6) [22]. Considering that Skade is a research prototype, and that this test was its first exposure to real trainees, we are satisfied with this mean rating. It suggests that Skade has the basic functions a trainee expects, and successfully integrates cyber range simulations. However, there is room for improvement.

The final questionnaire asked about problems that we anticipated trainees may have. For each potential problem, the 45 trainees answered if they had no problems, some problems, or serious problems. The results show that serious problems were rare, apart from getting needed help, e.g., using the hints (N = 8 had serious problems with getting needed help, N = 11 had some problems). Trainees experienced (some, or rarely, serious) problems with the purpose of the challenge questions (N = 16), access permissions within machines (N = 14), unresponsive systems (N = 11), interaction within machines (N = 9), challenge background information (N = 9), and low performance of the platform (N = 8).

The written feedback provided by the trainees largely aligns with these numbers. While most of this feedback pointed to problems, some were also clearly positive. For example, "Very good learning experience!" and "Tasks were fun. Hope platform evolves with more threat hunting. Liked it a lot."

6.2 Limitations

Our experiment was executed according to plan. Yet, the two hypotheses concerning the utility of clues, which we based on previous research, were not supported in the data. We performed some post-hoc analysis to try to identify why, and tested several potential explanations.

First, one might suspect that the clues themselves were poor. Our answer is that they were not. We designed the clues to provide a direct investigation path, with explicit scaffolding on how queries are structured in SIEM and related log analytics tools, and with context tied specifically to the observed attack and its correct details (e.g., IP addresses). Trainee feedback was in line with this, e.g., one trainee commenting: "Include example search syntax for OpenSearch, Wazuh etc. to improve experience." Thus, the clues we provided were already focused on the type of information that trainees wanted more of.

Second, the challenges may have been either too difficult or too easy for the clues to have a meaningful effect. This is plausible, as both *perceived learning* and *difficulty satisfaction* exhibit ceiling effects and therefore violate the assumption of normality. This is common in scales like these, were respondents use the end of the scales. To assess whether this violation affects the results, we compared Pearson correlations with non-parametric alternatives (Spearman and Kendall's Tau) and conducted bootstrapped tests with 1000 iterations. Due to the ceiling effects, the Spearman correlation between the two non-normal variables is 0.17 lower than the Pearson correlation we reported. However, all relationships remain robust and retain their statistical significance. This also applies to the non-significant relationship involving clues, even when outliers are removed. For example, the effect does not become statistically significant if we exclude trainees with the lowest ratings (< 2), the highest ratings (> 5), or both.

Third, the outcome measures could have been chosen differently. For instance, the concept of flow is known to relate to learning. We used *perceived learning* and *difficulty satisfaction* because they are the main targets of the intervention. The most established measure of flow is the Flow State Scale [16], which consists of 36 items. Some of these overlap with our measures (e.g., "My abilities matched the high challenge of the situation"), while others are not expected to be influenced by clues (e.g., "I had total concentration"). Thus, although other measures could be tested with clues and used as quality indicators for systems such as Skade, we believe the scales used in this test are reasonable and meaningful.

Fourth, one may suspect that the results were biased because some trainees dropped out after not receiving clues in the first challenge, or because of biased responding (e.g., social desirability). Dropouts are unlikely to bias the results, since roughly half of those who dropped out had been assigned to each condition: four received clues in the first challenge and three in the second challenge. We see no indications of other response biases. In addition, participation was voluntary and uncompensated, so there is little reason to suspect that participants answered in a particular way because they believed it was expected.

Fourth, one may suspect that trainees missed the availability of clues. Indeed, the results show that trainees do not always use available clues. This could lead to clue availability sometimes lacking an impact on perceived learning or difficulty satisfaction. However, the percent of clues used by trainees is not related to their difficulty satisfaction ($r = -0.079$, $p = 0.606$) or perceived learning ($r = -0.014$, $p = 0.926$). Thus, any missed clues did not alter the results.

Fifth, the sample and tests are limited to junior threat hunters and system administrators with limited cybersecurity competence. It is possible that clues would have a clearer effect in other sample frames. For instance, it may be that the individuals who needed clues in this study were overall too inexperienced to benefit from them, whereas the availability of clues could be the factor that tips the scale for a more competent audience.

Sixth, it could be that the randomization procedure has produced some bias. We randomized for each trainee, and the result was not perfectly even, as shown in Table 3. In combination with the results of our mixed model, suggesting that the challenge package and order were somewhat important, this could lead to our dependent variables being biased by the randomization procedure. Further research, with other randomization procedures, will be needed to clarify the impact of this bias. However, the effects of clues are insignificant when the mixed model is set to control for the effects of the randomization procedure.

Altogether, we believe it is unlikely that these potential issues have biased the result and hidden a true relationship. On the contrary, we believe that the availability of clues is not an effective way to control the difficulty satisfaction or the perceived learning. This may be, for example, because the trainees get stuck on very different things and need more personalized help than what can be offered with predefined clues.

6.3 The Prospect of Automating Cybersecurity Training

The need to close the competence gap is real. Consequently, there are many efforts on how to provide training and exercises cheaper and more effectively.

Table 3. The result of the randomization for each challenge package (pair of challenges).

Package	Clues first challenge	Clues second challenge	Total
A	4	10	14
B	6	6	12
C	11	8	19
Total	21	24	45

Some of these efforts were described in Sect. 3, and they share the design principles of Skade. For instance, most of the platforms employ structured learning tasks and clearly defined learning objectives (often related to MITRE ATT&CK). Accordingly, Skade is not the first documented attempt to automate exercise management of challenges in cyber ranges. For instance, the progression management described in [3] is along the same lines.

In our research, Skade received an acceptable rating according to the System Usability Scale, despite being only an initial prototype. However, our main focus was on clues, and their impact on trainee satisfaction with difficulty and their perceived learning. Unfortunately, our implementations of clues did not have much success in this regard. We believe that there are other ways of supporting trainees with clues or similar:

- Provide more dynamic clues by integrating large language models that target what the particular trainee needs help with better.
- Provide an "I give up"-button, which displays a step by step solution to the question and gives access to the next question.
- Differentiate clues of different types, e.g., separate clues on technical details and syntax from clues on mindset or threat intelligence.
- Adapt the difficulty level to the trainee, e.g., by asking for introductory self-assessments or by predicting proficiency based on past performance.

Thus, apart from our static clues, there are others ways that can be explored in balancing the difficulty levels of challenges.

7 Conclusions

Adapting the difficulty to trainees' proficiency levels is critical in hands-on cybersecurity training, such as threat hunting. However, the technical complexity of many threat hunting tasks makes it hard to find and implement the right difficulty levels.

Our tests indicate that the usability of our threat hunting training platform, Skade, is shaped by the trainees' satisfaction with the task difficulty, and by whether they believe that the system improved their threat hunting skills. While many trainees reported that the challenges were too difficult or that they lacked guidance, providing relevant clues did not produce measurable improvements in the satisfaction of the difficulty or in the perceived learning. These findings suggest that difficulty management cannot be solved by adding clues alone.

Acknowledgments. This study was funded by MSB (grant number 2022-1213). This study was also supported by DHS S&T and CISA pursuant to Technical Annex 07 of Project Arrangement 016-2011 between USA and Sweden.

Disclosure of Interests. The authors have no competing interests to declare that are relevant to the content of this article.

A Questionnaires

Table 4. The questions in the questionnaires answered after each challenge.

After challenge survey
Part 1 - Overall, I am satisfied with the...
Completely disagree - Completely agree
1... ease of completing the tasks in this challenge.
2... amount of time it took to complete the tasks in this challenge.
3... support information (on-line help, messages, documentation) when completing the tasks.
Part 2 - After this challenge...
Completely disagree - Completely agree
1... I am more ready to do threat hunting than before.
2... I have changed the way I think about how to investigate threats.
3... I feel better prepared to proactively search for threats in our operational computer networks.
4... my threat hunting skills have improved.

Table 5. The questions in the final questionnaire.

Evaluation of the app	
Part 1	
Strongly disagree - Strongly agree	
1	I think that I would like to use this system frequently.
2	I found the system unnecessarily complex.
3	I thought the system was easy to use.
4	I think that I would need the support of a technical person to use this.
5	I found the various functions in this system were well integrated.
6	I thought there was too much inconsistency in this system.
7	I would imagine that most people would learn to use this system very quickly.
8	I found the system very cumbersome to use.
9	I felt very confident using the system.
10	I needed to learn a lot of things before I could get going with this system.
Part 2	
No problem at all - Serious problems	
1	Background information, e.g., the purpose of the challenge.
2	Permissions to machines and data, e.g., accessing logs.
3	Purpose of the questions, e.g., the learning objectives that questions are linked to.
4	Getting the help you needed, e.g. using the hints.
5	Interaction with machines, e.g., remote desktops in the browser.
6	Unresponsive systems, e.g., downtime in the cyber range and 404 errors.
7	Low performance, e.g., because the web page is slow.

References

1. Alessi, S.M.: Fidelity in the design of instructional simulations. J. Comput.-Based Instr. **15**(2), 40–47 (1988)
2. Beason, R.E., Phelan, M., Devine, S., Aiken, M., Orban, J.: Evaluation of hands-on cybersecurity skills development. Technical report, Idaho National Laboratory (INL), Idaho Falls, ID (United States) (2021)
3. Beuran, R., Inoue, T., Tan, Y., Shinoda, Y.: Realistic cybersecurity training via scenario progression management. In: 2019 IEEE European Symposium on Security and Privacy Workshops (EuroS&PW), pp. 67–76. IEEE (2019)
4. Beuran, R., Tang, D., Tan, Z., Hasegawa, S., Tan, Y., Shinoda, Y.: Supporting cybersecurity education and training via LMS integration: CyLMS. Educ. Inf. Technol. **24**(6), 3619–3643 (2019)
5. Carnegie Mellon University.: TopoMojo: A VM Topology Manager (2019). https://www.sei.cmu.edu/library/topomojo-a-vm-topology-manager/
6. Chowdhury, N., Gkioulos, V.: Cyber security training for critical infrastructure protection: a literature review. Comput. Sci. Rev. **40**, 100361 (2021)
7. Crumpler, W., Lewis, J.A.: The cybersecurity workforce gap. Report, Center for Strategic and International Studies, Washington, DC (2019). https://www.csis.org/analysis/cybersecurity-workforce-gap
8. Cyberbit: Cyberbit range: Assessing cyber skills to improve soc team performance. Technical report, Cyberbit (2020). https://www.cyberbit.com/wp-content/uploads/2020/11/Assesing-Cyber-Skills-to-Improve-SOC-Team-Performance.pdf
9. Dobson, G., Podnar, T., Cerini, A., Osterritter, L.: R-EACTR: a framework for designing realistic cyber warfare exercises. Technical report CMU/SEI-2017-TR-005, Carnegie Mellon University, Software Engineering Institute's Digital Library (2017). https://insights.sei.cmu.edu/library/r-eactr-a-framework-for-designing-realistic-cyber-warfare-exercises/. Accessed 14 Aug 2024
10. ENISA: 2024 report on the state of cybersecurity in the union panel series. Technical report, ENISA (2024). https://www.enisa.europa.eu/publications/2024-report-on-the-state-of-the-cybersecurity-in-the-union
11. Floros, E., et al.: Towards the design of cyber range training programs for enhanced preparedness: investigating the training needs in critical infrastructures. In: 2025 IEEE Global Engineering Education Conference (EDUCON), pp. 1–10. IEEE (2025)
12. Gustafsson, T., Almroth, J.: Cyber range automation overview with a case study of CRATE. In: Asplund, M., Nadjm-Tehrani, S. (eds.) NordSec 2020. LNCS, vol. 12556, pp. 192–209. Springer, Cham (2021). https://doi.org/10.1007/978-3-030-70852-8_12
13. Holm, H.: Lore a red team emulation tool. IEEE Trans. Dependable Secure Comput. (2022)
14. Holm, H., Sommestad, T.: SVED: scanning, vulnerabilities, exploits and detection. In: MILCOM 2016-2016 IEEE Military Communications Conference, pp. 976–981. IEEE (2016)
15. Holm, H., Sommestad, T.: Realistic and balanced automated threat emulation. Comput. Secur. **151**, 104351 (2025)
16. Jackson, S.A., Marsh, H.W.: Development and validation of a scale to measure optimal experience: the flow state scale. J. Sport Exerc. Psychol. **18**(1), 17–35 (1996)

17. Jamovi: Jamovi (version 2.7.12) [computer software] (2025). https://www.jamovi.org
18. Kebande, V.R.: The impact of virtual laboratories on active learning and engagement in cybersecurity distance education. arXiv preprint (2024). https://arxiv.org/abs/2404.04952
19. Lazarov, W., et al.: Lessons learned from using cyber range to teach cybersecurity at different levels of education. Technology, Knowledge and Learning, pp. 1–36 (2025)
20. Lee, D., Kim, D., Lee, C., Ahn, M.K., Lee, W.: Icstasy: an integrated cybersecurity training system for military personnel. IEEE Access **10**, 62232–62246 (2022)
21. Lewis, J.R.: Psychometric evaluation of an after-scenario questionnaire for computer usability studies: the ASQ. SIGCHI Bull. **23**, 78–81 (1991). https://doi.org/10.1145/122672.122692
22. Lewis, J.R.: The system usability scale: past, present, and future. Int. J. Hum.-Comput. Interact. **34**, 577–590 (2018). https://doi.org/10.1080/10447318.2018.1455307
23. Mayes, J.T., Fowler, C.J.: Learning technology and usability: a framework for understanding courseware. Interact. Comput. **11**(5), 485–497 (1999)
24. Miazi, M.N.S., Pritom, M.M.A., Shehab, M., Chu, B., Wei, J.: The design of cyber threat hunting games: a case study. In: 2017 26th International Conference on Computer Communication and Networks (ICCCN), pp. 1–6. IEEE (2017)
25. Noble, C.: The relationship between fidelity and learning in aviation training and assessment. J. Air Transp. World Wide **7**(2-3) (2002)
26. OECD: Building a Skilled Cyber Security Workforce in Europe. OECD Publishing (2024). https://doi.org/10.1787/3673cd60-en. https://www.oecd.org/en/publications/building-a-skilled-cyber-security-workforce-in-europe_3673cd60-en.html
27. Office of the national cyber director: 2024 report on the cybersecurity posture of the united states office of the national cyber director executive office of the president. Technical report, The White House (2024). https://bidenwhitehouse.archives.gov/wp-content/uploads/2024/05/2024-Report-on-the-Cybersecurity-Posture-of-the-United-States.pdf
28. Pfaller, T., Skopik, F., Smith, P., Leitner, M.: Towards customized cyber exercises using a process-based lifecycle model. In: Proceedings of the 2024 European Interdisciplinary Cybersecurity Conference, pp. 37–45 (2024)
29. Rovai, A.P., Wighting, M.J., Baker, J.D., Grooms, L.D.: Development of an instrument to measure perceived cognitive, affective, and psychomotor learning in traditional and virtual classroom higher education settings. Internet High. Educ. **12**, 7–13 (2009). https://doi.org/10.1016/j.iheduc.2008.10.002
30. Ruhe, E., Mcnamee, M., Du, G., Malmborg, J.: Tools for improving student motivation used during online teaching in higher education due to covid-19. Pedagogiska inspirationskonferensen-Genombrottet **11** (2021)
31. Russo, E., Costa, G., Armando, A.: Building next generation cyber ranges with crack. Comput. Secur. **95**, 101837 (2020)
32. SANS Institute: New 2025 SANS Threat Hunting Survey Reveals 61% of Organizations Struggle with Staffing Shortages. Press release, GlobeNewswire (2025), led by SANS Principal Instructor Josh Lemon; Bethesda, MD; full text available from GlobeNewswire
33. Sommestad, T., Karlzén, H., Kvist, H., Gustafsson, H.: Skade – a challenge management system for cyber threat hunting. In: 9th Workshop On The Security Of Industrial Control Systems & Of Cyber-Physical Systems. Haag, Netherlands (2023)

34. Thuv, A., et al.: Ainception scenario and storylines. Technical report, Norwegian Defence Research Establishment FFI (2025)
35. Turner, J., Paris, S.G.: How literacy tasks influence children's motivation for literacy. Read. Teach. **48**(8), 662–673 (1995)
36. Vlachogianni, P., Tselios, N.: Perceived usability evaluation of educational technology using the system usability scale (SUS): a systematic review. J. Res. Technol. Educ. **54**(3), 392–409 (2022). https://doi.org/10.1080/15391523.2020.1867938
37. Vykopal, J., Vizvary, M., Oslejsek, R., Celeda, P., Tovarnak, D.: Lessons learned from complex hands-on defence exercises in a cyber range. In: Proceedings - Frontiers in Education Conference, FIE, vol. 2017-October, pp. 1–8. Institute of Electrical and Electronics Engineers Inc. (2017). https://doi.org/10.1109/FIE.2017.8190713
38. Vykopal, J., Seda, P., Švábenský, V., Čeleda, P.: Smart environment for adaptive learning of cybersecurity skills. IEEE Trans. Learn. Technol. **16**(3), 443–456 (2022)
39. Wei, J., Chu, B.T., Cranford-Wesley, D., Brown, J.: A laboratory for hands-on cyber threat hunting education. In: Journal of the Colloquium for Information Systems Security Education, vol. 7 (2020)

When Trust Backfires: How Manipulative Cookie Designs Shape User Consent Under GDPR

Florian Mercks[1(✉)], Sandra Kostic[2], and Maija Poikela[3]

[1] Berlin, Germany
mercksflorian@hotmail.com
[2] Fraunhofer AISEC, Berlin, Germany
sandra.kostic@aisec.fraunhofer.de
[3] Berlin Institute of Health at Charité, Berlin, Germany
maija.poikela@bih-charite.de

Abstract. Cookie consent interfaces are a central mechanism through which websites communicate data practices and request permission for non-essential tracking under the General Data Protection Regulation (GDPR). Yet many websites present these dialogs in ways that nudge or pressure users into granting higher levels of consent—for example through visually dominant consent options, hidden rejection paths, or pre-enabled tracking categories. While prior research has examined the usability and legal implications of such manipulative cookie consent designs, less is known about how they interact with trust in the underlying brand.

We address this gap through a mixed-methods study combining an in-person lab study ($N = 40$) and a remote survey experiment ($N = 252$). Using a 2×2 between-subjects design that varies brand trust (high vs. low) and interface design (GDPR-compliant vs. manipulative), we examine how usability perceptions, brand trust, and involvement relate to consent decisions. Our results show that manipulative cookie consent interfaces are associated with substantially higher consent, even when perceived usability is lower. This pattern is especially pronounced for the higher-trust brand, suggesting that trust can backfire in privacy-critical contexts by reducing scrutiny of consent choices.

We discuss implications for trustworthy interface design, privacy regulation, and the ethical risks of exploiting brand trust in consent flows.

Keywords: GDPR · cookie consent · brand trust · manipulative designs · user consent · privacy · trust · usability · user study

1 Introduction

With the introduction of the General Data Protection Regulation (GDPR), organizations operating in the European Union became subject to stricter requirements regarding the processing of personal data [17,56]. In the context of cookies

F. Mercks—Independent Researcher.

© The Author(s), under exclusive license to Springer Nature Switzerland AG 2027
K. Renaud et al. (Eds.): EICC 2026, CCIS 3012, pp. 22–37, 2027.
https://doi.org/10.1007/978-3-032-28957-5_2

and similar tracking technologies, consent requirements are additionally shaped by the ePrivacy framework [53,56]. In practice, cookie consent dialogs have become a key interface through which websites communicate data practices and request user permission for non-essential tracking.

While these interfaces should support meaningful privacy decisions, many designs highlight consent-maximizing choices, obscure privacy-protective alternatives, or introduce friction when users attempt to reject tracking. Such *manipulative cookie consent designs* are widely documented in the literature on consent banners and dark patterns [10,33,58,71] and raise concerns about fairness, transparency, and the validity of consent [12,30,57].

Prior research has examined usability issues in cookie dialogs [14,22,68], the prevalence and evolution of manipulative consent mechanisms [6,8,13,31], and user preferences regarding consent notices [7,9,70]. However, one important dimension remains underexplored: **brand trust**. Users may assume that highly trusted organizations—especially banks handling sensitive financial data—behave responsibly and will not misuse their data. This assumption may lead users to accept data collection more readily, even when they encounter opaque or manipulative cookie consent interfaces [24,44,46].

Privacy research has long documented discrepancies between stated privacy attitudes and actual behavior, often referred to as the *privacy paradox* [2–4]. Users report valuing privacy, yet often accept extensive tracking when decisions are made under time pressure, interface complexity, or trust cues.

Against this background, we investigate how trust in a brand relates to users' vulnerability to manipulative cookie consent designs in a mixed-methods study that systematically varies (1) *brand trust* (high vs. low) and (2) *interface design* (GDPR-compliant vs. manipulative) in a 2×2 between-subjects experiment, combining a lab study and a larger remote experiment.

Accordingly, we address the following research questions:

- **RQ1:** How do manipulative versus GDPR-compliant cookie consent interfaces affect users' consent decisions?
- **RQ2:** How does brand trust relate to users' consent decisions in cookie consent interfaces?
- **RQ3:** To what extent do usability and involvement help explain consent decisions across interface conditions?

This paper provides empirical evidence that manipulative cookie consent interfaces are associated with higher consent in a high-stakes banking context, identifies a trust backfire effect in which the higher-trust brand appears to benefit disproportionately from manipulative designs, and combines quantitative and qualitative analyses to relate consent decisions to usability, trust, involvement, and recurring privacy personas.

2 Related Work

2.1 Manipulative Cookie Designs in Consent Interfaces

Manipulative designs describe interface strategies that steer users toward decisions that primarily benefit the service provider rather than reflecting users' own preferences [58]. In the context of cookie consent banners, this includes visually dominant "Accept All" buttons, hidden or de-emphasized rejection options, preselected tracking categories, and complex multi-step opt-out flows [10–12]. We use the term "manipulative" to refer to interface designs that systematically steer users toward consent decisions, consistent with prior work on dark patterns.

Large-scale measurements have shown that such designs are widespread and often conflict with legal requirements for valid consent [8,13,57]. Experimental work demonstrates that manipulative cookie consent designs significantly increase consent rates and can nudge users into accepting tracking that they might otherwise avoid [9,12,22]. Recent work has also investigated how design variations affect present and future consent decisions [30] and how browser- or system-level tools interact with banner behavior [31,32].

Our work adds to this literature by examining how manipulative cookie consent designs interact with brand trust rather than treating consent banners as brand-neutral interface components.

2.2 GDPR, Consent, and Usability

The GDPR requires that consent, where used as the legal basis, be freely given, specific, informed, and unambiguous [17,54]. From a usability perspective, this implies that users must be able to efficiently understand what is being collected and exercise meaningful control over data categories [61,62]. Studies show that consent banners with clear labels, balanced button prominence, and low interaction complexity better support informed decisions and reduce accidental consent [14,22,57].

However, usability and compliance often compete with organizational incentives to maximize data collection. Many banners create friction for privacy-protective actions while making high-consent choices effortless [12,57]. Our work confirms that even objectively low-usability interfaces can be associated with higher consent, and that this pattern is strengthened when users trust the organization behind the interface.

2.3 Brand Trust and Privacy-Related Decision-Making

Brand trust refers to a consumer's belief that a brand is reliable, competent, and benevolent [16,40,50]. It is particularly important in high-stakes domains such as online banking, health, and e-commerce, where users must share sensitive information and depend on the organization's integrity [44,45,49]. Prior work in financial services and digital channels shows that trust can increase willingness to disclose personal data and adopt new interactive services such as chatbots [23,24,43].

In privacy research, trust is often discussed as a key factor that can either mitigate or exacerbate privacy concerns [4,46]. Some studies suggest that high trust can reduce scrutiny of privacy policies and interfaces, functioning as a heuristic that shortcuts more effortful evaluations. Our study builds on this line of work by explicitly manipulating brand trust in the context of cookie consent interfaces and observing how it interacts with manipulative design patterns.

3 Method

3.1 Study Design

We conducted a 2×2 between-subjects experiment with the following factors:

1. **Brand trust:** a banking brand with high perceived trust versus a banking brand with lower perceived trust (based on pretesting and prior studies on trusted brands [24,67]).
2. **Interface design:** a GDPR-compliant interface versus a manipulative interface using design patterns known to influence consent [12,57].

Participants were randomly assigned to one of the four resulting conditions. Random assignment was implemented in a between-subjects design to ensure independence of observations and avoid carry-over effects. We combined data from an in-person moderated study ($N = 40$) with a larger remote experiment ($N = 252$), yielding a final sample of $N = 292$ after quality checks; see Fig. 1. The study can be understood as a controlled investigation of how brand trust and manipulative design jointly shape consent.

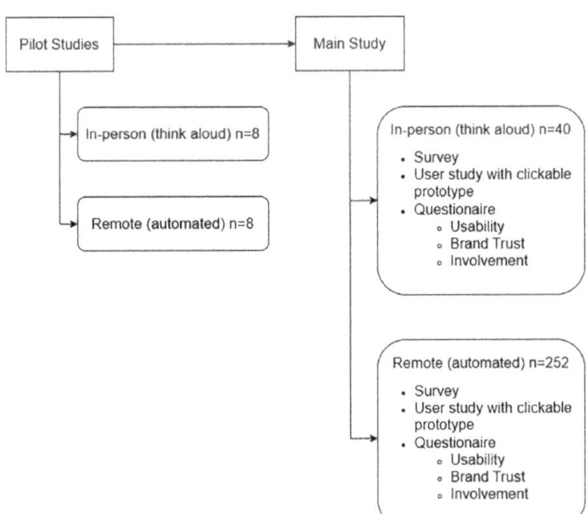

Fig. 1. Overview of pilot studies and the main study.

3.2 Materials

We implemented two cookie consent interfaces as interactive prototypes in Quant-UX [69]. The layout, color scheme, and textual content were kept as constant as possible across conditions. Only the brand name/logo and consent layout varied.

GDPR-Compliant Interface. The GDPR-compliant version followed common recommendations for usable and lawful consent banners [14,54,57]; see Fig. 2. It used equally prominent "Accept All" and "Reject All" buttons, enabled only strictly necessary cookies by default, and provided clearly labeled categories with straightforward access to detailed settings.

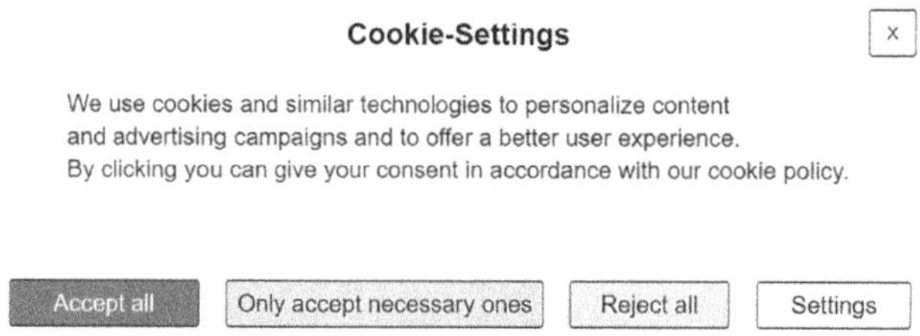

Fig. 2. GDPR-compliant interface for the high-trust bank condition.

Manipulative Interface. The manipulative version was inspired by patterns reported in prior work [10–12,57]; see Fig. 3. It used a visually dominant "Accept and continue" button, pre-enabled all cookie categories, hid "Reject All" behind secondary actions and additional steps, and employed more ambiguous wording for granular controls.

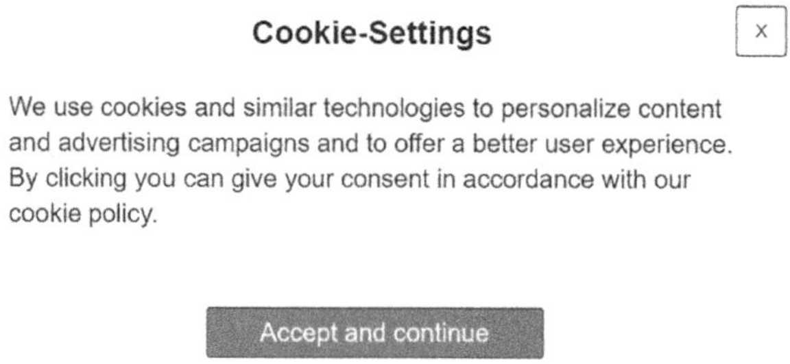

Fig. 3. Manipulative interface for the high-trust bank condition.

3.3 Participants and Recruitment

Participants were recruited using a combination of convenience and snowball sampling. Recruitment strategies differed between the in-person and remote study components. For the in-person pilot and main study, participants were recruited from the researchers' extended social networks and from the university environment. Additional participants were approached on site at a university cafeteria using an information stand describing the study purpose and duration.

For the remote main study, participants were recruited via student-oriented online channels, including mailing lists, survey platforms, and social media groups dedicated to academic study participation.

3.4 Variables

In addition to the usability of the cookie designs, we measured brand trust and involvement to better understand which factors are associated with different levels of consent. This was especially relevant because consent decisions can be shaped not only by interface characteristics, but also by emotional, contextual, and heuristic influences.

The main variables considered in this study were usability, measured using the System Usability Scale (SUS) [18,68]; brand trust, adapted from established brand trust scales [16,24,40]; involvement, measured via the Personal Involvement Inventory [38,39]; and consent level, an ordinal measure of the final cookie choice (0 = no consent, 1 = partial consent, 2 = full consent), inspired by prior consent studies [12,14,22].

Demographics included age, gender, and occupation. We also collected self-reported privacy attitudes and prior familiarity with cookie banners, but these were not central to the analyses reported here.

3.5 Procedure

In the **in-person study**, participants were welcomed, briefed, and asked to interact with the assigned cookie interface while thinking aloud [41]. After completing the task of deciding on a cookie configuration for the bank website, they completed the questionnaires and took part in a short semi-structured interview about their decision, privacy attitudes, and trust in the brand.

In the **remote study**, participants received a link to the prototype and the survey. They completed the consent task on their own device, followed by the same core questionnaires and open-ended questions capturing their reasoning and perceived trade-offs.

After data collection, we removed incomplete or obviously invalid responses and merged both datasets.

3.6 Quantitative Analysis

Quantitative data were analyzed using descriptive and inferential statistical methods. Consent level served as the primary dependent variable. Because consent level was measured on an ordinal three-level scale, we employ linear regression models as an exploratory approximation for the ordinal outcome. While an ordered logistic regression would provide a more appropriate specification, we focus on linear models to obtain interpretable estimates of direction and relative strength within the scope of this study. This exploratory approximation is intended to support interpretable comparisons of direction and relative strength and is consistent with prior exploratory analyses in similar HCI research contexts. The regression analyses reported here should therefore be interpreted conservatively as exploratory associations within the experimental setting rather than as formal causal estimates. Usability (SUS), brand trust, and involvement were treated as the main predictors, while age and gender were included as control variables. Statistical significance was assessed at the 0.05 level.

3.7 Qualitative Analysis

Open-ended responses and interview transcripts were analyzed using qualitative content analysis inspired by Kuckartz [65] and prior work on privacy behaviors [4]. Open-ended responses were initially prepared and organized in Microsoft Word and subsequently imported into MAXQDA for systematic coding and category management. We developed an initial coding scheme based on privacy strategy categories (e.g., accept everything, selectively accept, actively reject, use technical tools), iteratively refined it on a subset of responses, and then applied it to the full dataset. This process yielded four recurring privacy personas, described in Sect. 4.6. These categories emerged through iterative coding across both interview transcripts and open-ended responses and represent recurring behavioral tendencies rather than strictly mutually exclusive participant groups.

4 Results

The final sample consisted of 292 participants (40 in-person, 252 remote). Table 1 summarizes the demographic and occupational characteristics of the sample.

4.1 Consent Levels Across Conditions

Consent behavior differed across interface conditions (see Fig. 4). Across the full sample, manipulative cookie consent interfaces were associated with higher levels of consent than GDPR-compliant interfaces. For the high-trust brand (Trusted Bank), the proportion of participants granting full consent (level 2) was substantially higher in the manipulative condition than in the GDPR-compliant condition. For the low-trust brand (Untrusted Bank), manipulative cookie consent interfaces also increased consent levels, but the difference between conditions

Table 1. Demographic distribution of the sample.

Category	Item	%
Age	18–29 years	54.34
	30–49 years	34.25
	50+ years	11.41
Gender	Female	47.9
	Male	48.9
	Other/no answer	3.2
Employment	Education/studies	53.9
	Employed (full/part time)	40.6
	Other	5.5
Field	IT-related	23.2
	Other fields	76.8

was smaller. This descriptive pattern suggests that manipulative cookie consent interfaces appear particularly effective when presented by a highly trusted brand. The observed pattern is consistent with a potential interaction between brand trust and interface design, where the effect of manipulative interfaces appears stronger for high-trust brands.

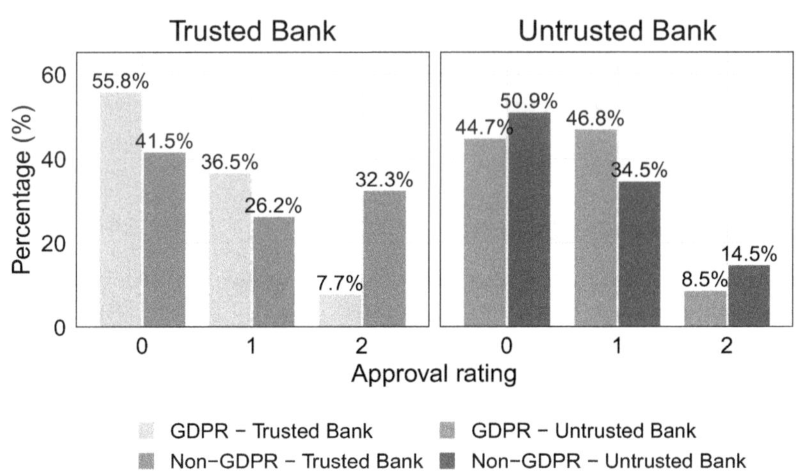

Fig. 4. Consent levels by scenario. Level 2 indicates full consent, level 1 partial consent, and level 0 no consent.

4.2 Usability and Consent

Usability scores differed markedly between interface conditions (see Fig. 5). GDPR-compliant interfaces achieved higher System Usability Scale (SUS) scores than manipulative cookie consent interfaces across both brands. A regression analysis showed a significant positive association between usability and consent level ($R^2 = 0.12$, $F(1, 198) = 28.49$, $p < .001$). This indicates a moderate association in the context of behavioral data. Within the limits of the analysis, this result should be interpreted as an observed association rather than a causal effect.

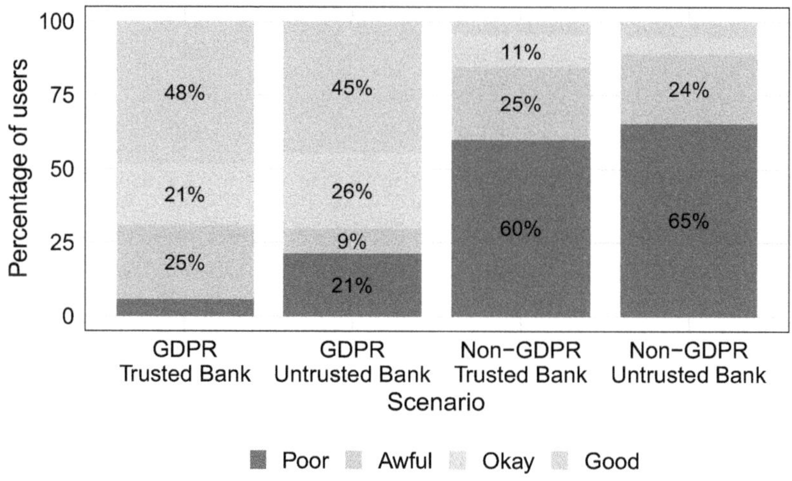

Fig. 5. Relative distribution of usability ratings by scenario.

4.3 Brand Trust and Consent

Brand trust scores differed across brands and interface conditions (see Fig. 6). Participants interacting with the high-trust brand (Trusted Bank) reported higher brand trust ratings than participants interacting with the low-trust brand (Untrusted Bank). Regression analysis showed that brand trust was significantly associated with consent level ($R^2 = 0.34$, $F(1, 228) = 140.9$, $p < .001$). This indicates a substantial proportion of variance explained for behavioral research contexts. Within the scope of this paper, this finding is interpreted as evidence of a strong association between trust and consent decisions rather than as a formal causal estimate.

4.4 Involvement and Consent

Levels of user involvement varied across brands and interface conditions (see Fig. 7). Regression analysis indicated a weak negative association between

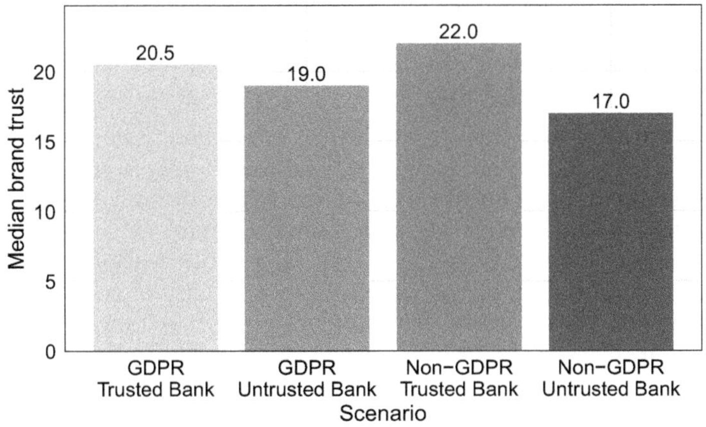

Fig. 6. Median brand trust ratings across conditions.

involvement and consent level ($R^2 = 0.02$, $F(1, 217) = 4.58$, $p = .034$), with higher involvement being associated with slightly lower consent decisions.

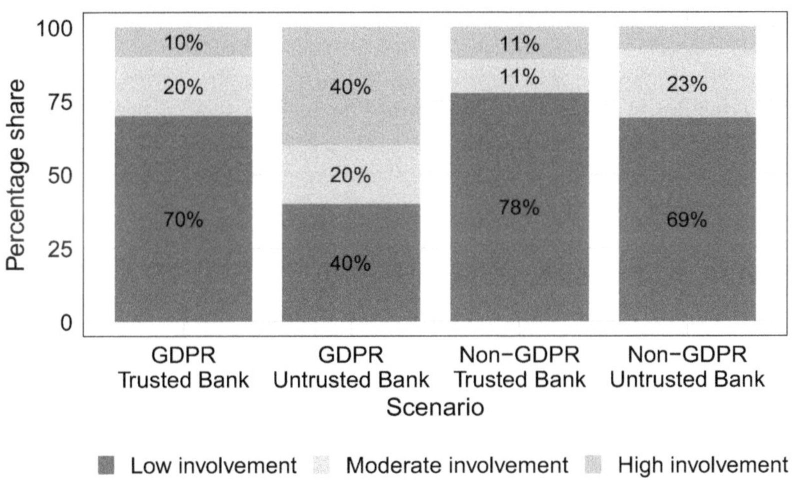

Fig. 7. Relative distribution of user involvement levels across scenarios.

4.5 Control Variables

Age and gender were included as control variables in the regression models. Neither variable showed a statistically significant association with consent level ($p > .05$).

4.6 Qualitative Findings

The qualitative analysis identified four recurring privacy personas based on participants' reported strategies when interacting with cookie consent interfaces:

- **Privacy Fundamentalists:** routinely reject tracking, use privacy tools (e.g., VPN, ad blockers), and invest time to find rejection options.
- **Pragmatists:** weigh benefits against privacy costs and sometimes accept tracking for convenience or access to desired content.
- **Unconcerned Users:** typically accept all cookies without much thought, often citing habituation or the perceived inevitability of tracking.
- **Trust-Driven Users:** make decisions primarily based on their trust in the organization; for trusted brands, they readily accept tracking, whereas for unknown brands, they are more cautious.

Participants' statements frequently referenced time pressure, perceived inevitability of tracking, and trust in the organization as factors influencing consent choices.

5 Discussion

Our results show that manipulative cookie consent designs are associated with higher consent levels than GDPR-compliant designs. Crucially, this pattern depends on brand trust: manipulative cookie consent interfaces appear to be particularly effective when used by a highly trusted brand.

This pattern can be interpreted as a *trust backfire effect*, where trust—typically considered a positive attribute—reduces users' scrutiny of privacy-relevant interface elements. In such contexts, users may rely on trust as a heuristic shortcut when making complex or time-constrained decisions rather than engaging in a detailed evaluation of consent options.

5.1 The Role of Usability and Trust

While GDPR-compliant interfaces achieved higher usability scores, this alone was not sufficient to prevent high consent in manipulative conditions. In our data, brand trust appeared to be more strongly associated with consent decisions than perceived usability, suggesting that social-psychological factors play a central role in privacy-related decision-making.

This finding relates to prior work on trust-based decision heuristics and the privacy paradox, which shows that users' stated privacy preferences can diverge from their actual behavior when convenience or trust cues are present [4,50].

5.2 Qualitative Interpretation

The qualitative personas provide additional explanatory depth for these findings. In particular, Trust-Driven Users frequently justified high-consent decisions by referring to their confidence in the organization, even when interface designs were perceived as inconvenient or opaque. This helps explain why high-trust brands benefit disproportionately from manipulative cookie consent designs.

5.3 Practical and Regulatory Implications

For practitioners, our findings highlight the ethical risk of leveraging trust to maximize data collection. Organizations with strong brand reputations, particularly in finance and health, should avoid manipulative designs in consent flows, as they may exploit users' trust in ways that undermine autonomy.

For regulators and standardization bodies, our study underscores the need to evaluate not only the formal compliance of cookie banners but also their contextual effects, including how trust and design interact [30,55]. Automated auditing tools for consent banners [31,32,57] should incorporate patterns that specifically capitalize on trust, such as familiar branding that softens or obscures aggressive tracking defaults.

6 Limitations

Our study has several limitations. First, the sample relied on convenience recruitment, with an overrepresentation of students and technology-savvy participants, which may limit generalizability to the broader population [66]. Second, we used interactive prototypes rather than live banking websites. This reduced ecological validity, but allowed precise control over the design factors of interest.

Third, our operationalization of brand trust and the difference between high- and low-trust brands reflect one particular national context and time period [24,67]. Other countries, industries, or cultural contexts might exhibit different trust profiles and responses to manipulative cookie designs.

Finally, the study combines in-person and remote data collection. While this increased the sample size and improved coverage, the two settings may also differ in attention, context, and task engagement. Future work could examine these settings separately and further investigate trust–interface relationships using formal interaction models.

7 Conclusion

This paper examined how manipulative cookie consent designs and brand trust jointly shape users' consent decisions under the GDPR. In a 2×2 experimental design with banking brands and GDPR-compliant versus manipulative interfaces, we found that manipulative designs were associated with higher consent and that trusted brands appeared to benefit disproportionately from these patterns.

These findings highlight a trust backfire effect in privacy-critical interfaces and underscore the need for guidelines and enforcement mechanisms that discourage the exploitation of user trust through manipulative cookie designs. Future work could extend our approach to other domains (e.g., health or e-commerce), explore long-term behavioral effects of different consent patterns, and develop design patterns that better reconcile organizational needs with user autonomy and data protection.

References

1. Lee, Y., Kozar, K.A., Larsen, K.R.: The technology acceptance model: past, present, and future. Commun. Assoc. Inf. Syst. **12**(1), 50 (2003)
2. Kumaraguru, P., Cranor, L.F.: Privacy indexes: a survey of Westin's studies (2005)
3. Westin, A.F.: Privacy and freedom. Washington Lee Law Rev. **25**(1), 166 (1968)
4. Norberg, P.A., Horne, D.R., Horne, D.A.: The privacy paradox: personal information disclosure intentions versus behaviors. J. Consum. Aff. **41**(1), 100–126 (2007)
5. Wein, T.: Data protection, cookie consent, and prices. Economies **10**(12), 307 (2022)
6. Sanchez-Rola, I., et al.: Can I opt out yet? GDPR and the global illusion of cookie control. In: Proceedings of the 2019 ACM Asia Conference on Computer and Communications Security, pp. 340–351 (2019)
7. Singh, A.K., Upadhyaya, N., Seth, A., Hu, X., Sastry, N., Mondal, M.: What cookie consent notices do users prefer: a study in the wild. In: Proceedings of the 2022 European Symposium on Usable Security, pp. 28–39 (2022)
8. Degeling, M., Utz, C., Lentzsch, C., Hosseini, H., Schaub, F., Holz, T.: We value your privacy... now take some cookies: measuring the GDPR's impact on web privacy. In: NDSS Symposium (2019)
9. Bermejo Fernandez, C., Chatzopoulos, D., Papadopoulos, D., Hui, P.: This website uses nudging: MTurk workers' behaviour on cookie consent notices. Proc. ACM Hum.-Comput. Interact. **5**(CSCW2), 1–22 (2021)
10. Graßl, P., Schraffenberger, H., Zuiderveen Borgesius, F., Buijzen, M.: Dark and bright patterns in cookie consent requests. J. Digit. Soc. Res. **3** (2021)
11. Soe, T.H., Nordberg, O.E., Guribye, F., Slavkovik, M.: Circumvention by design: dark patterns in cookie consent for online news outlets. In: Proceedings of the 11th Nordic Conference on Human-Computer Interaction: Shaping Experiences, Shaping Society, pp. 1–12 (2020)
12. Nouwens, M., Liccardi, I., Veale, M., Karger, D., Kagal, L.: Dark patterns after the GDPR: scraping consent pop-ups and demonstrating their influence. In: Proceedings of the 2020 CHI Conference on Human Factors in Computing Systems, pp. 1–13 (2020)
13. Kretschmer, M., Pennekamp, J., Wehrle, K.: Cookie banners and privacy policies: measuring the impact of the GDPR on the web. ACM Trans. Web (TWEB) **15**(4), 1–42 (2021)
14. Utz, C., Degeling, M., Fahl, S., Schaub, F., Holz, T.: (Un)informed consent: studying GDPR consent notices in the field. In: Proceedings of the 2019 ACM SIGSAC Conference on Computer and Communications Security, pp. 973–990 (2019)
15. Zaichkowsky, J.L.: Conceptualizing involvement. J. Advert. **15**(2), 4–34 (1986)
16. Delgado-Ballester, E., Munuera-Aleman, J.L., Yague-Guillen, M.J.: Development and validation of a brand trust scale. Int. J. Mark. Res. **45**(1), 35–54 (2003)
17. GDPR: General data protection regulation. Regulation (EU), vol. 679 (2016)
18. Brooke, J.: SUS–a quick and dirty usability scale. Usability Eval. Ind. **189**(194), 4–7 (1996)
19. Mathur, A., Kshirsagar, M., Mayer, J.: What makes a dark pattern... dark? Design attributes, normative considerations, and measurement methods. In: Proceedings of the 2021 CHI Conference on Human Factors in Computing Systems (2021)
20. Greenberg, S., Boring, S., Vermeulen, J., Dostal, J.: Dark patterns in proxemic interactions: a critical perspective. In: Proceedings of the 2014 Conference on Designing Interactive Systems, pp. 523–532 (2014)

21. Habib, H., Cranor, L. F.: Evaluating the usability of privacy choice mechanisms. In: Eighteenth Symposium on Usable Privacy and Security (SOUPS 2022), pp. 273–289 (2022)
22. Habib, H., Li, M., Young, E., Cranor, L.: "Okay, whatever": an evaluation of cookie consent interfaces. In: Proceedings of the 2022 CHI Conference on Human Factors in Computing Systems, pp. 1–27 (2022)
23. Alzaidi, M.S., Agag, G.: The role of trust and privacy concerns in using social media for e-retail services: the moderating role of COVID-19. J. Retail. Consum. Serv. **68**, 103042 (2022)
24. Lappeman, J., Marlie, S., Johnson, T., Poggenpoel, S.: Trust and digital privacy: willingness to disclose personal information to banking chatbot services. J. Finan. Serv. Mark. (2022)
25. Glaser, B.G., Strauss, A.L.: The Discovery of Grounded Theory: Strategies for Qualitative Research. Aldine, New York (1967)
26. Hevner, A.R., March, S.T., Park, J., Ram, S.: Design science in information systems research. MIS Q. **28**(1), 75–105 (2004)
27. Peffers, K., Tuunanen, T., Rothenberger, M., Chatterjee, S.: A design science research methodology for information systems research. J. Manag. Inf. Syst. **24**(3), 45–77 (2008)
28. Wilde, T., Hess, T.: Forschungsmethoden der Wirtschaftsinformatik. Wirtschaftsinformatik **49**(4), 280–287 (2007)
29. Yin, R.K.: Case Study Research: Design and Methods. Sage Publications, Thousand Oaks (2009)
30. Bielova, N., Litvine, L., Nguyen, A., Chammat, M., Toubiana, V., Hary, E.: The effect of design patterns on (present and future) cookie consent decisions. In: USENIX Security Symposium. USENIX Association. Accepted for publication (2024)
31. Demir, N., Urban, T., Pohlmann, N., Wressnegger, C.: A large-scale study of cookie banner interaction tools and their impact on users' privacy. Proc. Priv. Enhancing Technol. **1**, 1–16 (2024)
32. Zimmeck, S., Kuller, E., Ma, C., Tassone, B., Champeau, J.: Generalizable active privacy choice: designing a graphical user interface for global privacy control. Proc. Priv. Enhancing Technol. **1**, 258–279 (2024)
33. Gray, C.M., Santos, C., Bielova, N., Toth, M., Clifford, D.: Dark patterns and the legal requirements of consent banners: an interaction criticism perspective. In: Proceedings of the 2021 CHI Conference on Human Factors in Computing Systems (2021)
34. Toth, M., Bielova, N., Roca, V.: On dark patterns and manipulation of website publishers by CMPs. In: PETS 2022–22nd Privacy Enhancing Technologies Symposium (2022)
35. Martini, M., Drews, C., Seeliger, P., Weinzierl, Q.: Dark patterns: Phänomenologie und Antworten der Rechtsordnung. Zeitschrift für Digitalisierung und Recht **1**(2021), 47–74 (2021)
36. Gefen, D., Karahanna, E., Straub, D.W.: Inexperience and experience with online stores: the importance of TAM and trust. IEEE Trans. Eng. Manage. **50**(3), 307–321 (2003)
37. Kim, J.B.: An empirical study on consumer first purchase intention in online shopping: integrating initial trust and TAM. Electron. Commer. Res. **12**, 125–150 (2012)
38. Zaichkowsky, J.L.: Measuring the involvement construct. J. Consum. Res. **12**(3), 341–352 (1985)

39. Zaichkowsky, J.L.: The personal involvement inventory: reduction, revision, and application to advertising. J. Advert. **23**(4), 59–70 (1994)
40. Gurviez, P., Korchia, M.: Proposal for a multidimensional brand trust scale. In: 32nd EMAC Conference, pp. 438–452 (2003)
41. Sarodnick, F., Brau, H.: Methoden der Usability Evaluation: Wissenschaftliche Grundlagen und praktische Anwendung. Huber (2011)
42. Rousseau, D., Sitkin, S., Burt, R., Camerer, C.: Not so different after all: a cross-discipline view of trust. Acad. Manag. Rev. **23**(3), 393–404 (1998)
43. Hyun, H., Thavisay, T., Lee, S.H.: Enhancing the role of flow experience in social media usage and its impact on shopping. J. Retail. Consum. Serv. **65**, 102492 (2022)
44. Jaspers, E.D.T., Pearson, E.: Consumers' acceptance of domestic internet-of-things: the role of trust and privacy concerns. J. Bus. Res. **142**, 255–265 (2022)
45. Oghazi, P., Karlsson, S., Hellström, D., Hjort, K.: Online purchase return policy leniency and purchase decision: mediating role of consumer trust. J. Retail. Consum. Serv. **41**, 190–200 (2018)
46. Ayaburi, E.W., Treku, D.N.: Effect of penitence on social media trust and privacy concerns: the case of Facebook. Int. J. Inf. Manage. **50**, 171–181 (2020)
47. AlSuwaidi, M., Eid, R., Agag, G.: Understanding the link between CSR and employee green behaviour. J. Hosp. Tour. Manag. **46**, 50–61 (2021)
48. Bishop, L., Holmes, B.J., Kelley, C.M., et al.: National consumer health privacy survey 2005: executive summary (2005)
49. Li, H., Gupta, A., Zhang, J., Sarathy, R.: Examining the decision to use standalone personal health record systems as a trust-enabled fair social contract. Decis. Support Syst. **57**, 376–386 (2014)
50. Mayer, R.C., Davis, J.H., Schoorman, F.D.: An integrative model of organizational trust. Acad. Manag. Rev. **20**(3), 709–734 (1995)
51. Suh, B., Han, I.: Effect of trust on customer acceptance of Internet banking. Electron. Commer. Res. Appl. **1**(3–4), 247–263 (2002)
52. Davis, F.D.: Perceived usefulness, perceived ease of use, and user acceptance of information technology. MIS Q. **13**(3), 319–340 (1989)
53. European Parliament and Council of the European Union: Directive 2002/58/EC concerning the processing of personal data and the protection of privacy in the electronic communications sector (ePrivacy Directive) (2002)
54. EU GDPR: Principles relating to processing of personal data. https://gdpr-info.eu/art-5-gdpr/. Accessed 23 Mar 2024
55. Schade, F.: Dark sides of data transparency: organized immaturity after GDPR? Bus. Ethics Q. **33**(3), 473–501 (2023)
56. European Union: Regulation (EU) 2016/679 of the European Parliament and of the Council on the protection of natural persons with regard to the processing of personal data and on the free movement of such data (General Data Protection Regulation) (2016)
57. Matte, C., Bielova, N., Santos, C.: Do cookie banners respect my choice?: measuring legal compliance of banners from IAB Europe's transparency and consent framework. In: 2020 IEEE Symposium on Security and Privacy (SP), pp. 791–809 (2020)
58. Narayanan, A., Mathur, A., Chetty, M., Kshirsagar, M.: Dark patterns: past, present, and future: the evolution of tricky user interfaces. Queue **18**(2), 67–92 (2020)
59. Koch, S., Wessels, M., Altpeter, B., Olvermann, M., Johns, M.: Keeping privacy labels honest. Proc. Priv. Enhancing Technol. (2022)

60. Zimmeck, S., Wang, O., Alicki, K., Wang, J., Eng, S.: Usability and enforceability of global privacy control. Proc. Priv. Enhancing Technol. (2023)
61. Nielsen, J.: Usability Engineering. Morgan Kaufmann (1994)
62. Dumas, J.S., Redish, J.: A Practical Guide to Usability Testing. Intellect (1999)
63. Gediga, G., Hamborg, K.: Evaluation in der Software-Ergonomie. Zeitschrift für Psychologie **210**, 40–57 (2002)
64. Faulkner, L.: Beyond the five-user assumption: benefits of increased sample sizes in usability testing. Behav. Res. Methods Instrum. Comput. **35**, 379–383 (2003)
65. Kuckartz, U.: Qualitative Inhaltsanalyse: Methoden. Beltz Juventa (2022)
66. (DeStatis), Federal Statistical Office of Germany: Bevölkerung nach Altersgruppen 2011 bis 2023. https://www.destatis.de/DE/Themen/Gesellschaft-Umwelt/Bevoelkerung/Bevoelkerungsstand/Tabellen/bevoelkerung-altersgruppen-deutschland-absulut-basis-2022.html. Accessed 07 Apr 2025
67. Readers Digest Deutschland: Trusted Brand Studie 2022: Marken sind Vertrauensanker in unsicheren Zeiten. https://www.readersdigest.de/blogs/markengut/trusted-brand-studie-2022-marken-sind-vertrauensanker-in-unsicheren-zeiten. Accessed 25 Mar 2024
68. Klug, B.: An overview of the system usability scale in library website and system usability testing. Weave J. Libr. User Exp. **1** (2017)
69. QuantUX. https://quant-ux.com/
70. Mejtoft, T., Vejbrink Starbrink, N., Roos Morales, C., Norberg, O., Andersson, M., Söderström, U.: Cookies and trust: trust in organizations and the design of cookie consent prompts. In: Proceedings of the European Conference on Cognitive Ergonomics 2023, pp. 1–6 (2023)
71. Gray, C.M., Kou, Y., Battles, B., Hoggatt, J., Toombs, A.L.: The dark (patterns) side of UX design. In: Proceedings of the 2018 CHI Conference on Human Factors in Computing Systems, pp. 1–14 (2018)

Cybersecurity Guidance for Smart Homes: A Cross-National Review of Government Sources

Victor Jüttner(✉) and Erik Buchmann

Center for Scalable Data Analytics and Artificial Intelligence (ScaDS.AI) Dresden/Leipzig, Leipzig, Germany
{victor.juettner,erik.buchmann}@uni-leipzig.de

Abstract. Smart homes are increasingly targeted by cyberattacks, yet residents often lack guidance when incidents occur. Since affected residents are likely to seek help from trustworthy sources, this paper asks: What actionable cybersecurity guidance do governments provide to smart home users whose systems have been compromised? To answer this question, we conduct an exploratory, user-centered review of governmental cybersecurity guidance for smart homes across eleven countries to identify and characterize the types of guidance governments provide and to systematize their content. Using a standardized search and screening process, we derive three emergent clusters: incident reporting, general security recommendations, and incident response. Our findings show that governments provide abundant general security advice and accessible reporting channels, but structured incident response guidance tailored to smart homes is rare. Only two sources offer step-by-step recovery guidance for non-expert users, highlighting a gap between preventive advice and post-incident support.

Keywords: Smart Home · Cybersecurity Advice · Incident Response · Government Policy · Usable Security

1 Introduction

Home networks experience on average 8 attacks against connected devices per day, based on telemetry from 2.6 million smart homes worldwide [11]. Yet residents often lack clear, actionable guidance on how to respond once a compromise has occurred. Domestic smart homes are typically operated by non-expert users who must make security decisions under uncertainty, time pressure, and limited technical insight. When incidents occur, affected residents are therefore likely to seek external guidance to understand what actions to take and how to recover their systems.

Governmental agencies represent an important source of such guidance. Official government resources are non-commercial and likely to be perceived as trustworthy, and they are frequently referenced in public cybersecurity awareness

© The Author(s), under exclusive license to Springer Nature Switzerland AG 2027
K. Renaud et al. (Eds.): EICC 2026, CCIS 3012, pp. 38–54, 2027.
https://doi.org/10.1007/978-3-032-28957-5_3

efforts such as Canada's *Get Cyber Safe* [25], France's *cybermalveillance* [9], Japan's *NOTICE* [36], or Australia's *Protect Yourself* [4]. At the same time, it remains unclear what actionable guidance governments actually provide to smart home users following a cybersecurity incident, how this guidance is structured, and whether it supports users beyond general preventive advice.

What actionable cybersecurity guidance do governments provide to smart home users whose systems have been compromised? This paper addresses this gap in the field of incident response by empirically examining governmental cybersecurity guidance for smart home users. We adopt the perspective of a smart home resident seeking help after an incident and focus on official governmental online resources that such users are likely to encounter.

To answer this question, we conduct an exploratory, user-centered review of governmental cybersecurity resources across eleven countries. Our goal is to identify and characterize the types of guidance governments provide and to systematize their content. To this end, we use a standardized search approach that reflects how affected residents might seek help to identify relevant official sources. All identified materials are manually screened and inductively clustered based on their primary type of guidance. This process leads to three categories: incident reporting, describing how users can notify authorities or relevant agencies; general security recommendations, focusing on preventive and hardening practices; and incident response, providing guidance on recovery after a compromise. We then systematize and analyze the content of each cluster.

Our analysis shows that governmental guidance for smart home users is both abundant and unevenly distributed. Across all countries, we find accessible channels for incident reporting and a wide range of general security recommendations, with substantial cross-agency agreement on baseline practices such as securing routers, updating devices, and strengthening authentication. In contrast, structured incident response guidance tailored to smart home environments is rare. Only two sources in our dataset provide step-by-step recovery guidance aimed at non-expert users.

By systematizing existing governmental recommendations and identifying where guidance is lacking, this paper contributes an empirical overview of official smart home cybersecurity advice and highlights the need for user-centered incident response frameworks for domestic environments.

2 Related Work

Related work addresses smart-home security constraints, online cybersecurity advice for end users, and government guidance for consumer IoT security.

2.1 Smart Home Security in Practice

Smart-home security is shaped by home-specific constraints such as limited expertise, time, and tooling. Nthala and Flechais argue that home network security must be conceptualized around households' lived realities; consequently, guidance must be feasible for non-experts and robust to uncertainty [37]. Ye et al. show that home routers ship with heterogeneous and sometimes risky default settings, and that resulting exposures can persist because users rarely revisit configuration unless prompted by an incident [48]. These constraints show that smart home security needs clear, low-effort guidance that residents can execute under stress [28].

2.2 Online Cybersecurity Advice for End Users

Redmiles et al. show that end users learn security practices from heterogeneous sources, and that advice sources are associated with whether recommended behaviors are adopted [43]. The quality of sources also varies substantially: Redmiles et al. also highlight the differences of clarity and actionability of security advice on the web and show that end-users are left on their own to prioritize which actions they need to take [44]. Quinlan et al. have shown an increase in vocabulary complexity and reading difficulty of security advice, which limits their practicability, especially for vulnerable users [40]. Focusing on smart-home contexts, Turner et al. show that searching the web yields a broad mix of sources with differing credibility and device-specific relevance, making it difficult for users to identify appropriate guidance when issues arise [47]. Taken together, the online security advice is fragmented: guidance varies in quality and specificity, and users are frequently left to judge credibility and relevance and to translate generic recommendations into context-appropriate actions.

2.3 Government Guidance for Smart Home Security

Governments bodies promote baseline consumer IoT security via awareness campaigns, public guidance portals, regulation, and labeling. Analyses of government-led cybersecurity awareness materials suggest that many initiatives rely heavily on education and assume that knowledge translates into behavior change [46]; ENISA likewise highlights the centrality of awareness in national strategies while noting persistent challenges in targeting and impact [23]. Rama and Keevy compare good practices across government-led cybersecurity awareness websites, illustrating variation in topic coverage and presentation [41]. These initiatives provide smart security hygiene and inform purchasing.

2.4 Research Gap

Prior work shows that smart home security advice is widely available online, but varies substantially in clarity and actionability, often leaving users to judge relevance and translate generic recommendations into context-appropriate actions [40,43,44,47]. Governments represent a particularly important source of such advice, as their portals are likely to be perceived as trustworthy and commonly promote baseline consumer IoT security practices through public awareness and guidance materials [23,41,46].

Despite this, existing literature provides limited empirical insight into the content and structure of official governmental guidance for smart home users. In particular, it remains unclear what types of actionable guidance governments make available to users following a cybersecurity incident and how this guidance is organized across different contexts. Addressing this gap requires a systematic examination of governmental online resources beyond individual campaigns or isolated examples.

3 Review Procedure

Our review procedure consists of four stages: purposive country selection, a standardized user-centered search for official sources, manual screening for relevance, and qualitative clustering of the identified sources (Fig. 1).

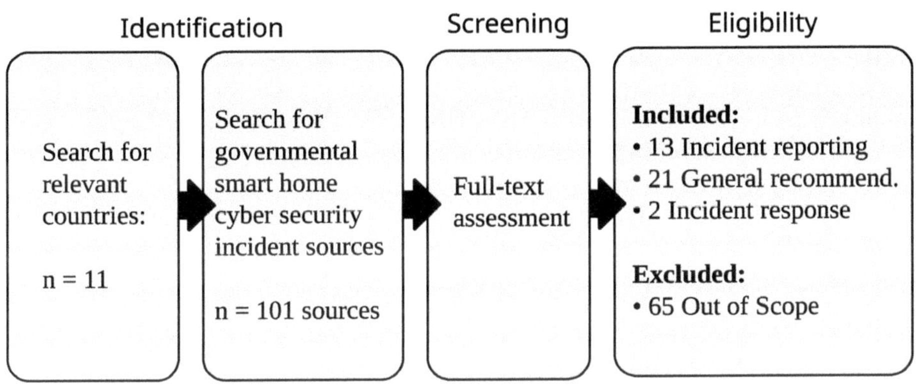

Fig. 1. Data Collection

3.1 Country Selection

We employed purposive sampling to select countries for an exploratory, cross-national review of governmental smart home cybersecurity guidance. To balance cross-national coverage with feasibility, we targeted a sample size of approximately 10–13 countries. We considered countries eligible for inclusion if national authorities provided publicly accessible cybersecurity guidance for private individuals. We identified candidate countries through an exploratory search targeting official government sources. To avoid restricting the analysis to a single market or regional context, we further aimed to include countries that varied in consumer smart home adoption levels and geographic location.

We assessed differences in consumer smart home adoption using available international statistics and national surveys [21,38]. These data confirmed that the selected countries span different levels of market maturity rather than uniformly high adoption. We achieved geographic diversity by including countries from North America, Europe, and the Asia–Pacific region, enabling comparison across different regulatory and governance contexts. Based on these criteria, we included eleven countries in the analysis: Australia, Austria, Canada, Finland, France, Germany, Japan, New Zealand, Singapore, the United Kingdom, and the United States.

3.2 Data Collection

For each selected country, we conducted a standardized, user-centered exploratory search. We used OpenAI's ChatGPT (gpt-5.2-2025-12-11) via the WebUI as a search engine. For each country, we issued the prompt *"My smart home devices have been hacked. I live in {COUNTRY} in english. Could you please provide me with official sources that offer advice on what I should do now?"*. This prompt reflects how a smart home resident might seek advice after an incident.

From ChatGPT's responses, we collected only the referenced sources; we did not analyze the generated text itself. To ensure consistency across countries, we started a new chat session for each search. When ChatGPT directed us to a page that did not contain guidance but linked to another page that did, we included the linked page. We translated sources that were not available in English using the built-in translation function of the Firefox browser.

We initiated the search process on 29 December 2025. Across the eleven countries, we identified a total of 101 distinct websites. These websites originated from 49 distinct governmental institutions, including national cybersecurity agencies, consumer protection authorities, and police.

3.3 Source Screening

We manually verified all identified sources to ensure relevance and consistency with the scope of this study. We included sources only if they met all of the following criteria:

1. The source originates from an official national authority.
2. The source explicitly targets private individuals or households.
3. The source offers actionable guidance, such as concrete steps for prevention, mitigation, or incident reporting.
4. The source addresses smart homes, IoT in domestic settings, or home network security.

We excluded sources that did not meet all criteria from further analysis.

3.4 Clustering

After screening, we analyzed the sources inductively to identify recurring patterns in their primary purpose and type of actionable guidance. We treated each source as a unit of analysis and assigned it to a cluster based on its dominant user-facing function (e.g., reporting, preventive guidance, or recovery support).

We refined clusters iteratively as we reviewed additional sources, until no new categories emerged. After clustering, we analyzed each cluster using a qualitative content analysis approach tailored to the content type and scope of the respective cluster.

4 Systematization

After screening, we included 35 sources in the final dataset (Table 1) and excluded the remaining sources as out of scope. We qualitatively analyzed the included sources and grouped them into three clusters based on their primary type of guidance: incident reporting, general security recommendations, and incident response.

Table 1. Distribution of guidance clusters by country

Country	Reporting	Recommendations	Response
Australia	2	4	0
Austria	1	2	0
Canada	2	3	0
Finland	2	1	0
France	0	2	1
Germany	0	1	0
Japan	0	1	0
New Zealand	2	1	0
Singapore	1	1	1
UK	1	1	0
USA	1	4	0
Total	12	21	2

We then analyzed each cluster using a qualitative content analysis approach tailored to its scope. For incident reporting sources, we examined the reporting channels offered (e.g., online forms, telephone hotlines, and email) and how sources framed incidents. For incident response sources, we analyzed the presence and structure of recovery guidance. For general security recommendation sources, we synthesized actionable steps mentioned across multiple sources and iteratively grouped them by their primary target, such as home routers, smart devices, online accounts, or general practices.

4.1 Incident Reporting

In our dataset, we identified 13 incident-reporting websites spanning 9 countries and operated by 11 distinct agencies. Across this incident-reporting cluster, agencies consistently offer a limited set of recurring reporting channels (Table 2).

These include guided online reporting tools, dedicated telephone hotlines, and email-based reporting. In addition, several agencies provide referral pages that help users identify and navigate to the most appropriate reporting destination.

Across the incident-reporting sources, reporting is generally framed broadly as reporting cybercrime or cybersecurity incidents affecting individuals and small businesses. The sources accept reports of common compromise scenarios (e.g., hacking/account compromise, scams/fraud, malicious software). We did not find any smart home or home network specific incident reporting website.

Table 2. Incident reporting channels and which public agencies provide them.

Reporting channel	ACSC (AU) [5,6]	BKA (AT) [12]	CCCS (CA) [15]	RCMP (CA) [42]	NCSC-FI(FI) [32,33]	MBIE (NZ) [30]	NCSC-NZ (NZ) [34]	CSA (SG) [20]	CoLP (UK) [19]	IC3 (US) [26]	Total
Online reporting tool	2x		x		2x		x			x	7
Telephone reporting	2x		x		2x		x				4
Email reporting		x				x	x				3
Referral page	x		x	x				x	x		5

4.2 Incident Response

Two sources went beyond general cybersecurity recommendations by providing explicit guidance on recovery following a compromise: GIP ACYMA from France [10] and the CSA in Singapore [22]. As only two sources in our dataset offer incident response guidance specifically for smart home environments, we do not systematize them further but instead present them individually.

GIP ACYMA provides a 12-step recovery plan that is structurally similar to industry incident response playbooks, but is tailored to non-expert users. It offers step-by-step instructions for mitigating an ongoing attack and re-securing the affected device or network. The guidance is intentionally generic and applies to a broad range of consumer systems, including computers, tablets, mobile phones, connected objects, servers, and home networks.

The CSA, by contrast, presents an extensive set of general cybersecurity recommendations alongside three recovery steps specifically targeted at smart home devices: *Disconnect the Device from the Internet*, *Change Credentials and/or Perform a Factory Reset*, and *Contact the Manufacturer for Assistance*.

4.3 General Recommendations

For each of the eleven countries we could find at least one source that provided general cyber security recommendations to use smart home devices, in total we could identify 21 sources providing 46 distinct recommendations, provided by 17 government agencies Table 4. For an explanation of all recommendations see Table 5. Our results show that recommendations are device specific, they are aimed at the router, smart device, online-account, or general recommendations. Most recommendation we found are either aimed at router, smart device or both. All recommendations in Table 3 were referenced at least twice among all sources. Recommendations that were only mentioned once are reported for completeness in Table 6 in the Appendix.

Several recommendations show broad cross-agency consensus, suggesting a widely accepted "baseline" for consumer cybersecurity. Across device categories, guidance most consistently emphasizes strengthening authentication (e.g., changing default credentials and using strong passwords), keeping software updated (including enabling automatic updates where available), and reducing attack surface and lateral movement risk (e.g., disabling unnecessary features and isolating smart devices via guest networks/VLANs). In contrast, a long tail of recommendations appears in only a small number of sources, indicating fragmentation or context-specific emphasis (e.g., monitoring internet usage, rebooting regularly, or advice about specific interfaces/communications).

Table 3. General recommendations by public agencies.

Recommendation	ACSC (AU) [3,4,7,8]	A-SIT (AT) [1,2]	CCCS (CA) [14]	GCS (CA) [25]	OPC (CA) [39]	NCSC-FI (FI) [31]	GIP ACYMA (FR) [9]	MEFSIN (FR) [29]	BSI (DE) [13]	NOTICE (JP) [36]	CERT NZ (NZ) [16]	SPF (SG) [45]	NCSC-UK (UK) [35]	CISA (US) [17,18]	FTC (US) [24]	IC3 (US) [27]	Total
Router																	
Change admin credentials	3x	x			x	x		x	x			x		x	x		11
Change SSID+Wi-Fi passw.	4x	x		x	x	x	x	x						x	x		12
Update regularly	2x	x				x								x	x	x	7
Enable automatic updates		2x		x	x	x			x		x		x				8
Disable remote management	2x		x			x			x					x			6
Replace after EOL	2x									x							3
Enable firewall						x			x					x			3
Use guest Wi-Fi	3x	x	x	x	x	x	x	x	x		x	x					13
Use WPA2/WPA3	3x	x			x	x	x					x		x	x		10
Check connected devices	x													x			2
Monitor internet usage	3x															x	4
Access network over VPN		x							x								2

(*continued*)

Table 3. (*continued*)

Recommendation	ACSC (AU) [3,4,7,8]	A-SIT (AT) [1,2]	CCCS (CA) [14]	GCS (CA) [25]	OPC (CA) [39]	NCSC-FI (FI) [31]	GIP ACYMA (FR) [9]	MEFSIN (FR) [29]	BSI (DE) [13]	NOTICE (JP) [36]	CERT NZ (NZ) [16]	SPF (SG) [45]	NCSC-UK (UK) [35]	CISA (US) [17,18]	FTC (US) [24]	IC3 (US) [27]	Total
Smart device																	
Change default credentials	3x	x	x	x	x		x		x	x	x	x	x	x	x		15
Update regularly	3x	x	x	x	x		x	x	x	x	x	x	x	2x	x	x	18
Enable automatic updates		2x		x	x	x			x		x		x				8
Disable remote management	x	x			x				x				x				5
Replace after EOL									x	x	x						3
Check reviews / ratings	3x		x		x		x	x	x		x	x	x				11
Avoid unnecessary internet	3x				x				x		x				x		7
Disable unused features	2x	x		x	x					x		x		x	x		9
Turn off when not in use	3x	x		x	x		x										7
Reboot regularly	3x																3
Secure physical access	3x								x								4
Encrypted communication		x	x						x								3
Avoid insecure interfaces					x									2x	x	x	5
Online																	
Limit social media sharing		x		x													2
Enable MFA	x	x	x						x			x	x		x		7
Protect digital identity		x			x		x								x		4
General																	
Use strong passwords	x	x	x	x	x		x	x	x	x		x	x	2x			13
Use a password manager	x	x							x								3

5 Discussion

Our findings reveal a clear imbalance in the type of cybersecurity guidance governments provide to smart home users. While general security recommendations and incident reporting mechanisms are widely available and relatively consistent across countries, structured incident response guidance tailored to smart home environments remains rare.

5.1 Implications

I1: General Recommendations. The abundance of general recommendations indicates an emerging consensus on baseline smart home security practices, such as changing default credentials, keeping devices updated, and segmenting home networks. However, frequent mention of certain practices should not be interpreted as evidence that these measures are sufficient, optimal, or appropriate in all incident scenarios.

I2: Incident Response. The lack of structured incident response guidance is perhaps unsurprising. Unlike organizational IT environments, smart homes lack established response frameworks or reliable ways to assess the security status. Without such foundations, it is difficult for authorities to prescribe recovery steps that are both actionable and verifiable for non-expert users. As a result, most guidance stops short of helping users understand when an incident has been resolved.

I3: Incident Reporting. At the same time, the ease of accessing reporting channels across countries is a positive finding. Governments have largely succeeded in making incident reporting visible and accessible, even if reporting is framed broadly as cybercrime rather than smart home–specific compromise.

5.2 Design and Practice Recommenations

R1: Step-by-Step Workflows. Existing recommendations should be reorganized into structured, step-by-step workflows that users can follow under stress. Rather than presenting long lists of preventive measures, guidance could be ordered into phases (e.g., containment, remediation, hardening), allowing users to progress incrementally from immediate response to longer-term security improvements.

R2: Validation Mechanisms. Second, current guidance lacks mechanisms for validation. Users are rarely given ways to assess whether their smart home is secure again after taking recommended actions. Designing lightweight validation cues, such as checks for unknown devices, confirmation of update status, or indicators of restored normal behavior could significantly improve user confidence and reduce premature termination of recovery efforts.

For practitioners and policymakers, this suggests that improving smart home incident response does not necessarily require inventing new recommendations, but rather systematizing existing advice and presenting it in a form that supports decision-making and reassurance.

5.3 Limitations

This study is exploratory and subject to several limitations. Most notably, source identification relied on a ChatGPT-assisted, web-based search strategy rather than a predefined registry of governmental resources. As a result, it is possible that relevant agencies, countries, or guidance documents were not identified. The country sample was purposive rather than exhaustive, and the findings do not claim global representativeness.

In addition, the analysis focused on the presence and structure of guidance rather than its real-world effectiveness. We did not evaluate whether following the identified recommendations would successfully resolve incidents, nor how users interpret or apply this guidance in practice.

6 Conclusion

Smart home residents increasingly face cybersecurity incidents but often lack clear, actionable guidance tailored to domestic environments. In this exploratory review, we examined governmental cybersecurity resources across eleven countries to assess what official guidance is available to users whose smart homes have been compromised. While governments widely provide general security recommendations and accessible incident reporting channels, structured incident response guidance specific to smart home contexts remains rare.

By systematically characterizing and synthesizing existing governmental guidance, this work highlights a gap between preventive advice and post-incident support for non-expert users. Rather than proposing new response frameworks, our findings provide an empirical foundation for future efforts aimed at translating existing advice into clearer, more structured recovery guidance. Strengthening post-incident support has the potential to better equip smart home residents to navigate cybersecurity incidents and assess recovery in domestic settings.

Acknowledgments. The authors acknowledge the financial support by the Federal Ministry of Research, Technology and Space of Germany and by Sächsische Staatsministerium für Wissenschaft, Kultur und Tourismus in the programme Center of Excellence for AI-research "Center for Scalable Data Analytics and Artificial Intelligence Dresden/Leipzig", project identification number: ScaDS.AI.

Disclosure of Interests. The authors have no competing interests to declare that are relevant to the content of this article.

A Appendix

Table 4. Public agencies included in the dataset.

Country	Agency	Abbrev.
AU	Australian Cyber Security Centre	ACSC
AT	Austrian Secure Information Technology Centre	A-SIT
AT	Federal Criminal Police Office	BKA
CA	Canadian Centre for Cyber Security	CCCS
CA	National public awareness campaign "Get Cyber Safe"	GCS
CA	Office of the Privacy Commissioner of Canada	OPC
CA	Royal Canadian Mounted Police	RCMP
FI	National Cyber Security Centre	NCSC-FI
FR	Public Action Interest Grouping against Cybermalware	GIP ACYMA
FR	Ministry of Economy and Finance	MEFSIN
DE	Federal Office for Information Security	BSI
JP	NOTICE project (botnet mitigation initiative)	NOTICE
NZ	Ministry of Business, Innovation and Employment	MBIE
NZ	National Cyber Security Centre	NCSC-NZ
NZ	New Zealand Computer Emergency Response Team	CERT NZ
SG	Cyber Security Agency of Singapore	CSA
SG	Singapore Police Force	SPF
UK	City of London Police	CoLP
UK	National Cyber Security Centre	NCSC-UK
US	Cybersecurity and Infrastructure Security Agency	CISA
US	Federal Trade Commission	FTC
US	Internet Crime Complaint Center (FBI)	IC3

Table 5. Explanations of recommendation labels.

Recommendation label	Explanation
Replace after EOL	Replace routers or smart devices once vendors no longer provide firmware or security updates.
Disable remote management	Disable administrative access to routers or devices from outside the local network (e.g., via the public Internet).
Enable automatic updates	Configure devices or routers to automatically install security and firmware updates.
Use guest Wi-Fi	Place smart home devices on a logically separate network (e.g., guest network or VLAN) to reduce exposure to other household devices.
Use WPA2/WPA3	Configure the Wi-Fi network to use modern encryption standards (WPA2 or WPA3).
Access network over VPN	Use a virtual private network to securely access the home network remotely rather than exposing management interfaces to the Internet.
Check connected devices	Review the list of devices connected to the home router to identify unknown connections.
Monitor internet usage	Observe network traffic (e.g. data volume) that could indicate compromise or misuse.
Avoid unnecessary internet	Limit or disable Internet connectivity for smart devices when external connectivity is not required for their core functionality.
Disable unused features	Turn off device capabilities (e.g., microphones, cameras, Bluetooth, remote access) that are not actively needed, thereby reducing attack surface.
Encrypted communication	Ensure that communication between devices, apps, and cloud services is protected using encryption rather than transmitted in plaintext.
Avoid insecure interfaces	Secure or disable device interfaces and protocols known to lack sufficient security guarantees (e.g., unencrypted web interfaces or legacy protocols).
Protect digital identity	Safeguard personal online accounts and identity information associated with smart home services, including account recovery mechanisms.
Use a password manager	Use dedicated software to generate and store unique, strong passwords across devices and online accounts.
Check reviews / ratings	Evaluate third-party assessments and user reviews before purchasing smart home devices.
Limit social media sharing	Minimize public disclosure of smart home usage, device types, or household routines on social media platforms.

Table 6. Cybersecurity recommendations mentioned by exactly one governmental agency.

Agency	Recommendation
ACSC (AU) [7]	Turn off router when on holiday
A-SIT (AT) [2]	Use antivirus software
	Child safety
	Manage access of smartphone apps to phone
	Use VPN to access Internet
A-SIT (AT) [1]	Use separate router for IoT
	Adjust privacy settings in smart device app
CCCS (CA) [14]	Monitor, detect, and correct security issues
OPC (CA) [39]	Advise guests about smart devices
NCSC-FI (FI) [31]	Reboot router regularly
	Check visibility of your IP from the Internet
CISA (US) [17]	Disable WPS on the router
	Reduce Wi-Fi signal strength
IC3 (US) [27]	Watch out for suspicious behavior
	Use only trusted apps
GIP ACYMA (FR) [9]	Keep phone apps updated

References

1. A-SIT: Sicherer einsatz von smart home systemen. https://www.onlinesicherheit.gv.at/Services/Technologie-Schwerpunkte/Smart-Home/Sicherer-Einsatz-von-Smart-Home-Systemen.html. Accessed 03 Jan 2026
2. A-SIT: Sicherheit im smart home: So schützen sie ihre geräte. https://www.onlinesicherheit.gv.at/Services/News/Sicherheit-im-Smart-Home.html. Accessed 03 Jan 2026
3. ACSC: Internet things devices. https://www.cyber.gov.au/protect-yourself/securing-your-devices/how-secure-your-devices/internet-things-devices. Accessed 03 Jan 2026
4. ACSC: Personal cyber security: Advanced steps guide. https://www.cyber.gov.au/protect-yourself/resources-protect-yourself/personal-cyber-security-guides/personal-cybersecurity-advanced-steps. Accessed 03 Jan 2026
5. ACSC: Report a cybercrime, incident or vulnerability. https://www.cyber.gov.au/report-and-recover/report. Accessed 11 Jan 2026
6. ACSC: Report and recover from hacking. https://www.cyber.gov.au/report-and-recover/recover-from/hacking. Accessed 11 Jan 2026
7. ACSC: Secure your Wi-Fi and router. https://www.cyber.gov.au/protect-yourself/staying-secure-online/secure-your-wifi-and-router. Accessed 03 Jan 2026
8. ACSC: Tips to secure your internet of things device (2020). https://www.cyber.gov.au/sites/default/files/2023-03/Tips%20to%20secure%20your%20Internet%20of%20Things%20device%20%28AUG%202020%29.pdf. Accessed 03 Jan 2026

9. Acyma, G.: La sécurité des objets connectés (IoT). https://www.cybermalveillance.gouv.fr/tous-nos-contenus/bonnes-pratiques/securite-objets-connectes-iot. Accessed 03 Jan 2026
10. Acyma, G.: Piratage d'un système informatique de particulier, que faire? https://www.cybermalveillance.gouv.fr/tous-nos-contenus/fiches-reflexes/piratage-systeme-informatique-particuliers. Accessed 03 Jan 2026
11. Bitdefender: The 2023 IoT security landscape report. Technical report, Bitdefender (2023). https://www.bitdefender.com/files/News/CaseStudies/study/429/2023-IoT-Security-Landscape-Report.pdf. Accessed 23 Mar 2026
12. BKA-AT: Delikte & Ermittlungen. https://www.bundeskriminalamt.at/306/start.aspx. Accessed 11 Jan 2026
13. BSI: Smarthome – den wohnraum sicher vernetzen. https://www.bsi.bund.de/DE/Themen/Verbraucherinnen-und-Verbraucher/Informationen-und-Empfehlungen/Internet-der-Dinge-Smart-leben/Smart-Home/smart-home_node.html. Accessed 03 Jan 2026
14. CCCS: Internet of things (IoT) security - itsap.00.012. https://www.cyber.gc.ca/en/guidance/internet-things-iot-security-itsap00012. Accessed 03 Jan 2026
15. CCCS: Report a cyber incident: Individuals. https://www.cyber.gc.ca/en/incident-management/report-cyber-incident-individuals. Accessed 11 Jan 2026
16. CERT NZ: Use the internet of things securely. https://www.ownyouronline.govt.nz/personal/get-protected/guides/using-the-internet-of-things/. Accessed 03 Jan 2026
17. CISA: Home network security. https://www.cisa.gov/news-events/news/home-network-security. Accessed 03 Jan 2026
18. CISA: Securing the internet of things (IoT). https://www.cisa.gov/news-events/news/securing-internet-things-iot. Accessed 03 Jan 2026
19. City of London Police (UK): Guide to reporting cyber crime & fraud (Report Fraud). https://www.reportfraud.police.uk/guide-to-reporting/. Accessed 11 Jan 2026
20. CSA: Cyber Aid. https://www.csa.gov.sg/resources/singcert/cyber-aid/. Accessed 11 Jan 2026
21. CSA: Cybersecurity awareness survey 2022 (2023)
22. CSA: Protecting your IoT devices (2024). https://www.csa.gov.sg/alerts-and-advisories/advisories/ad-2024-012. Accessed 15 Sept 2025
23. ENISA: Raising awareness of cybersecurity: A key element of national cybersecurity strategies (2021). https://ega.ee/wp-content/uploads/2021/12/ENISA-Report-Raising-awareness-a-key-element-of-national-cybersecurity-strategies.pdf
24. FTC: Securing your internet connected devices home. https://consumer.ftc.gov/articles/securing-your-internet-connected-devices-home. Accessed 03 Jan 2026
25. Get Cyber Safe: TVS and smart devices. https://www.getcybersafe.gc.ca/en/secure-your-devices/tvs-and-smart-devices. Accessed 03 Jan 2026
26. IC3: Complaint Form. https://complaint.ic3.gov/. Accessed 11 Jan 2026
27. IC3: Home internet connected devices facilitate criminal activity. https://www.ic3.gov/PSA/2025/PSA250605. Accessed 03 Jan 2026
28. Jüttner, V., Löffler, C.S., Buchmann, E.: Actionable cybersecurity notifications for smart homes: a user study on the role of length and complexity. In: Proceedings of the Future Technologies Conference (FTC) 2025, vol. 2, pp. 298–319. Springer, Cham (2026)
29. MEFSIN: Objets connectés : les risques à connaître. https://www.economie.gouv.fr/dgccrf/les-fiches-pratiques/objets-connectes-les-risques-connaitre. Accessed 03 Jan 2026

30. Ministry of Business, Innovation and Employment (MBIE), New Zealand: If you've been hacked. https://www.business.govt.nz/operations/it-risks-and-scams/if-youve-been-hacked. Accessed 11 Jan 2026
31. NCSC-FI: Home network and router security. https://www.kyberturvallisuuskeskus.fi/en/ncsc-news/instructions-and-guides/home-network-and-router-security. Accessed 03 Jan 2026
32. NCSC-FI: Report an information security incident. https://www.kyberturvallisuuskeskus.fi/en/report. Accessed 11 Jan 2026
33. NCSC-FI: What to do if you detect an information security incident. https://www.kyberturvallisuuskeskus.fi/en/ncsc-news/instructions-and-guides/what-do-if-you-detect-information-security-incident. Accessed 11 Jan 2026
34. NCSC-NZ: Pūrongotia (Report it). https://www.ncsc.govt.nz/report/. Accessed 11 Jan 2026
35. NCSC-UK: Smart devices: using them safely in your home. https://www.ncsc.gov.uk/guidance/smart-devices-in-the-home. Accessed 03 Jan 2026
36. NOTICE: Notice | let's protect IoT devices. https://notice.go.jp/en. Accessed 03 Jan 2026
37. Nthala, N., Flechais, I.: Rethinking home network security. In: Proceedings of the 2018 European Workshop on Usable Security (EuroUSEC 2018). Internet Society (2018). https://ora.ox.ac.uk/objects/uuid:44fd30ab-a51e-4e9f-a7aa-9947950f2110
38. OECD: Measuring the internet of things (2023). https://doi.org/10.1787/021333b7-en
39. OPC: Smart devices and your privacy. https://www.priv.gc.ca/en/privacy-topics/technology/02_05_d_72_iot/. Accessed 03 Jan 2026
40. Quinlan, M., Ceross, A., Simpson, A.: The efficacy potential of cyber security advice as presented in news articles. Interact. Comput. **37**(1), 30–48 (2025). https://doi.org/10.1093/iwc/iwae048
41. Rama, P., Keevy, M.: Public cybersecurity awareness good practices on government-led websites. Int. J. Res. Bus. Soc. Sci. **12**(7), 94–104 (2023). https://doi.org/10.20525/ijrbs.v12i7.2840. https://ssbfnet.com/ojs/index.php/ijrbs/article/view/2840
42. RCMP: Report cybercrime and fraud. https://rcmp.ca/en/federal-policing/cybercrime/national-cybercrime-coordination-centre/report-cybercrime-and-fraud. Accessed 11 Jan 2026
43. Redmiles, E.M., Kross, S., Mazurek, M.L.: How I learned to be secure: a census-representative survey of security advice sources and behavior. In: Proceedings of the 2016 ACM SIGSAC Conference on Computer and Communications Security (CCS 2016). Association for Computing Machinery (2016). https://doi.org/10.1145/2976749.2978307
44. Redmiles, E.M., et al.: A comprehensive quality evaluation of security and privacy advice on the web. In: 29th USENIX Security Symposium (USENIX Security 2020), pp. 89–108. USENIX Association (2020). https://doi.org/10.5555/3489212.3489218
45. SPF: Police advisory - securing of smart home devices including IP cameras. https://www.police.gov.sg/media-hub/news/2021/20210125_police_advisory_securing_of_smart_home_devices_including_ip_cameras. Accessed 03 Jan 2026
46. van Steen, T., Norris, E., Atha, K., Joinson, A.: What (if any) behaviour change techniques do government-led cybersecurity awareness campaigns use? J. Cybersecurity **6**(1) (2020). https://doi.org/10.1093/cybsec/tyaa019

47. Turner, S., Nurse, J.R.C., Li, S.: When googling it doesn't work: the challenge of finding security advice for smart home devices. In: Human Aspects of Information Security and Assurance: 15th IFIP WG 11.12 International Symposium, HAISA 2021. Springer (2021). https://www.hooklee.com/Papers/HAISA2021.pdf
48. Ye, J., de Carné de Carnavalet, X., Zhao, L., Zhang, M., Wu, L., Zhang, W.: Exposed by default: a security analysis of home router default settings. In: Proceedings of the 19th ACM Asia Conference on Computer and Communications Security (ASIA CCS 2024). Association for Computing Machinery (2024). https://doi.org/10.1145/3634737.3637671

Security Governance and Media

Negative by Design: Media Frames of Hackers in Germany

Raphael Morisco(✉)

Karlsruhe Institute of Technology (KIT), Karlsruhe, Germany
EduSecAI@proton.me

Abstract. This paper examines how the term 'hacker' is framed in German mainstream media. It presents a longitudinal content analysis of 301 publicly accessible articles from four widely circulated German news outlets (*Die Zeit*, *Süddeutsche Zeitung*, *Bild*, and *Der Spiegel*), sampled from January 2017, 2018, 2019, and 2020. The analysis shows a strong predominance of negative connotations and recurring frames that associate hackers with criminality, geopolitical aggression, and digital attack. Drawing on framing research, scandalization theory, and the concept of media reality, the paper interprets these representations as patterned constructions rather than isolated lexical choices. The contribution of the study lies in empirically documenting how a socially salient cybersecurity figure is represented across a bounded longitudinal sample of German reporting. The paper concludes by discussing implications for public understandings of cybersecurity and by outlining the methodological limits of the selected corpus.

Keywords: Hacker · media framing · German media · content analysis · cybersecurity · public discourse

1 Introduction

The term 'hacker' occupies a prominent place in public discourse on digital technology, cybercrime, and cybersecurity. In journalistic reporting, however, the concept often functions not merely as a descriptive label but as a socially charged signifier associated with danger, deviance, and technical opacity. Such representations matter because media do not simply reflect social reality; they also participate in constructing it through recurring frames, semantic associations, and patterns of emphasis.

Previous scholarship has shown that hackers are frequently portrayed through dramatized and stereotypical narratives, especially in popular media and cultural discourse [1,2]. These portrayals often reveal broader societal fears surrounding technology, crime, and national security rather than offering differentiated accounts of hacker cultures and practices [1]. More broadly, research in media and communication studies has shown that media coverage contributes to the social construction of risks, public issues, and collective interpretations [9,10,13].

© The Author(s), under exclusive license to Springer Nature Switzerland AG 2027
K. Renaud et al. (Eds.): EICC 2026, CCIS 3012, pp. 57–66, 2027.
https://doi.org/10.1007/978-3-032-28957-5_4

Despite the cultural and political significance of cybersecurity discourse, however, the representation of hackers in German mainstream media has received comparatively little systematic empirical attention.

This paper addresses that gap through a longitudinal content analysis of 301 publicly accessible articles from four widely circulated German news outlets: *Die Zeit*, *Süddeutsche Zeitung*, *Bild*, and *Der Spiegel*. The corpus is based on articles published in January 2017, 2018, 2019, and 2020 that contain the keyword 'hacker'. January was selected as a fixed sampling month to ensure comparability across years and to avoid seasonal distortions associated with the German *Sommerloch* ("summer news drought"). The aim is not to claim representativeness for all German media discourse, but to identify recurring patterns in connotation and framing within a clearly delimited and comparable sample.

The paper makes two main contributions. First, it provides an empirical account of how the term 'hacker' is connoted in selected German mainstream media coverage. Second, it identifies recurring framing mechanisms through which hackers are associated with criminality, geopolitical threat, and digital attack.

To guide the analysis, the study addresses the following research questions:

- **RQ1:** How is the term 'hacker' connoted in selected German mainstream media coverage?
- **RQ2:** Which recurring media frames can be identified in the representation of hackers?
- **RQ3:** To what extent are these connotations and frames stable across the sampled period from 2017 to 2020?

The remainder of the paper is structured as follows. Section 2 outlines the theoretical background on media framing, scandalization, and media reality. Section 3 presents the dataset, sampling strategy, coding procedure, and empirical findings. Section 4 discusses the results, implications, and limitations. Section 5 concludes the paper.

2 Theoretical Background

2.1 Media Framing and Scandalization

Media framing research examines how issues are not only selected and prioritized but also presented from particular interpretive perspectives [3,20]. In media-effects research, such framing processes are closely related to broader mechanisms through which communication shapes salience, interpretation, and public understanding [9,10]. In the context of cybersecurity reporting, this is particularly relevant because terms such as 'hacker' can function as condensed public labels that evoke threat, deviance, and criminality beyond their technical meaning.

Earlier work has shown that media portrayals of hackers often reveal broader social anxieties surrounding digital technology rather than accurately reflecting hacker cultures themselves [1]. In this sense, representations of hackers can be understood as part of wider public narratives about technological risk, social order, and national security.

To analyze these representations more closely, this paper also draws on scandalization theory. In addition to Kepplinger's work on dramatization mechanisms [4], broader scandal research has emphasized that public scandals are communicatively constructed and amplified through media-specific dynamics [5]. Kepplinger identifies several forms of dramatization in public communication, including horror labels, criminal association, defamation, catastrophe suggestion, catastrophe collage, and serial scandalization [4]. These mechanisms are useful heuristics for examining how journalistic reporting amplifies specific threat perceptions and stabilizes recurring interpretive patterns.

2.2 Media Reality

The concept of media reality (*Medienwirklichkeit*) refers to the mediated construction of socially shared interpretations of the world [6–8,12]. From a constructivist perspective, media representations are not treated as transparent reproductions of reality but as structured selections shaped by institutional routines, narrative conventions, and culturally embedded categories [6,8,11]. Therefore, such mediated constructions are linked to broader processes of public knowledge formation and issue interpretation [13]. In the present context, this perspective is useful because it allows the representation of hackers to be analyzed not simply as a matter of word choice, but as part of a broader process through which digital threat actors become publicly intelligible.

Taken together, framing research, scandalization theory, and the concept of media reality provide the analytical basis for examining how the figure of the hacker is constructed in German mainstream reporting.

3 Dataset and Method

This section presents the corpus construction, sampling strategy, and coding procedure used for the longitudinal content analysis. The aim is to identify recurring connotations and frames associated with the term 'hacker' in a bounded sample of German mainstream reporting.

3.1 Dataset and Sampling Strategy

To construct a transparent and comparable corpus, circulation and reach data of major German-language print and online media were consulted using IVW rankings [15]. Based on these data and the practical requirement of searchable and sufficiently accessible archives, four outlets were selected for the final sample: *Süddeutsche Zeitung*, *Die Zeit*, *Bild*, and *Der Spiegel*. These outlets represent prominent segments of the German news landscape while also allowing sufficiently consistent retrieval of articles across the sampled years.

The analysis focuses on articles published in January 2017, 2018, 2019, and 2020 that contain the keyword 'hacker'. January was selected as a fixed recurring sampling window in order to ensure comparability across years while keeping the corpus manageable for manual coding. The choice of the same month in each year provides a consistent basis for longitudinal comparison, although it does not capture possible seasonal variation in reporting.

The sampled period from 2017 to 2020 should be understood as a bounded pre-pandemic baseline rather than as a representation of the most recent state of media discourse. The study therefore does not claim to capture later developments in cybersecurity reporting after 2020; instead, it identifies patterns within a clearly delimited period that can serve as a point of comparison for future research.

Only publicly accessible articles were included. This improves transparency and reproducibility, but it may also introduce bias, since openly accessible content can differ from paywalled reporting in style, topic selection, or degree of sensationalism. The findings should therefore be interpreted as referring to a specific and bounded corpus rather than to German media discourse as a whole.

The target sample consisted of up to 320 articles. After applying the inclusion criteria and excluding inaccessible items, the final corpus comprised 301 articles. The corpus construction and initial analytical framework are described in detail in [16]. For each article, the following attributes were documented: outlet, title, publication date, semantic field, connotation, keywords, topic relevance to cybersecurity, access date, and URL.

3.2 Coding and Analytical Procedure

The analysis combines descriptive content analysis with an interpretive framing perspective. Connotation was coded as *negative*, *positive*, or *none/off-topic*. A negative connotation was assigned when the term 'hacker' was explicitly associated with criminality, threat, attack, warfare, fear, or comparable semantic fields. A positive connotation was assigned when the term referred to constructive, protective, or socially valued forms of technical expertise. Articles were coded as *none/off-topic* when the keyword appeared without substantive relevance to cybersecurity-related reporting.

In addition to connotation, the analysis identified recurring framing elements, including terminological inaccuracy, geopolitical attribution, semantic association with attack-related vocabulary, and forms of dramatization derived from scandalization theory [4]. These categories were used as analytical heuristics to identify patterns across the corpus rather than as an attempt to claim exhaustive representativeness.

The coding was conducted manually in order to preserve contextual sensitivity across heterogeneous article formats and topics (Figs. 1, 2 and Table 1).

Across the full corpus of 301 articles, the term 'hacker' is predominantly associated with negative connotations. In total, 241 articles (80%) were coded as negative, 10 articles (3%) as positive, and 50 articles (17%) as none/off-topic (see Fig. 3).

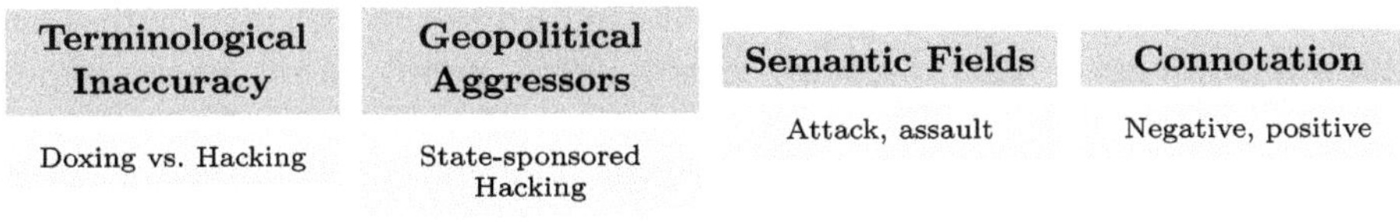

Fig. 1. Linguistic components of media frames.

Table 1. Detailed listing of the connotation of the term 'hacker' in the month of January (2017 to 2020). Abbreviations for newspapers are as follows: DZ: Die Zeit; S: Süddeutsche; B: Bild; DS: Der Spiegel.

Newspapers	Jan 2017				Jan 2018				Jan 2019				Jan 2020			
	DZ	S	B	DS	DZ	S	B	DS	DZ	S	B	DS	DZ	S	B	DS
Negative	16	13	18	17	13	13	17	14	19	19	20	19	10	10	6	17
Positive	2	0	2	1	0	1	0	2	1	0	0	0	0	0	1	0
None	2	7	0	2	7	6	0	4	0	1	0	1	5	10	2	3
Total Number of Articles	20	20	20	20	20	20	17	20	20	20	20	20	15	10	9	20

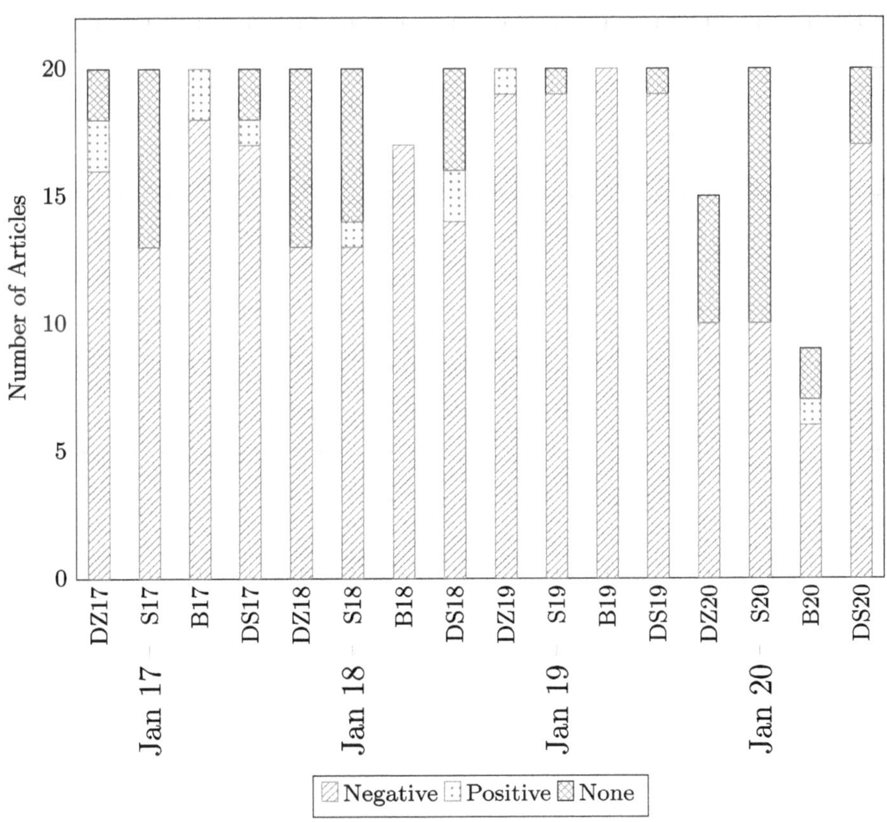

Fig. 2. Connotation of the term 'hacker' in the month of January from 2017 to 2020.

This pattern remains visible across all sampled years, although with some variation. In January 2017, 64 of 80 articles (80%) were coded as negative. In January 2018, the number of negative articles was 57 of 77 (74%). The strongest concentration of negative connotations appears in January 2019, where 77 of 80 articles (96%) were coded as negative. In January 2020, the share of negative articles remained dominant at 43 of 64 articles (67%), although the proportion of off-topic items increased.

Overall, the longitudinal pattern suggests that negative connotations are not isolated outliers but a stable feature of the sampled reporting.

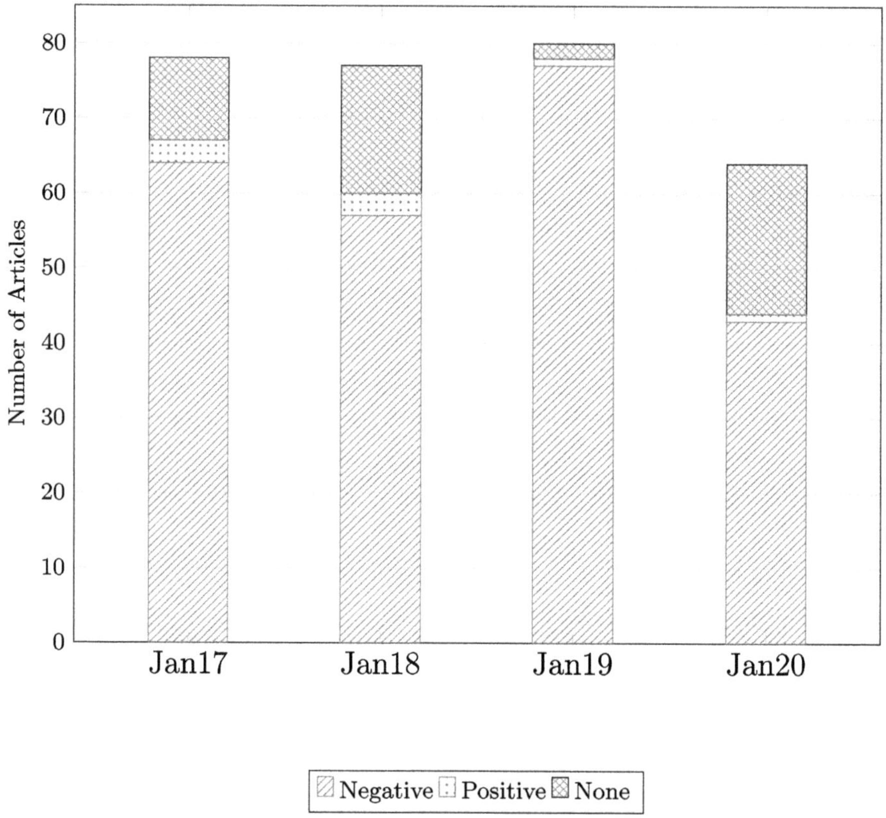

Fig. 3. Overall connotation of the term ‘hacker’ for January in the years 2017–2020.

3.3 Recurring Media Frames

Beyond the numerical distribution of connotations, the corpus reveals several recurring framing patterns.

First, the data show repeated *terminological inaccuracy*. In some cases, incidents were described as 'hacker attacks' although the underlying activity was more accurately described using other concepts. A prominent example is the public discussion of the 2019 publication of politicians' personal data, which was initially framed in parts of the media as a hacking case although it is more precisely discussed as doxing [17–19]. Such cases indicate that the term 'hacker' often functions as a generalized threat label. This pattern is relevant not only terminologically but also from a framing perspective, since recurring labels help stabilize particular interpretations of digital events [20].

Second, the corpus contains recurring *geopolitical attributions*. Across the sampled years, references to Russia, Iran, and later Saudi Arabia repeatedly appear in connection with hacking and cyberattack narratives. This contributes to a frame in which hackers are linked not only to criminal individuals but also to hostile state actors and national security threats.

Third, the term 'hacker' is frequently embedded in *attack-centered semantic fields*. Particularly frequent lexical associations include terms such as 'attack', 'assault', 'cyberwar', 'fear', and 'data'. These semantic patterns reinforce an interpretive environment in which the hacker is positioned primarily as a source of threat and disruption.

Fourth, the corpus shows multiple forms of *dramatization* in the sense of scandalization theory [4,5]. Excluding visual exaggeration, which was not part of the linguistic analysis, all remaining dramatization types could be identified in the material. The most salient patterns include criminal association, catastrophe suggestion, and serial scandalization. In practical terms, this means that the term 'hacker' is repeatedly linked to language that intensifies danger, amplifies uncertainty, and foregrounds harmful outcomes.

Taken together, these patterns suggest that the media representation of hackers in the sampled corpus is shaped not only by negative connotation in a narrow lexical sense, but also by recurring narrative and semantic structures that stabilize the hacker as a figure of digital threat.

4 Discussion

The findings indicate that the term 'hacker' is represented in the sampled corpus predominantly through negative connotations and recurring threat-oriented frames. This confirms the broader observation from earlier scholarship that media portrayals of hackers often reveal more about social anxieties surrounding digital technology than about hacker cultures themselves [1]. In the present study, this tendency becomes visible in three especially persistent patterns: criminalization, geopolitical threat attribution, and semantic linkage to attack and warfare.

From a media-theoretical perspective, these patterns can be understood as elements of media reality [12]. The reporting does not merely label isolated incidents but helps stabilize a socially intelligible figure of the hacker as a condensed symbol of digital threat. This matters because public understandings of cybersecurity are shaped not only by technical incidents themselves but also by the ways in which those incidents are narrated, typified, and repeated. In this respect, the findings are also consistent with broader work on media effects and public knowledge formation [9,10,13].

These observations also have implications for cybersecurity education. If media reporting repeatedly presents hackers primarily as threatening or criminal figures, this may narrow public understanding of cybersecurity and obscure constructive roles such as ethical hacking, vulnerability research, and responsible disclosure. From an educational perspective, this suggests the value of linking security literacy with critical media competence, so that learners can better contextualize journalistic portrayals of digital threats.

At the same time, the findings should not be overstated. The study is limited to four German news outlets, one recurring month per year, and a corpus ending in 2020. It therefore does not claim representativeness for all German media discourse and does not capture later developments in cybersecurity reporting, including post-pandemic changes, intensified cyberwar narratives [21,22], and the growing visibility of AI-related threat narratives [23]. In addition, the restriction to publicly accessible articles may introduce selection bias.

The present study focuses exclusively on linguistic framing and therefore does not analyse accompanying images. Visual representations were not part of the coding procedure and were consequently excluded from the analytical framework. The findings of this study therefore refer specifically to textual and semantic patterns in journalistic reporting.

Even with these limitations, the findings suggest that a narrow public narrative centered on the hacker as an individualized adversary may obscure more structural dimensions of contemporary cybersecurity, including organizational vulnerability, socio-technical interdependence, cyberwarfare dynamics [21,22], and preventive security practices. Comparative research further indicates that such constructions are not unique to the German case but can also be observed across different national media contexts [14]. A corpus-based comparison of *China Daily* and *The New York Times*, for example, finds that hacker narratives often reflect national political contexts: U.S. coverage frequently associates hackers with foreign adversaries such as Russia or China, while Chinese reporting largely responds to U.S. accusations. The authors therefore conclude that media portrayals of hackers are shaped by national ideological contexts and may contribute to reinforcing geopolitical divisions in cybersecurity discourse.

Taken together, these observations suggest that the patterns identified in the German news landscape are relevant not only for media analysis but also for broader discussions about cybersecurity communication and public understanding. If public discourse continues to frame cybersecurity primarily through the figure of the malicious hacker, broader systemic dimensions of digital security may remain underrepresented in public debate and policy discussions.

5 Conclusion

This paper examined how the term 'hacker' is represented in a selected sample of German mainstream media articles published in January 2017, 2018, 2019, and 2020. Across 301 articles, the analysis identified a strong predominance of negative connotations and recurring frames that associate hackers with criminality, geopolitical hostility, and digital attack. These findings suggest that the figure of the hacker functions in the sampled reporting less as a differentiated socio-technical actor and more as a condensed symbol of digital threat.

Beyond the descriptive result itself, the study contributes to cybersecurity research by showing how media framing participates in shaping public understandings of cybersecurity-related actors and risks [9,10,20]. This is relevant because public discourse does not merely reflect technical realities; it also influences how risks are perceived, communicated, and socially organized.

At the same time, the paper is limited by its focus on four outlets, one recurring month per year, and a corpus ending in 2020. Future research could extend the corpus to more recent years, compare different national media systems, or combine content analysis with audience-oriented research in order to examine how such frames are received and interpreted.

Taken together, the findings indicate that media portrayals of hackers are not merely terminological accidents. They are patterned communicative constructions that contribute to a broader media reality of cybersecurity.

Acknowledgments. The author thanks the anonymous reviewers for their constructive comments and suggestions, which helped to improve the clarity and structure of the paper. The author is also grateful to Karen Renaud for her guidance during the shepherding process.

Parts of the theoretical perspective of this study were influenced by research on constructivist media theory and media scandalization, particularly the work of Bernhard Pörksen and Hans Mathias Kepplinger.

References

1. Thomas, D.: Hacker Culture. University of Minnesota Press, Minneapolis (2003)
2. Moebius, S., Schroer, M.: Einleitung. In: Moebius, S., Schroer, M. (eds.) Diven, Hacker, Spekulanten: Sozialfiguren der Gegenwart, pp. 7–11. Suhrkamp, Berlin (2010)
3. Bonfadelli, H., Friemel, T.N.: Medienwirkungsforschung, 6th edn. UVK Verlagsgesellschaft, Konstanz, München (2017)
4. Kepplinger, H.M.: Die Mechanismen der Skandalisierung: Warum man den Medien gerade dann nicht vertrauen kann, wenn es darauf ankommt, 4th edn. Lau Verlag, Reinbek (2018)
5. Pflügler, S., Baugut, P.: Normverstoß, Medienkonstrukt, Machtkampf oder Internethysterie? Publizistik **60**(3), 305–324 (2015)
6. Pörksen, B.: Schlüsselwerke des Konstruktivismus. In: Pörksen, B. (ed.) Schlüsselwerke des Konstruktivismus, 2nd edn, pp. 3–18. Springer VS, Wiesbaden (2015)

7. Pörksen, B.: Konstruktivismus: Medienethische Konsequenzen. Springer VS, Wiesbaden (2014)
8. Hickethier, K.: Einführung in die Medienwissenschaft, 2nd edn. Metzler, Stuttgart, Weimar (2010)
9. Merten, K.: Wirkungen von Kommunikation. In: Merten, K., Schmidt, S.J., Weischenberg, S. (eds.) Die Wirklichkeit der Medien, pp. 291–328. Springer Fachmedien, Wiesbaden (1994)
10. Jäckel, M.: Medienwirkungen: Ein Studienbuch zur Einführung, 5th edn. VS Verlag für Sozialwissenschaften, Wiesbaden (2011)
11. Schweiger, W.: Der (des)informierte Bürger im Netz. Springer, Wiesbaden (2017)
12. Bentele, G.: Medienwirklichkeit. In: Bentele, G., Brosius, H.-B., Jarren, O. (eds.) Lexikon Kommunikations- und Medienwissenschaft, pp. 224–226 (2013)
13. Beck, K.: Kommunikationswissenschaft, 5th edn. UVK, Konstanz, München (2017)
14. Pei, J., Li, D., Cheng, L.: Media portrayal of hackers in China Daily and The New York Times. Discourse Commun. **16**(5), 598–618 (2022). https://doi.org/10.1177/17504813221099190
15. Informationsgemeinschaft zur Feststellung der Verbreitung von Werbeträgern e.V.: IVW Quartalsauflagen. Accessed 7 Nov 2020
16. Morisco, R.: Framing the hacker: media representations and public discourse in Germany. arXiv preprint: arXiv:2511.09608 (2025)
17. Douglas, D.M.: Doxing: a conceptual analysis. Ethics Inf. Technol. **18**(3), 199–210 (2016). https://doi.org/10.1007/s10676-016-9406-0
18. Schneier, B.: We Have Root. Wiley, Hoboken (2019)
19. Kühl, E.: Privatsphäre: Eine Waffe namens Doxing. Die Zeit (2019). https://www.zeit.de/digital/datenschutz/2019-01/privatsphaere-doxing-daten-sammeln-datensicherheit-politiker. Accessed 5 Feb 2021
20. Scheufele, B.: Framing. In: Bentele et al. (eds.) Lexikon Kommunikations- und Medienwissenschaft, p. 96 (2013)
21. DiMaggio, J.: The Art of Cyberwarfare. No Starch Press, San Francisco (2022)
22. Kurz, C., Rieger, F.: Cyberwar - Die Gefahr aus dem Netz. C. Bertelsmann, München (2018)
23. Vulpe, S.-N., et al.: AI and cybersecurity: a risk society perspective. Front. Comput. Sci. **6**, 1462250 (2024). https://doi.org/10.3389/fcomp.2024.1462250

Digital Control: National Tactics of Internet Censorship

Julia Lenz[1(✉)], Simon Volpert[1], Sebastian Zillien[1], Philip Rünz[2], Luca Caviglione[3], and Steffen Wendzel[1]

[1] University of Ulm, Ulm, Germany
{julia.lenz,simon.volpert,sebastian.zillien,steffen.wendzel}@uni-ulm.de
[2] FernUniversität in Hagen, Hagen, Germany
[3] CNR - IMATI, Genova, Italy
luca.caviglione@ge.imati.cnr.it
https://www.uni-ulm.de/en/in/omi/

Abstract. Internet censorship has evolved from simple content blocking to complex socio-technical control systems. While prior surveys primarily focus on techniques such as DNS poisoning or deep packet inspection, they rarely examine how national contexts shape deployments. This paper presents a comparative analysis of 20 countries, conceptualizing censorship as a four-dimensional architecture comprising legal regulation, technical filtering, information flooding and infrastructural denial. Synthesizing existing research across regime types, we identify a systematic "sliding scale" in the dominant locus of control: democratic systems rely mainly on regulatory mechanisms, hybrid regimes combine filtering with narrative manipulation, and highly restrictive states block physical access. Our findings show that censorship adapts to political and institutional contexts, often shifting its control across domains calling for broader measurement approaches beyond the network layer.

Keywords: Internet Censorship · Censorship Measurement · Free Internet Access · TCP/IP · DNS · Internet Technology

1 Introduction

The Internet enables large-scale connectivity and access to vast amounts of information and services. At the same time, governments, corporate, institutional, and non-governmental organizations may deem certain interactions or content inappropriate due to political, religious, cultural, legal, or security-related concerns. Internet censorship therefore aims to restrict access to and exchange of information considered undesirable by a censor [32].

Given the growing importance of digital communication infrastructures, a substantial body of research examines the technical mechanisms used to enforce censorship. Recent surveys such as [90] review network-level filtering techniques and detection methods, complementing previous research in [51], which surveyed censorship trends across 70 countries. The resulting snapshot is somewhat

© The Author(s), under exclusive license to Springer Nature Switzerland AG 2027
K. Renaud et al. (Eds.): EICC 2026, CCIS 3012, pp. 67–87, 2027.
https://doi.org/10.1007/978-3-032-28957-5_5

alarming, since advanced techniques ranging from deep packet inspection (DPI) to per-protocol manipulation are nowadays prevalent. Other notable research includes the seminal work in [6] providing a comprehensive but now outdated view of Internet censorship and the work in [87] analyzing real-world censorship behaviors. Together, these studies offer detailed information on *which* techniques are deployed and *where* they occur. However, existing research focuses on identifying the *presence* of specific filtering capabilities (e.g., DNS tampering or TLS filtering) [51] rather than examining how these mechanisms are embedded within broader political, legal, and infrastructural systems. Censorship is not merely a technical intervention, but a multi-dimensional system of control shaped by differences in institutional design, regulatory frameworks, and state capacity. Consequently, systematic comparison of these socio-political architectures across regime types remains limited [4,90,91].

This paper addresses this gap through a comparative and country-focused analysis of Internet censorship. Drawing on the annual "Freedom on the Net" (FOTN) report [32] and complementary sources, we examine how national actors implement and operate censorship. Rather than cataloging individual techniques, we identify distinct *models of control* and conceptualize censorship as a multidimensional architecture. Accordingly, this paper asks: *How do national Internet censorship architectures vary across regime types, and which control mechanisms emerge as dominant within different political contexts?* To answer this question, we identify recurring patterns, overlaps, and singularities across countries to synthesize evidence into a four-dimensional analytical framework capturing legal, technical, informational, and infrastructural dimensions of censorship. Our contributions are threefold: (i) we introduce a structured model for describing national censorship architectures beyond technique-level inventories; (ii) we provide a cross-country mapping of these dimensions; (iii) we propose a novel "sliding scale" model for liking dominant control mechanisms to political contexts. Our paper is accompanied by a project website: https://atlas.omi.uni-ulm.de/

The remainder of this paper is structured as follows. Section 2 describes the methodology and sampling strategy. Section 3 presents the country-specific findings. Section 4 discusses commonalities, trends, and shifts as well as overall implications, limitations and future research directions. Lastly, Sect. 5 concludes the paper.

2 Methodology

To enable structured cross-country comparison while preserving analytical depth, we do not cover all countries worldwide. Instead, we analyze 20 countries through a stratified random sampling approach based on the categorization of the 2025 FOTN report [32]. Four sampling strata were constructed. Three correspond to the FOTN categories: "not-free", "partially free", "free", from which five countries were randomly selected each. The fourth stratum consists of countries "not covered" by FOTN, from which five additional cases were drawn. Although these countries lack an Internet-specific FOTN rating, Freedom House assesses them in its separate annual report "Freedom in the World" [31], providing a broader "World" Freedom Score measuring political rights and civil liberties. Accordingly, Table 1 reports both the FOTN "Internet" score (where available) and the "World" score of our final sample, also highlighted in Fig. 1. For "not covered" countries, only the latter exists. Within each stratum, eligible countries were randomly sampled with a uniform probability. No geographic quotas were imposed, as the primary analytical variable is regime type rather than regional distribution (e.g., continent-based balancing). Furthermore, each country is analyzed using the following four-dimensional framework:

(i) **Legal/regulatory controls** (•): Restrictive laws and regulations making undesired topics illegal.
(ii) **Technical filtering** (•): Rule-based filtering of undesired content.
(iii) **Information flooding** (•): Propaganda through bot/troll armies, narrative manipulation, and fake news.
(iv) **Infrastructure denial** (•): (Near) total Internet shutdowns, infrastructure-level disconnects.

These dimensions are not mutually exclusive but provide a structured lens for comparing censorship architectures across national contexts. Throughout the paper, color-coded markers annotate text and tables: a solid bullet (•) denotes a primary censorship mechanism and an open circle (○) indicates a secondary or occasional application. Country assessments are based on qualitative synthesis of peer-reviewed studies, reports from established digital rights organizations, regulatory documents, and reputable news outlets. Sources were identified through academic database queries, citation snowballing, and web searches.

Fig. 1. Geographical distribution of the 20 selected countries. Highlighted regions indicate the nations included in our analysis across the various FOTN categories.

Table 1. Random selection of five countries per category. "World" and "Internet" denote Freedom in the World and Freedom on the Net scores, respectively. Cell colors map a continuous scale from 0 (Red/Worst) to 100 (Green/Best).

Country	World	Internet	Country	World	Internet
Not-Free			**Free**		
China	9	9.0	Canada	97	85.0
Iran	11	13.0	France	89	76.0
Thailand	34	39.0	Germany	95	74.0
Turkey	33	31.0	Italy	89	74.0
Vietnam	20	22.0	Taiwan	94	79.0
Partially Free			**Not Covered**		
India	63	51.0	Eritrea	3	N/A
Indonesia	56	48.0	North Korea	3	N/A
Mexico	59	61.0	Spain	90	N/A
Nigeria	44	59.0	Syria	5	N/A
Ukraine	51	62.0	Turkmenistan	1	N/A

3 Censorship Deployments in Selected Countries

This section presents country-specific analyses of Internet censorship. As most governments rarely disclose official blocklists or filtering infrastructure, our insights rely mostly on indirect evidence, such as long-term measurement studies that track blocking and traffic alternations across time and locations [90].

3.1 FOTN Non-free Countries

China. One of the most sophisticated censorship systems worldwide is operated by China, integrating technical censorship measures with a rigid legal framework. Through intermediary liability, the state delegates censorship duties to private platforms, compelling them to police content under threat of heavy fines or license revocation [24] •. The network-level censorship is primarily implemented through the *Great Firewall* (GFW), a large-scale filtering system designed to detect and block undesirable traffic, particularly encrypted protocols which are critical for censorship circumvention [92] •. The GFW also employs *residual censorship* [13], whereby once a connection attempt has been flagged, the same traffic tuple may remain blocked for 120 to 180 s, effectively suppressing repeated connection attempts. Beyond filtering, coordinated "information flooding" shapes online discourse •. Han highlights the role of the "Fifty Cent Army" (*wumaodang*), whose members engage in "discourse competition" by posting pro-government narratives and diverting discussions away from critical topics [38].

During politically sensitive periods, China occasionally resorts to infrastructural interventions such as temporary Internet shutdowns and DNS tampering [51] ○. Although less common, these measures illustrate its ability to escalate censorship through direct network disruptions.

Iran. Among the most repressive Internet environment worldwide is Iran ●. Authorities regularly intensify restrictions around elections, including criminalizing regime-critical content [10,32]. Censorship practices date back to at least 2002, when roughly 50% of the most visited websites were blocked and over 95% of adult content filtered [10] ●. Since then, DPI, DNS tampering, and HTTP keyword filtering are used to target political, social, and security-related content. [14,40]. In 2020, additional protocol-based filtering that inspects the first two client packets of DNS, HTTP, and HTTPS traffic were introduced. If protocol fingerprints do not match expected patterns, flows are dropped for 60 s [14]. Encrypted DNS, including DoT, is also blocked [11], while measurement studies show inconsistencies in protocol-level enforcement, such as the continued accessibility of unencrypted HTTP/2 traffic [47]. The state promotes a curated information environment through the "National Internet", prioritizing domestic services and enabling stronger narrative control ●. During nationwide protests in 2019, authorities used this infrastructure to substitute global connectivity and have since pursued a more isolated and potentially religiously framed "Halal Internet" [40]. Beyond filtering, throttling and connectivity disruptions have also been widely used ●. Mainly SSH and other encrypted flows have been degraded via TCP packet drops [29]. During the most recent protests, which began in late December 2025, authorities have blocked Internet access while also intensifying forceful suppression of demonstrations [5]. Following the escalation on 28 Feb 2026, Iran re-imposed a near-total wartime "international kill-switch and whitelisting" model. External connectivity dropped to near-zero ($\sim$ 98% traffic loss) [22,58], while domestic intranet access persisted and politically vetted "white-SIM / tiered access" channels expanded alongside coercive enforcement (SMS prosecution threats targeting circumvention attempts) [30].

Thailand. The censorship system of Thailand relies on *lèse-majesté* laws and the expanded 2017 Computer Crime Act, which are used as primary tools to prosecute online speech ●. Enforcement is bolstered by the 2010 Cyber Scout program and the "Social Sanction" network mobilizing citizens to report dissent. Together, these mechanisms create a climate of self-censorship, illustrated by the 2024 arrest of two Prachatai journalists covering anti-monarchy graffiti [32]. Technical filtering is less documented, although studies report blocking related to religious debates and pornography [1] ●. Proposals for a centralized "single gateway" architecture (resembling the GFW) have been discussed, but doubts remain regarding their political feasibility and implementation [48]. Monitoring also occurs at the ISP level, where active probing techniques are used to identify infected user machines and suspicious traffic patterns, which can subsequently inform blocking decisions or the detection of circumvention tools [15]. Studies of circumvention practices further indicate that most users adopt standard VPNs or Tor, while a minority resorts to manual workarounds like RSS feeds [33]. Evidence also points to coordinated information manipulation, including large-scale fabricated Tweets to make the regime appear more popular [79] ●. The government has additionally employed targeted infrastructural controls ○. In mid-2025, authorities ordered telecom providers to shutdown broadband and mobile Internet links to Cambodia near the border region, officially citing cybersecurity concerns [83]. Although framed as an anti-scammer measure, this restriction reflects the willingness to enforce disruptions during geopolitical disputes.

Turkey. A robust legal censorship regime is maintained by Turkey, frequently issuing content takedown requests under vague laws, restricting criticism of political leadership [81,88] ●. For example, large volumes of such requests were directed at Twitter in 2015, and internationally operating platforms often comply with such country-specific removal mechanisms [81,88]. Moreover, in 2014, the government attempted to block access to both Twitter and YouTube by acting over DNS. To prevent users from circumventing these blocks through alternative DNS services, Turkish ISPs began advertising a "bogus" BGP route to divert traffic intended for OpenDNS and Google DNS servers [81]. At the technical level, censorship includes DNS tampering, BGP hijacking, TLS-based filtering and DPI deployed by ISPs to enable user profiling and the blocking of selected keywords or destinations [51] ●. Turkey also relies on a decentralized online propaganda network known as "AKTrolls", who harasses dissidents and floods platforms with pro-regime content [72] ●. Furthermore, authorities have occasionally resorted to infrastructural interventions ○. Internet throttling and regional shutdowns have been observed during politically sensitive periods, including protests and the 2016 coup attempt [51].

Vietnam. Internet censorship in Vietnam follows a "bamboo diplomacy" approach characterized by flexibility and pragmatism [60] ●. Rather than deploying a centralized firewall, the Communist Party allows Western platforms like Facebook and Google to operate while simultaneously pressuring them to comply with domestic content removal requests [49]. This strategy of "platform coercion" allows the state to harness the economic benefits of global connectivity while maintaining control over "toxic" content. Under fear-based control, effectively summarized by the proverb "kill one to warn one hundred" [19], high-profile arrests target users who cross vague "red lines", such as criticizing key leaders, reinforcing a climate of self-censorship [49]. Technically, censorship relies on a combination of DNS tampering, IP blocking, and firewall-based filtering, with inconsistencies observed across ISPs (e.g., FPT vs VNPT), suggesting that enforcement varies across the national network infrastructure [49] ●. Complementing these technical measures, "Force 47", a military cyber-unit of over 10,000 agents, and "public opinion shapers" ( ) flood social media with pro-state narratives and counter critical voices [60] ●. Moreover, authorities utilize "prior restraint" as cultural control by vetting musical and artistic works before release, reinforcing broader patterns of self-censorship [62]. In addition to account-based access restrictions, bloggers and activists have reported being cut off from Internet services, placed under surveillance, or denied access to mobile networks, showcasing the selective use of connectivity restrictions [32] ○.

3.2 FOTN Partially Free Countries

India. Censorship in India is legally grounded in Section 69A of the Information Technology Act, allowing the government to order takedowns of online content and block websites deemed a threat to sovereignty, public order or national security [43] ●. These provisions enable the restriction of access to selected Internet destinations [61]. At the technical level, blocking is typically implemented through DNS and HTTP filtering, although enforcement varies considerably across ISPs [78] ●. Some providers block only subsets of targeted sites or apply their own filtering policies, while others leave content accessible, highlighting a decentralized and sometimes arbitrary system [78]. The network topology further shapes censorship: roughly $\sim 1\%$ of autonomous systems (AS) carry around 95% of the traffic to blocked sites, creating a centralized structure that could support large-scale filtering [36]. For mobile apps, blocking is typically enforced directly by service providers rather than through network-level interception [36]. These technical efforts are reinforced by extensive social manipulation ◐. Party-affiliated IT cells and "cyber troops" use social media platforms such as WhatsApp to spread disinformation, deploy bot networks, and amplify partisan narratives, particularly during elections [20]. Simultaneously, studies report growing levels of self-censorship among content creators on streaming platforms, who may voluntarily restrict expression to maintain access to the domestic large "unfettered" digital market [53]. Uncommonly among democracies, the state frequently resorts to regional Internet shutdowns to control unrest, as seen during the Kashmir lockdown and protests surrounding the Citizenship Amendment Act [56] ○.

Indonesia. The censorship regime of Indonesia is described as an emerging "digital authoritarianism" [77], and grounded in the Electronic Information and Transactions Law and Ministerial Regulation 5 ●. These regulations require platforms to register with the government and grant authorities broad powers to order content takedowns deemed illegal or harmful [50]. Enforcement is decentralized: the Ministry of Communication and Informatics maintains the *TrustPositif* blocklist, while implementation is carried out by individual ISPs. The censorship is mainly implemented through DNS hijacking, where ISPs redirect requests for blacklisted domains to warning pages such as "Internet Positif" ●. The decentralized enforcement model often leads to inconsistent blocking and collateral damage [74]. Monitoring technologies such as DPI systems and automated "Cyber Crawling" programs complement these filtering mechanisms by detecting and tracking undesirable content, thereby enabling more targeted blocking [69]. Coordinated information manipulation also occurs through government-aligned influencers and so-called "buzzers" who amplify pro-government narratives and harass critics on social media [77] ◐. Instead of total connectivity blackouts, the government utilizes tactical bandwidth throttling during crises ○. During the 2019 post-election riots, authorities slowed upload speeds on platforms like WhatsApp and Facebook to curb viral image and video sharing while preserving text functionality [74].

Mexico. While Mexico formally prohibits state-mandated Internet shutdowns under the 2021 IFT Guidelines, Article 190 of the 2014 Telecommunications Law allows authorities to suspend telecommunication services to prevent criminal activity, granting the state conditional power over connectivity infrastructure [2] ○. Technical censorship is relatively rare and often driven by economic motivations ○. In 2006, a major national provider blocked access to www.skype.com to protect revenue from long-distance calls [28]. Measurements of Tor connectivity between 2010 and 2019 also suggest that some ISPs intermittently blocked routes to Tor directory authorities, although disruptions may also stem from misconfigurations, NAT behavior, or DPI-based filtering [44]. Some content control also occurs outside the network layer. For instance, a recent work addresses how distributed media such as anime is often "sanitized" before release, though users frequently bypass restrictions through P2P file sharing of uncensored versions [46]. The authors highlight the difficulty of enforcing blockages due to the ability of P2P to route data beyond national borders [46]. Lastly, coordinated information campaigns have been observed ◐. As an example, the government shaped the discussion on the COVID-19 pandemic performed via Twitter [66].

Nigeria. Online censorship in Nigeria is primarily enforced through legal instruments such as the 2015 Cybercrime Act and various regulatory provisions, including anti-blasphemy statutes ●. These laws have been used to prosecute journalists and social media users, encouraging self-censorship and fostering a chilling effect online. Authorities have also invoked Sect. 146 of the 2003 Communications Act, to compel ISPs to block websites, often without transparent procedures [32]. Although studies document the proliferation of laws restricting online contents as well as movies and games, a comprehensive technical evaluation of the censorship infrastructure remains limited [37]. Technical filtering approaches include DNS manipulation, TCP injection, and blockpage delivery, as observed during the 2017–2018 measurement campaign by ICLab [61] ●. Nonetheless, these mechanisms appear to be implemented inconsistently and with limited oversight. Political actors also employ troll networks and disinformation campaigns to shape public opinion during elections and protests, using bots, fake accounts, and state-linked influencers [3] ◐. Authorities have additionally resorted to infrastructural restrictions during periods of political tension, including regional Internet and mobile network shutdowns in 2021 and a nationwide Twitter-ban after the platform removed a presidential tweet [8] ○.

Ukraine. The censorship model in Ukraine reflects the country's defensive response to Russian "hybrid warfare" and military aggression conducted through digital disinformation on the "online battlefield" [17]. Rather than suppressing domestic dissent, restrictions focus on countering foreign influence and safeguarding the national information space. Prior to the invasion, the 2017 Law "On Sanctions" enabled the blocking of major Russian platforms (e.g., Odnoklassniki, Yandex), creating what Golovchenko [35] terms a "friction-based" model that increased the "cost" of access (e.g., forcing VPN use) rather than imposing a complete blockade, encouraging migration toward domestic and Western alternatives ●. Since 2022, censorship has intensified under martial law through the 2022 Law "On Media" and amendments to the Criminal Code (Article 436-2), criminalizing pro-Russian narratives and effectively merging network-level filtering with punitive legal restrictions [17]. Technically, the National Centre for Operations and Technology Management of Telecommunications Networks issues direct blocking orders to network operators, resulting in the blocking of over 48 million Russian IP addresses and numerous AS associated with Russian infrastructure [17] ●. Wartime information management also includes centralized broadcasting through the nationwide "Unified News" telethon, designed to counter disinformation and maintain a consistent public narrative [70] ○. Systematic Internet shutdowns have not been used as a censorship tool. Most connectivity disruptions during the war result from Russian attacks on infrastructure and ISP-level rerouting in occupied territories [70]. Internationally, similar restrictions have been adapted by the EU against Russian state media such as RT and Sputnik [73].

3.3 FOTN Free Countries

Canada. Internet governance in Canada is increasingly shifting toward a model of "soft censorship" based on regulatory control rather than large-scale technical filtering. Judicial site-blocking orders and platform mandates have introduced limited forms of network-level intervention [26] ●. In the *GoldTV* case, courts authorized ISPs to block IPTV piracy services, establishing a precedent for court-mandated blocking and filtering under copyright enforcement [25]. Recent legislation further expands regulatory influence. Bill C-11 (Online Streaming Act), grants authorities indirect influence over platform algorithmic "discoverability", manipulating what users see in their feeds based on national criteria rather than user preference [39]. Similarly, the *Online News Act* (Bill C-18), intending to support the Canadian news industry by forcing dominant digital platforms to pay the domestic organizations, triggered major platform responses. Meta decided to block news content for users, significantly altering the country's information ecosystem, affecting accessibility [82]. Additional proposals such as Bill C-63 aim to regulate "legal but harmful" speech, intensifying debates about platform liability and potential self-censorship [23,42]. While such filtering remains limited in scope, the government has also used DNS tampering in exceptional cases, such as blocking COVID-19–related content during the pandemic [45] ●.

France. The regulation of online content in France primarily relies on legal and regulatory mechanisms. A notable example is the "graduated response" system introduced through the HADOPI framework, which allows authorities to identify users distributing copyrighted material via peer-to-peer networks and report their IP addresses to law enforcement agencies [54] ●. These measures focus on deterrence and enforcement of intellectual property law rather than broad content blocking. Although relatively limited in scope, technical filtering mechanisms have also been documented ○. Measurement studies indicate the use of DNS and HTTP/keyword filtering [51], mainly to mitigate child pornography or as a reaction against the 2001 terrorism wave [16]. However, even if some works hint at the use of filtering techniques, to the best of our knowledge there is no prior literature analyzing used techniques or their pervasiveness.

Germany. In Germany, online regulation relies primarily on legal enforcement and content removal rather than systematic network-level filtering [32]. The state briefly employed technical blocking through the *Zugangserschwerungsgesetz* (2010–2011), targeting websites hosting child exploitation material via DNS-based filtering [51] ● ○. After its repeal, policy shifted toward content removal and platform liability. A central pillar is the *Network Enforcement Act* (NetzDG), which requires large social media platforms to remove illegal content within strict timeframes, strengthening enforcement of existing criminal law while raising concerns about potential overblocking and chilling effects on expression [85]. Today, enforcement focuses on removing illegal content, including extremist material (e.g., Holocaust denial) and copyright violations. Takedown procedures are coordinated through the *Clearingstelle Urheberrecht im Internet*, facilitating cooperation between rights holders, regulators, and ISPs.

Italy. State-mandated content filtering has increased in Italy, primarily targeting copyright violations ●. The national telecom regulator AGCOM operates the "Piracy Shield" system, which allows domain and IP-level blocking of services associated with illegal streaming. Internet and VPN providers must enforce these blocking orders, typically triggered through a ticket-based reporting system that flags IPv4/v6 addresses or domain names responsible for copyright violations or illegal activities. Critics argue that the system lacks sufficient oversight and may lead to overblocking and collateral damage. In fact, endpoints are blocked after a ticket is submitted, but "Piracy Shield" cannot handle the presence of shared IP addresses or DNS entries. A notable incident in October 2024 showed how a misclassification led to the blocking of `drive.usercontent.google.com`, disrupting access to Google Drive and YouTube for many users [7] ●. Measurement studies also show inconsistent enforcement and frequent circumvention via open or foreign DNS resolvers [64]. In addition to DNS filtering, some operators deploy TCP and HTTP tampering, for instance, Vodafone applies HTTP proxy inspection for more targeted filtering [4].

Taiwan. As a digitally open democracy, Taiwan faces asymmetric threats from China, including attempts at "extraterritorial censorship" and cognitive warfare campaigns targeting its public discourse [75]. Censorship often arises indirectly through economic and political pressure rather than direct state control ●. Media outlets with business ties to China often self-censor, a form of "outsourced" control described as part of the "3 Is" of China's strategy: propaganda *Initiation*, media *Investment* in local media outlets, and *Ideological* polarization [75,80]. As a direct countermeasure to the "Investment" strategy, the National Communications Commission revoked the broadcast license of the pro-China television channel *CtiTV* in 2020, citing repeated misinformation violations [75]. Rather than relying on extensive technical filtering, Taiwan focuses on countering disinformation campaigns ○. Malicious foreign-linked accounts amplify polarizing narratives online [59], while a "whole-of-society" response involving journalists, civil society, and academic institutions works to collaboratively detect and debunk disinformation [71]. Technical blocking remains minimal ○. In a few national security cases, authorities have restricted certain Chinese streaming platforms (OTT services). The state also promotes "digital democracy" through platforms such as *vTaiwan*, which use algorithms to visualize public consensus and facilitate deliberative policymaking rather than manipulate information visibility [55].

3.4 FOTN Non-covered Countries

Eritrea. As an extreme case of "stable authoritarianism", Eritrea enforces censorship primarily through infrastructure denial and systemic information control [65]. Control is embedded in the broader political and institutional system: from physical surveillance to an educational curriculum that restricts digital literacy ○. Some argue that the state employs a strategy of "mis-education", in which the curriculum is designed to produce docile students to secure obedience and discourage political mobilization [76]. Information control is reinforced by a state monopoly on media ○. Since banning independent outlets in 2001, domestic news has been tightly controlled through government-driven channels [18], with occasional shutdowns during unrest [65]. Traditional network-level filtering is minimal, as the regime largely prevents widespread Internet use ○. Thus, Internet penetration remains extremely low (approx. 1.3%),with access tightly controlled by the state-owned monopoly *EriTel* [76] ●. Mobile data services are virtually nonexistent, and most users rely on state-monitored Internet cafés with slow speeds, high costs and physical surveillance [34].

North Korea. Near-total isolation from the global Internet characterizes information control in North Korea, which is largely enforced via strict legal and punitive measures. Unauthorized access to foreign information sources is criminalized, and citizens attempting to connect to the Internet risk severe penalties, including imprisonment [9] ○. Traditional filtering is largely unnecessary because only a few government-approved institutions and elites can access the outside Internet ○. Consequently, domestic information flows are tightly controlled through state-run media and the highly curated national intranet "Kwangmyong", which hosts only government-approved websites and services ○. Thus, propaganda exists but is embedded in infrastructural isolation. By preventing physical connectivity to the global Internet, the regime avoids needing sophisticated filtering systems ●. This isolation hinders Internet measurement efforts, as there is almost no traffic to or from the outside world.

Spain. A recent analysis identified multiple censorship measures in response to political conflicts [88], most notably during the 2017 Catalan independence referendum ●. Authorities ordered the blocking of websites and applications associated with the referendum, providing a legal basis for subsequent network-level interventions. ISPs used technical filtering such as DNS manipulation, CDN node seizures, HTTP blocking with DPI, and additional TLS-layer censorship ●. For example, Vodafone deployed SNI blocking, which aimed to alter the TLS extension used to route data through HTTPS-reachable destinations hosted on the same physical machine. Other blockages against TLS traffic have been inferred by considering the presence of many certificate verification failures plaguing some AS, suggesting the presence of TLS intercept attempts [52].

Syria. Internet censorship in Syria has a long history closely tied to regime stability and political control. Since the late 1990s, authorities have restricted access to email, FTP, and blocked foreign political content using centralized infrastructure, proxies and imported filtering tools [41] ●. Technical censorship includes blocking IP subnets (e.g., those linked to Israel), keyword filtering, and disabling access to instant messaging services and Facebook plugins [21] ●. Analyses of leaked proxy logs reveal extensive HTTP filtering and countermeasures against VPNs and circumvention tools. By 2012, popular VPN protocols such as PPTP and L2TP had largely been rendered ineffective [27]. Despite these restrictions, users appear to be aware of the censorship tactics and frequently attempt to access blocked content via Tor, VPNs, and cached versions of blocked websites. In addition, pro-regime cyber groups such as the "Syrian Electronic Army", have conducted coordinated disinformation campaigns and harassment against opposition voices online [12] ●. Political changes following the leadership transition in December 2024 have introduced uncertainty regarding the future of the censorship system, with some reports suggesting a possible loosening of controls, although the long-term trajectory remains unclear [84] ○.

Turkmenistan. The Internet environment in Turkmenistan ranks among the most restrictive worldwide. Although official legislation promotes digital development, in practice the state tightly controls connectivity, monopolizes access, enforces pervasive censorship and opaque cybersecurity controls [67] ●. Most VPN use is prohibited, with violations punishable by imprisonment, reflecting the regimes punitive approach [89]. Authorities justify these restrictions as part of the "Golden Age" ideology, emphasizing national unity through informational isolation [68]. Technical filtering is extensive, including DNS tampering, HTTPS blocking, port-agnostic filtering, and large-scale IP/domain bans [57] ●. Measurement campaigns identified more than 122,000 blocked domains from a sample of

15.5 million, often due to overly broad filtering rules causing substantial collateral damage [63]. Port filtering is also severe: in one measured AS, only $\sim 1,150$ of $65,535$ ports remained accessible [63]. These policies vary across networks and IP prefixes, reflecting an inconsistent enforcement architecture. Authorities have also periodically deployed shutdowns for preemptive censorship, collective punishment, or protest suppression rather than genuine emergency response [86] •. In April 2023, access to roughly 75% of global IP addresses was reportedly blocked.

4 Discussion: The Sliding Scale of Control

This section synthesizes the cross-country findings and develops a comparative interpretation of the identified censorship architectures. Building on data presented in Table 2, we highlight recurring patterns across regime types and derive the proposed "sliding scale" model. Moreover, we discuss limitations and future directions.

Table 2. Global Spectrum of Control: Censorship Mechanisms by FOTN Status

Country	**FOTN Status**	Infrastructure (Shutdowns/Intranet)	**Tech Filtering** (DNS/IP/DPI)	Info Flooding (Trolls/Bots)	Legal/Reg (Fines/Orders)
China	Not Free	○	•	•	•
Iran	Not Free	•	•	•	•
Thailand	Not Free	○	•	•	•
Turkey	Not Free	○	•	•	•
Vietnam	Not Free	○	•	•	•
India	Partially Free	○	•	•	•
Indonesia	Partially Free	○	•	•	•
Mexico	Partially Free	-	○	•	○
Nigeria	Partially Free	○	•	•	•
Ukraine	Partially Free	-	•	○	•
Canada	Free	-	•	-	•
France	Free	-	○	-	•
Germany	Free	-	○	-	•
Italy	Free	-	•	-	•
Taiwan	Free	-	○	○	•
Eritrea	Not Covered	•	○	○	○
North Korea	Not Covered	•	○	○	○
Spain	Not Covered	-	•	-	•
Syria	Not Covered	○	•	○	•
Turkmenistan	Not Covered	•	•	-	•

• = Primary Mechanism, ○ = Secondary/Occasional, - = Not Observed/Evidence Missing

4.1 Commonalities: A Spectrum of Intervention

Table 2 categorizes the censorship mechanisms of our selected countries, sorted by their FOTN status. Rather than a binary divide between "Free" and "Not

Free," the data shows a diagonal progression in control strategies from the physical to the legal domain. We therefore combine the four intervention dimensions from our framework into three overarching models of censorship behavior, defined by their primary operating domain and freedom (Fig. 2).

Model 1: Regulatory Defense (The Legal Domain): Countries classified as "Free" (e.g., Canada, Taiwan, Germany) primarily exert control through the **Legal/Regulatory** dimension. Technical blocking is either absent or limited to specific judicial orders (e.g., Canada's *GoldTV* case or Italy's *Piracy Shield*). Here, the focus is on content regulation and platform liability rather than network interference.

Model 2: Filter & Flood (The Network Domain): As we move to "Partially Free" and "Not Free" nations like Indonesia and Vietnam, the primary mechanisms consolidate around **Technical Filtering** and **Information Flooding**. These regimes maintain connectivity for economic reasons but rely on network-level blocking (e.g., DNS poisoning) and state-sponsored narrative manipulation (e.g., Vietnam's Force 47) to curate the online environment.

Model 3: Infrastructural Denial (The Physical Domain): At the bottom of the spectrum, highly restrictive regimes like Eritrea and North Korea resort to the **Infrastructure** dimension as their primary tool. The main mechanism is preventing physical access to the global Internet entirely, creating a domestic intranet or maintaining a digital void.

Figure 2 visualizes the synthesis of the four dimensions into three models as a "diagonal shift" in censorship strategies. As countries move from "Free" to "Not Free", control shifts from the Legal/Application (Model 1) to the Network (Model 2) and Physical dimension (Model 3). While depicted as a progression, shifts may also occur in reverse as countries liberalize and reduce reliance on infrastructural or network-level controls. Suggesting that while techniques vary, the intent to control remains constant, adapting to technological and legal capacity.

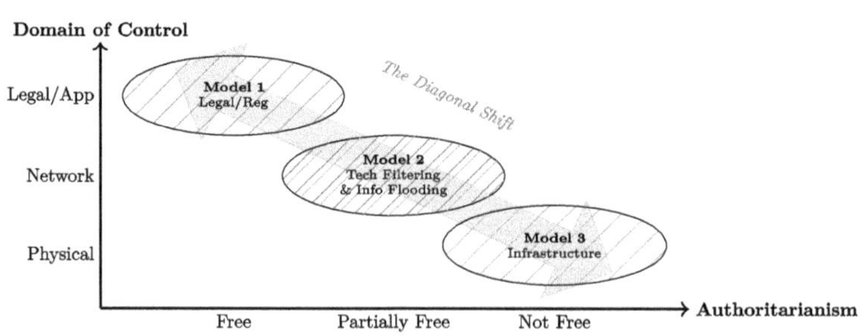

Fig. 2. Sliding Scale of Control: As authoritarianism grows, censorship shifts from legal tools to physical infrastructure.

4.2 Architectural Trade-Offs in Censorship Design

Beyond regime differences, our inductive analysis reveals recurring design trade-offs in how states balance costs, precision, and collateral damage.

High Cost, High Precision. Regimes with high state capacity pursue "Engineering Maximalism." **China** pairs advanced DPI with human flooding to maintain granular control while limiting economic impact [92], whereas **Iran** relies on centralized protocol filtering for "Ideological Whitelisting," often trading usability for strict compliance [14]. These high-investment models allow states to filter dissent without severing global connectivity.

Coarse-Grained Enforcement. Conversely, other nations rely on decentralized or automated blocking where collateral damage is an accepted operational cost. In **Indonesia**, filtering is inconsistent across regions due to fragmented ISP infrastructure, while in **Italy**, the "Piracy Shield" platform prioritizes speed over accuracy, causing automated, mistaken bans of legitimate services. As noted in the recent review in [90], these policy-driven or cost-saving shortcuts often result in significant inconsistency and overblocking.

Strategic Outsourcing. A middle path is observed in **Vietnam**, which employs "Platform Coercion" to outsource censorship to foreign tech giants [49]. By pressuring companies to remove content under threat of law, the state balances economic openness with political control without assuming the massive infrastructure costs of building a domestic firewall.

4.3 Limitations and Future Directions

Our study has several limitations. First, the sample includes only about 10% of global states; although stratified by regime type, it may miss distinctive architectures in unexamined regions such as South America. Second, the analysis relies on secondary rather than primary data, consistent with the paper's goal of synthesizing existing findings into a comparative framework, but this may reduce depth for closed regimes where reliable data are scarce. The proposed four-dimensional framework should therefore be understood as a foundation rather than a definitive taxonomy. Future studies can apply these dimensions to new countries, refine coding criteria, or empirically test the regime-dependent patterns. An accompanying interactive exhibit is being developed as a continuously updated resource to map and explain global Internet censorship practices: https://atlas.omi.uni-ulm.de/.

5 Conclusion

This paper compared Internet censorship in 20 countries, focusing on national control architectures rather than technical details. We identified three distinct models: infrastructural denial in closed autocracies, filter & flood in mixed regimes, and regulatory defense in democracies. Our findings show that censorship adapts to political and institutional contexts, shifting its primary locus of control. While authoritarian regimes invest in technical firewalls to block dissent, democratic nations increasingly adopt similar mechanics under the guise of copyright protection or national security. This convergence suggests that future censorship research should move beyond traffic metrics to auditing the legal and algorithmic systems that govern information visibility. The proposed framework provides a structured baseline for systematically comparing national approaches and guiding future empirical research on evolving censorship.

Acknowledgments. We like to thank Saskia Imhof for her supportive work regarding the investigation of multiple countries.

Conflict of Interest Disclosure. The authors have no competing interests to declare that are relevant to the content of this article.

References

1. Abdulmana, S., Saleh, B.: Coordinate negative content filtering and threat detection in Thailand on the Internet infrastructure. In: The 5th International Conference on Information and Communication Technology for the Muslim World (ICT4M). IEEE (2014)
2. Access Now: Mexico: New guidelines jeopardize net neutrality and empower surveillance (2021). https://www.accessnow.org/press-release/mexico-guidelines-jeopardize-net-neutrality/
3. Access Now: #KeepItOn: Authorities in Nigeria must reconnect people in Zamfara North now (2025). https://www.accessnow.org/press-release/keepiton-authorities-in-nigeria-must-reconnect-people-in-zamfara-north-now/
4. Aceto, G., Montieri, A., Pescapé, A.: Internet censorship in Italy: an analysis of 3G/4G networks. In: 2017 IEEE International Conference on Communications (ICC). IEEE (2017)
5. Aceto, G., Persico, V., Pescapè, A.: Iran's January 2026 internet shutdown (2026). https://arxiv.org/abs/2603.28753
6. Aceto, G., Pescapé, A.: Internet censorship detection: a survey. Comput. Netw. **83** (2015). https://doi.org/10.1016/j.comnet.2015.03.008
7. Agenzia Nazionale Stampa Associata (ANSA): Anti-piracy shield halts google drive (2024). https://www.ansa.it/english/news/general_news/2024/10/20/anti-piracy-shield-halts-google-drive_2e7b137e-361d-4026-a704-9895c798e65a.html
8. Al Jazeera: Twitter restricted in Nigeria after government decree (2021). https://www.aljazeera.com/news/2021/6/5/twitter-restricted-in-nigeria-after-government-decree
9. Amnesty International: North Korea, the surveillance state (2025). https://www.amnesty.org.uk/north-korea-surveillance-state-prison-camp-internet-phone-technology
10. Aryan, S., Aryan, H., Halderman, J.A.: Internet censorship in Iran: a first look. In: 3rd USENIX FOCI (2013)
11. Basso, S.: DNS over TLS blocked in Iran (2020). https://ooni.org/post/2020-iran-dot/

12. Bertram, S.K.: 'Close enough' – the link between the Syrian Electronic Army and the Bashar al-Assad regime, and implications for the future development of nation-state cyber counter-insurgency strategies. Contemp. Voices: St Andrews J. Int. Relat. **8**(1) (2017). https://doi.org/10.15664/jtr.1294
13. Bock, K., Bharadwaj, P., Singh, J., Levin, D.: Your censor is my censor: weaponizing censorship infrastructure for availability attacks. In: 2021 IEEE SPW. IEEE (2021). https://doi.org/10.1109/SPW53761.2021.00059
14. Bock, K., Fax, Y., Reese, K., Singh, J., Levin, D.: Detecting and evading censorship-in-depth: a case study of Iran's protocol whitelister. In: 10th USENIX Workshop on Free and Open Communications on the Internet (FOCI 2020) (2020)
15. Bou-Harb, E., Debbabi, M., Assi, C.: A novel cyber security capability: inferring internet-scale infections by correlating malware and probing activities. Comput. Netw. **94** (2016)
16. Breindl, Y., Wright, J.: Internet filtering trends in western liberal democracies: French and German regulatory debates. FOCI **12** (2012)
17. Burdiak, P.: Online battlefield: the legal landscape of Ukraine's internet censorship before and after the 2022 Russian invasion. Mod. Hist. Polit. Issues **47** (2023). https://doi.org/10.31861/mhpi2023.47.195-205
18. Murthy, C.S.H.N.: State-owned media and democratization in Eritrea: an analytical study. Glob. Media J. Afr. Ed. **6**(2) (2013). https://doi.org/10.5789/6-2-108
19. Cain, G.: Kill one to warn one hundred: the politics of press censorship in Vietnam. Int. J. Press/Polit. **19**(1) (2014). https://doi.org/10.1177/1940161213508814
20. Campbell-Smith, U., Bradshaw, S.: Global cyber troops country profile: India. Project on Computational Propaganda (2019)
21. Chaabane, A., Chen, T., Cunche, M., De Cristofaro, E., Friedman, A., Kaafar, M.A.: Censorship in the wild: analyzing internet filtering in Syria. In: Proceedings of 2014 Conference on Internet Measurement Conference ACM (2014). https://doi.org/10.1145/2663716.2663720
22. Cloudflare: Traffic volume in Iran (2026). https://radar.cloudflare.com/traffic/ir?dateRange=1d
23. Coe, P.: The draft online safety bill and the regulation of hate speech: have we opened Pandora's box? J. Media Law **14**(1) (2022). https://doi.org/10.1080/17577632.2022.2083870
24. Creemers, R.: China's social credit system: an evolving practice of control. Available at SSRN 3175792 (2018)
25. Crowne, E.: All that glitters: federal court of Canada issues Canada's first website blocking order against GoldTV. J. Intellect. Property Law Pract. **15**(2) (2020). https://doi.org/10.1093/jiplp/jpaa002
26. Deibert, R.: Canada and the challenges of cyberspace governance and security, vol. 6. The School of Public Policy Publications (2013). https://doi.org/10.55016/ojs/sppp.v6i1.42426
27. Eissa, T., Cho, G.: Internet anonymity in Syria, challenges and solution. In: Kim, K.J., Chung, K.-Y. (eds.) IT Convergence and Security 2012. LNEE, vol. 215, pp. 177–186. Springer, Dordrecht (2013). https://doi.org/10.1007/978-94-007-5860-5_21
28. Erixon, F., Hindley, B., Lee-Makiyama, H.: Protectionism online: internet censorship and international trade law. Technical report, ECIPE Working Paper (2009)
29. Fifield, D.: Threat Modeling and Circumvention of Internet Censorship. Ph.D. thesis, UC Berkeley EECS (2017)

30. Filter Watch: Wartime digital isolation: Iran's strategic internet shutdown (2026). https://filter.watch/english/2026/03/10/investigative-report-irans-strategic-internet-shutdown-of-february-28/
31. Freedom House: Freedom in the world 2025: The uphill battle to safeguard rights (2025). https://freedomhouse.org/report/freedom-world
32. Freedom House: Freedom of the net 2025 (2025). https://freedomhouse.org/report/freedom-net/2025/uncertain-future-global-internet
33. Gebhart, G., Kohno, T.: Internet censorship in Thailand: user practices and potential threats. In: 2017 IEEE European Symposium on Security and Privacy (EuroS&P). IEEE (2017). https://doi.org/10.1109/EUROSP.2017.50
34. Ghebregiorgis, F., Mihreteab, H.T.: Determinants of internet use and internet penetration in eritrea: evidences from the city of Asmara. J. Econ. Manag. Sci. **1**(1) (2018). https://doi.org/10.30560/jems.v1n1p28
35. Golovchenko, Y.: Fighting propaganda with censorship: a study of the Ukrainian ban on Russian social media. J. Polit. **84**(2) (2022). https://doi.org/10.1086/716949
36. Gosain, D., Singh, K., Sharma, R., Suresh Babu, J., Chakravaty, S.: Out in the open: on the implementation of mobile app filtering in India. In: International Conference on Passive and Active Network Measurement. Springer (2024)
37. Gumede, W.: Rise in censorship of the Internet and social media in Africa. J. Afr. Media Stud. **8**(3) (2016)
38. Han, R.: Defending the authoritarian regime online: China's "voluntary fifty-cent army". China Q. **224** (2015). https://doi.org/10.1017/S0305741015001216
39. Hannaford, N.: Death by a thousand clicks: the rise of internet censorship and control in Canada (2025). https://www.jccf.ca/wp-content/uploads/2025/12/Death-by-a-thousand-clicks_Final.pdf
40. Hashemzadegan, A., Gholami, A.: Internet censorship in Iran: an inside look. J. Cyberspace Stud. **6**(2), 183–204 (2022). https://doi.org/10.22059/JCSS.2022.349715.1080
41. Helwani, I.: Cyberactivism in [Syria: Emergence, transformation, potentials, and limitations. Güvenlik Stratejileri Dergisi **20**(48) (2024)
42. Hu, W., Barradas, D.: Work in progress: a glance at social media self-censorship in North America. In: 2023 IEEE European Symposium on Security and Privacy Workshops (EuroS&PW) (2023). https://doi.org/10.1109/EuroSPW59978.2023.00072
43. Indian Gov: Section 69a Of It Act 2 | Ministry of Electronics and Information Technology (2008). https://www.meity.gov.in/documents/act-and-policies/section-69a-of-it-act-2-YjNxgTMtQWa
44. Iszaevich, G.E.W.: Distributed detection of tor directory authorities censorship in Mexico. In: The Eighteenth International Conference on Networks (2019)
45. Jin, L., Hao, S., Wang, H., Cotton, C.: Understanding the practices of global censorship through accurate, end-to-end measurements. Proc. ACM Meas. Anal. Comput. Syst. **5**(3) (2021). https://doi.org/10.1145/3491055
46. Josephy-Hernández, D.E.: A brief history of anime censorship in the United States, México and Costa Rica: reception and adverse reactions. In: Routledge Handbook of East Asian Translation. Routledge (2025)
47. Lange, F., Niere, N., Jonathan von Niessen: censors ignore unencrypted HTTP/2 traffic (2024). https://upb-syssec.github.io/blog/2024/http2/
48. Laungaramsri, P.: Mass surveillance and the militarization of cyberspace in post-coup Thailand. Adv. Southeast Asian Stud. **9**(2) (2016)

49. Luong, D.N.A.: A study of Vietnam's control over online anti-state content. In: A Study of Vietnam's Control over Online Anti-state Content, pp. 1–27. Trends in Southeast Asia, ISEAS–Yusof Ishak Institute (2023)
50. Manullang, S.O.: The legality of devious cyber practices: readiness of Indonesia's cyber laws. Society **10**(2) (2022). https://doi.org/10.33019/society.v10i2.482
51. Master, A., Garman, C.: A worldwide view of nation-state internet censorship. In: Free and Open Communications on the Internet (FOCI'23) (2023)
52. Medina, M.: Governmental censorship of the internet: Spanish vs. Catalans case study. Libr. Trends **68**(4) (2020)
53. Mehta, S.N., Amit-Danhi, E.R.: The road to censorship: the case of digital audio-visual industries in India. Int. J. Cult. Policy **31**(7) (2025)
54. Meyer, T.: Graduated response in France: the clash of copyright and the Internet. J. Inf. Policy **2** (2012)
55. Moats, D., Tseng, Y.S.: Using quali-quantitative methods to interrogate the role of algorithms in digital democracy platforms. Inf. Commun. Soc. **27**(5) (2024). https://doi.org/10.1080/1369118X.2023.2230286
56. Momen, M.N., S., H., Das, D.: Mediated democracy and internet shutdown in India. J. Inf. Commun. Ethics Soc. **19**(2) (2021). https://doi.org/10.1108/JICES-07-2020-0075
57. Net4People: Bidirectional DNS, HTTPS, HTTP injection in Turkmenistan (2021). https://github.com/net4people/bbs/issues/80
58. NetBlocks: Mapping internet freedom in real time (2026). https://netblocks.org/
59. Nguyen, N.L., Wang, M.H., Dai, Y.C., Dow, C.R.: Understanding malicious accounts in online political discussions: a multilayer network approach. Sensors **21**(6) (2021). https://doi.org/10.3390/s21062183
60. Nguyen-Pochan, T.T.P.: State management of social media in Vietnam. Russian J. Vietnamese Stud. **5**(1S) (2021). https://doi.org/10.54631/VS.2021.S-23-33
61. Niaki, A.A., et al.: ICLab: a global, longitudinal Internet censorship measurement platform. In: 2020 IEEE Symposium on Security and Privacy (SP). IEEE (2020). https://doi.org/10.1109/SP.2020.00014
62. Norton, B.: Music and censorship in Vietnam since 1954. In: Hall, P. (ed.) The Oxford Handbook of Music Censorship. Oxford University Press (2017). https://doi.org/10.1093/oxfordhb/9780199733163.013.29
63. Nourin, S., et al.: Measuring and evading Turkmenistan's internet censorship: a case study in large-scale measurements of a low-penetration country. In: Proceedings of ACM Web Conference 2023. ACM (2023). https://doi.org/10.1145/3543507.3583189
64. Park, J., Jang, R., Mohaisen, M., Mohaisen, D.: A large-scale behavioral analysis of the open DNS resolvers on the Internet. IEEE/ACM Trans. Network. **30**(1) (2022). https://doi.org/10.1109/TNET.2021.3105599
65. Piskunova, N.: Small, stable, authoritarian? Assessing Eritrea's consistency in abusing media freedoms. In: Saqib, F., Wlodarczyk, M. (eds.) Authoritarian and Populist Influences in the New Media. IGI Global (2017). https://doi.org/10.4018/978-1-5225-2187-7.ch014
66. Piña-García, C.A., Espinoza, A.: Coordinated campaigns on Twitter during the coronavirus health crisis in Mexico. Tapuya Latin Am. Sci. Technol. Soc. **5**(1) (2022). https://doi.org/10.1080/25729861.2022.2035935
67. Progres: The regulation of the Internet in Turkmenistan (2024). https://progres.online/reports/internet-freedom/the-regulation-of-the-internet-in-turkmenistan/
68. Qurim: Turkmenistan and their "Golden DPI" (2019). https://www.qurium.org/alerts/turkmenistan-and-their-golden-dpi/

69. Rama, B.G.A.: Press freedom in the digital era in Indonesia: a human rights perspective. JIHAD: Jurnal Ilmu Hukum dan Administrasi **6**(4) (2024). https://doi.org/10.58258/jihad.v6i4.7543
70. Ramesh, R., et al.: Network responses to Russia's invasion of Ukraine in 2022: a cautionary tale for internet freedom. In: Proceedings of 32nd USENIX Conference on Security Symposium. USENIX Association (2023)
71. Rauchfleisch, A., Tseng, T.H., Kao, J.J., Liu, Y.T.: Taiwan's public discourse about disinformation: the role of journalism, academia, and politics. Journal. Pract. **17**(10) (2023). https://doi.org/10.1080/17512786.2022.2110928
72. Saka, E.: Social media in Turkey as a space for political battles: AKTrolls and other Politically motivated trolling. Middle East Critique **27**(2) (2018). https://doi.org/10.1080/19436149.2018.1439271
73. Santos Okholm, C., Ebrahimi Fard, A., Ten Thij, M.: Blocking the information war? Testing the effectiveness of the EU's censorship of Russian state propaganda among the fringe communities of Western Europe. Internet Policy Rev. **13**(3) (2024). https://doi.org/10.14763/2024.3.1788
74. Satriawan, I., Elven, T.M.A., Lailam, T.: Internet shutdown in Indonesia: an appropriate response or a threat to human rights? Sriwijaya Law Review (2023). https://doi.org/10.28946/slrev.Vol7.Iss1.1018.pp19-46
75. Shen, P.: How China initiates information operations against Taiwan. Taiwan Strategists **12** (2021)
76. Shiker, Z.R., Tsegay, S.M.: (Mis)Education in authoritarian regimes: the case of Eritrea. Educ. Sci. **15**(7) (2025). https://doi.org/10.3390/educsci15070801
77. Sihidi, I.T.: The rise of symptoms of digital authoritarianism: lesson from Indonesia. Int. J. Polit. Public Policy **2**(1) (2025). https://doi.org/10.70214/28vg6e57
78. Singh, K., Grover, G., Bansal, V.: How India censors the web. In: Proceedings of 12th ACM Conference on Web Science. ACM (2020). https://doi.org/10.1145/3394231.3397891
79. Sirikupt, C.: Drowning out dissent: the Thai military's quest to fabricate popular support on Twitter. Int. J. Press/Polit. **31**(1) (2026). https://doi.org/10.1177/19401612241279158
80. Skrzypczak, J.: The effect of the 'China Factor' on Taiwan's media system security as an example of the 'privatization and outsourcing' of censorship and propaganda in the digital age. Przegląd Strategiczny **9**(12) (2019). https://doi.org/10.14746/ps.2019.1.22
81. Tanash, R.S., Chen, Z., Thakur, T., Wallach, D.S., Subramanian, D.: Known unknowns: an analysis of twitter censorship in Turkey. In: Proceedings of 14th ACM WPES. ACM (2015). https://doi.org/10.1145/2808138.2808147
82. Tenenboim, O.: Under-the-radar engagement: how and why news users limit their public expression. J. Comput.-Mediated Commun. **30**(1) (2025). https://doi.org/10.1093/jcmc/zmae024
83. The Internet Monitor: Internet Monitor (2026). https://thenetmonitor.org/
84. The New Arab Staff: New Syria government unblocks websites banned by Assad regime (2025). https://www.newarab.com/news/new-syria-government-unblocks-websites-banned-assad-regime
85. Theil, S.: The German NetzDG: A Risk Worth Taking? Verfassungsblog: On Matters Constitutional (2018). https://doi.org/10.17176/20180208-184652
86. Thumfart, J.: Digital rights and the state of exception. Internet shutdowns from the perspective of just securitization theory. J. Glob. Secur. Stud. **9**(1) (2024). https://doi.org/10.1093/jogss/ogad024

87. Tschantz, M.C., Afroz, S., Paxson, V.: SoK: towards grounding censorship circumvention in empiricism. In: 2016 IEEE Symposium on Security and Privacy (S&P). IEEE (2016). https://doi.org/10.1109/SP.2016.59
88. Ververis, V., Marguel, S., Fabian, B.: Cross-country comparison of internet censorship: a literature review. Policy Internet **12**(4) (2020)
89. Watch, H.R.: Turkmenistan events 2023 (2024). https://www.hrw.org/world-report/2024/Country-chapters/turkmenistan
90. Wendzel, S., Volpert, S., Zillien, S., Lenz, J., Rünz, P., Caviglione, L.: A survey of internet censorship and its measurement: methodology, trends, and challenges. Comput. Secur. **164** (2025). https://doi.org/10.1016/j.cose.2025.104732
91. Winter, P.: Measuring and Circumventing Internet Censorship. Ph.D. thesis, Karlstads universitet (2014)
92. Wu, M., et al.: How the great firewall of China detects and blocks fully encrypted traffic. In: Proceedings of 32nd Usenix Security Symposium USENIX (2023)

Modelling the Stages of an Organisational Whistleblower's Journey

Karen Renaud[1,2,3(✉)], Karl van der Schyff[4], and Merrill Warkentin[5]

[1] University of Strathclyde, Glasgow, UK
karen.renaud@strath.ac.uk
[2] University of South Africa, Pretoria, South Africa
[3] Rhodes University, Grahamstown, South Africa
[4] Abertay University, Dundee, UK
k.vanderschyff@abertay.ac.uk
[5] Mississippi State University, Starkville, MS, USA
m.warkentin@msstate.edu

Abstract. Whistleblowers perform an essential service in revealing wrongdoing. Most feel compelled to highlight nefarious organisational activities. If their efforts within the organisation are ignored and they feel driven to "go public", disclosures can harm the organisation's cybersecurity and whistleblowers themselves usually pay a high price for their actions. We argue that, rather than vilifying whistleblowers, organisations 'harness' their propensity to keep the organisation on the straight and narrow. To achieve this, it is crucial to understand the whistleblower's journey from being triggered by some unethical organisational activity to the final external whistleblowing act. We reviewed the archival literature to produce a synthesis of whistleblower stages. We carried out a case study and interviewed two whistleblowers to inform derivation of a staged whistleblowing model. This makes it possible to identify pressure points for targeted interventions which can encourage internal whistleblowing and thereby prevent the tangential dismantling of organisational information security.

Keywords: Whistleblowing · Insider · Internal vs. External

1 Introduction

Whistleblowing is a topic of profound importance to the study of organisations, especially considering issues of power and resistance [5,16,49,66]. While '*everyone supports whistleblowers until they have one in their own organization*' [95, p iix], whistleblowers are rarely celebrated [82], even though whistleblowing has the power to deliver great benefit in many cases [7,12]. Such whistleblowing is often seen as a betrayal, so it is probably fair to say that organisations dread whistleblowers, considering them a threat [86]. Certainly, whistleblowers themselves pay a heavy price for acting on their corporate social responsibility [37,48,53].

© The Author(s), under exclusive license to Springer Nature Switzerland AG 2027
K. Renaud et al. (Eds.): EICC 2026, CCIS 3012, pp. 88–105, 2027.
https://doi.org/10.1007/978-3-032-28957-5_6

Whistleblowers, classified as an cybersecurity-related insider threat [83], are a threat to the organisation's cybersecurity because: (1) they always collect evidence to support their claims, and this can violate the confidentiality of the organisation's information; (2) if a bad agent within the organisation is exploiting an unreported vulnerability, the integrity or the availability of the information could well be compromised. It is thus worth encouraging internal whistleblowing [55] so that these events can be improve the organisation [60] and its cybersecurity, in the long run. Figure 1 outlines the structure of the paper.

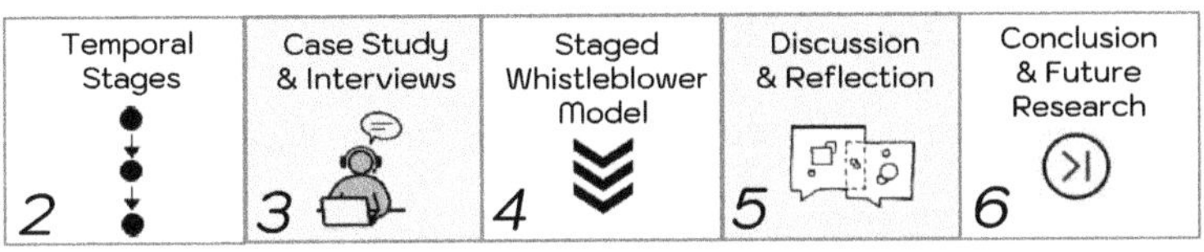

Fig. 1. Paper Structure & Contributions.

2 Delineating Whistleblowing

Whistleblowing is defined by Near and Miceli [67, p.4] as: "*The disclosure by organisation members (former or current) of illegal, immoral, or illegitimate practices under the control of their employers, to persons or organisations that may be able to effect action*". Gøtzsche [34] argues that modern whistleblowers perform great deeds for the common good, confirmed by Seaton [92]. Lamb [54] found that, according to her participants, "*morally, blowing the whistle was the right thing to do; to do nothing made them complicit; and, they would blow the whistle again* (p.ii). Duska [24] claims that there is no obligation for any employee to feel loyalty to an employer who is harming society. This is because, as argued by Dasgupta and Kesharwani [18], the whistleblower is demonstrating loyalty to the organisation as a whole and not to unethical managers. Kenny *et al.* [49] agree, claiming that whistleblowers are in need of support in carrying out their whistleblowing activities.

Some consider whistleblowing to be corrosive and impermissible [88] and many whistleblowers are treated as outcasts [6]. These authors believe that encouraging whistleblowing effectively incites disloyalty, and creates dissent and conflict. The actions of the whistleblower undoubtedly harm: (1) the organisation, who will suffer reputational damage, disruption which can result in costly litigation [43,44], and (2) the whistleblower him or herself, who often pays a heavy price for blowing the whistle [11] including losing their citizenship [94], incarceration [93] and vilification [26,56]. We conclude that whistleblowers do indeed have a role to play in deterring organisations from behaving unethically and *going rogue*. If they feel compelled to blow the whistle, they might well compromise the organisation's cybersecurity, at some cost to the organisation, on top of the reputation harm that comes from the whistleblowing event.

Culiberg and Mihelič [17, p.787] argue that '*more needs to be done in terms of explaining the process that occurs between the time a wrongdoing is witnessed and when actual whistleblowing occurs.*' This idea of a progression is confirmed by various authors [10,78,104]. Moreover, Near and Miceli [69] argue that '*every whistleblowing situation is a process that unfolds over time, not limited to a single event*' (p.109). To target interventions, it is helpful to identify key pressure points in the whistleblower's journey (Table 1).

Table 1. Temporal Stages of the Whistleblowing Process (WB; Int = Internal; Ext = External; Resp = Response; Eval = Evaluate)

<table>
<tr><th colspan="4">STAGE 1: Formulate Intention: [18,29,36,40,58,68,72,85]</th></tr>
<tr><td>Become Aware: [8,17,21,23,27, 31,33,54,75,79, 80,84,99,102,110, 110]</td><td>Consider Serious/Unethical/ Immoral: [17,21,23,31, 33,50,54,74,99,109]</td><td>Personal Responsibility: [17,23,33,46,99]</td><td>Make Decision: [2,8,21,23,33,45, 54,72,75,80,84,99, 110]</td></tr>
</table>

<table>
<tr><th>STAGE 2: Int WB</th><th>STAGE 3: Eval Org Resp</th><th>STAGE 4: Ext WB</th></tr>
<tr><td>[8,17,18,21,27,29,31,36,40, 58,68,69,75,79,80,84,102, 109,110]</td><td>[18,27,29,40,58,68,69,75, 80,84]</td><td>[2,18,36,40,58,68,69,80, 102]</td></tr>
<tr><th colspan="3">STAGE 5: Reflection: [27,41]</th></tr>
<tr><td>Org Response: [2,18,36,40]</td><td>Consequences: [2,21,36,76,79,85,102]</td><td>Plan Next Steps: [13,63,102]</td></tr>
</table>

3 Case Study and Interviews

3.1 Research Methodology

Watts and Buckley [106] point out that most whistleblower research relies on surveys, but also cite Miceli and Near [64] to outline the limitations of this kind of research, due to most survey responses being collected from 'clean' organisations or students. To avoid this limitation, we engaged directly with the experiences of two whistleblowers, using media reports for two, and interviews for another two.

Data Gathering: *Case Study.* We made use of a non-experimental qualitative methodology (i.e., data was not manipulated during evaluation) [81], because of the nature of the secondary data sources. To obtain a representative sample, we included public sources and media items that portrayed Snowden as a hero, but also those that did not. We also included documentaries and public debates. The following criteria were applied to choose suitable media items: (1) feature key government actors that formed a core part of the media campaigns surrounding Edward Snowden; (2) provide an holistic (and impartial) analysis of the events surrounding the Edward Snowden disclosures; (3) feature Edward Snowden himself; (4) feature a public debate between individuals with a vested interest in privacy matters; (5) be expert lectures by cybersecurity experts.

Data Gathering: ***Interviews.*** We carried out narrative based interviews with two whistleblowers: a National Health Service General Practice Doctor (UK), and a banker (South Africa). We used the 'Tell Me a Story' approach advised by Hull *et al.* [42] so that the whistleblower told the story in their own words. This ensured that we did not prime responses. We asked follow-up questions to enhance our understanding but listened, rather than spoke, so as to give them the freedom to speak in a narrative way. The interview protocol was approved by the first author's institutional ethics review board.

Analysis: ***Case Study:*** We interrogated each statement Snowden made, to uncover the underlying thought processes. We provide key quotes to support our progression.

Analysis: ***Interviews:*** We analysed the interview data inductively so as to gain the greatest insight possible. Such inductive analyses enables researchers to draw conclusions and identify patterns from unstructured data. In other words, without any preconceived ideas as to what theoretical frameworks or constructs we should be considering. Such inductive analyses enables researchers to draw conclusions and identify patterns from unstructured data – in our case the interview data. We argue that this method is suitable given its ability to find fresh viewpoints and managing the complexity of numerous phenomena of interest [14]. We started our analysis by reading (and re-reading) the transcripts. This first phase guarantees objectivity, allowing themes to emerge spontaneously [97]. Once familiar with the transcripts, we coded them – an essential component of inductive analysis. Initially, smaller sentence fragments were coded to get a sense of the initial themes and patterns. Eventually, we were merged these codes on a conceptual level as the links between the codes become clearer. This, in turn, enabled us to identify core overarching themes. Identifying themes in this manner also improved our understanding how these themes interacted with other themes. We did this until we reached a point of theoretical saturation [30]. We did not carry out our inductive analysis in isolation to comply with the concept of triangulation. In this regard, our literature review informed the interview process and the choice of subjects to interview. This strategy produces a well-rounded perspective of the facts, which enhanced our analysis [22]. We gave our interviewees a chance to view and comment on our findings before continuing with our analysis. Such checking strengthens the validity of our findings; allowing them to engage further with the researcher [57]. From the above process, a narrative emerged that confirmed the stages and factors we had already identified from the research literature.

3.2 Case Study: Edward Snowden (NSA Whistleblower)

Edward Snowden is one of the most (in)famous whistleblowers. He revealed widespread electronic surveillance, by the NSA, of both USA and foreign citizens, including telephonic and Internet data [73]. He leaked exfiltrated information via press releases through The Guardian and Washington Post newspapers in early June, 2013 [32]. Video quotes attributed as follows: (source video number, time in video) – see Appendix for Sources.

STAGE 1—Formulate Intention to Blow Whistle. Trigger event & (i) Becoming Aware: '*when I was sitting at my desk at the NSA working with tools of mass surveillance every day I saw that all of our communications are being intercepted all of the time in the absence of any suspicion of wrongdoing.*' (3:9). (ii) Feeling that situation is serious: '*over time, that awareness of wrongdoing sort of builds up and you feel compelled to talk about it and the more you talk about it the more you're ignored the more you told it's not a problem until eventually you realise that these things need to be determined by the public not by somebody who is simply hired by the government.*' (7,3). (iii) Feeling that it is his personal responsibility to do this: '*I think anyone who opposes that sort of world has an obligation to act.*'

STAGE 2—Blow Whistle Internally: He claims that his attempts to blow the whistle internally were ignored [87] and proceeds directly to contemplate external whistleblowing.

STAGE 3—Evaluate Outcomes: N/A

STAGE 4—Blow Whistle Externally: He shared his materials with Barton Gellman, formerly of The Washington Post, filmmaker Laura Poitras and Glenn Greenwald of The Guardian [20]. '*And that's why when I worked with journalists, I made sure that I never had to make that decision. I wanted to remove from myself the authority, the sort of positional capability to publish any document on my own. And that's the reason why now, we're sitting in 2016, I've never made public a single classified document.*' (4,30). '*When I really came to struggle with these issues, I thought to myself, how can I do this in the most responsible way, that maximises the public benefit while minimising the risks?*' (15,8).

Snowden rationalises the whistleblowing act: '*I came in and I took the Oath of Service. Now I will point out, it's not an oath of secrecy. In fact, no oath of secrecy exists within the CIA or the NSA. The Oath of Service is to support and defend the Constitution of the United States of America against all enemies, foreign and domestic. It's not a pledge of loyalty to the President, to an official, to a policy, or anything else. It's to defend the Constitution, specifically. And not just against those outside, but also against those inside.*' (4,11).

STAGE 5—Reflection: *Evaluating Outcomes & Consequences:* Snowden was charged with crimes against the United States of America, two of which fall under the Espionage Act of 1917 [4]. He sought sanctuary in Russia and had his passport withdrawn by the US government [94], despite the UN Convention of 1961, which requires: '*that states establish safeguards in their nationality laws to prevent statelessness at birth and later in life.*' [100, p. 3]. The following statement suggests that he felt he has been unfairly disparaged: '*They included espionage charges. Now, they knew that I wasn't working for any foreign government. That* *was clear from the beginning. They knew that I was working with journalists and that the recipient of the information was the public at large.*' (6,36).

3.3 Interview 1: UK National Health Service (NHS) Whistleblower

Rost [90] reviewed 233 healthcare whistleblowers and found that 90% were fired or demoted and 27% faced lawsuits. However, only 16% said that they would *not* blow the whistle again, which suggests a mindset of doing what they deem to be right, despite the personal cost. There is evidence that UK NHS whistleblowers pay a heavy price for blowing the whistle e.g., [9,51, p.119], with the courts tending to support the health service rather than the whistleblower (e.g., Dr Chris Day [19]).

Interviewing a healthcare whistleblower was thus important to help verify our process model We refer to the Doctor we interviewed as 'W1'. W1 had a very successful GP practice in a building that housed a number of independent practices. He had amassed a wealth of experience both in the UK and in other countries, in a variety of medical specialisms, before establishing his GP practice.

STAGE 1—Formulate Intention to Blow Whistle (Motivation)
W1 became aware of the fact that some of the other local practices' general practitioners (GPs) were regularly working a full day in their GP practices and then going to the local hospital to work a full night shift before returning to their practice to work the following day (*trigger*). This is not in violation of current UK regulations because, whilst the totality of the GPs hours worked included employed work in their role as employees at the hospitals, and the self-employed work at their practices, it is only *employed* work that is regulated by UK regulations. As a consequence, the totality of hours is not regulated. However, it *is* in violation of common sense and self and others' preservation. W1 became aware of this, and other GPs and staff members also mentioned concerns about this practice to him. All felt that this was compromising patient safety and care and, in some cases, could link specific patient incidents to this practice (*serious*). As other staff were uncomfortable at the prospect of speaking out, W1 thus felt he had to do something to address the situation (*personally responsible*). This confirms what Kenny [47] argues, that whistleblowing is an '*intrinsically collective phenomenon, even when it appears as though only one person is speaking out*' (p. 212)

STAGE 2—Blow Whistle Internally: (1) *Takes Personal Action:* W1 went to speak to one of the doctors who was engaging in this practice (*confront wrong-doer)*. He was angrily dismissed. (2) *Gathers Evidence:* W1 gathered evidence in the form of rotas[1] from the hospital. W1 has an entire room of his house filled with documents he gathered during his experiences (*retains records*). W1 then decided to blow the whistle internally, but the channels for doing so were not widely publicised. So, he reported the wrong-doing to the clinical director of the GP collective where the wrongdoers were doing the night shifts.

[1] Work schedules.

STAGE 3—Evaluate Outcomes: The overworking reduced, but was re-established within three months. W1 then discovered, by submitting a Freedom of Information request, that another hospital and Health Board were allowing the same kind of overworking by the same GPs.

STAGE 4—Blow Whistle Externally: W1 then attempted to blow the whistle externally but, perhaps due to the affection and loyalty the British public had for their NHS [25] at the time, and because the correspondent thought the issue would be too complex for the paper's readership, the nearest city's newspaper declined to print the story. The local newspaper did print part of the story but it did not go any further as further publication was suppressed by the Editor, who, W1 was informed, had been pressurised by external sources. W1 argued that he had attempted to prevent the unethical behaviours both by speaking to the medics themselves and also by speaking to managers. He felt that he had no option but to go public because he believed that patients were receiving substandard care.

STAGE 5—Reflection: *Consequences:* W1 was ostracised by the medical staff of the other practices in his building. Those who had the same concerns about the wrongdoing did not publicly support him. A number of spurious complaints were made about him (including criticism of his appearance, his demeanour, and his patient care), all of which lacked substance. The complaints were constructed or encouraged by those about whom his complaint of dangerously excessive overworking was made. W1 was eventually forced out of his practice and had not been able to establish a new practice since then. When asked if he thought his whistleblowing was preventing him from establishing a new GP practice, he said he could sense the fear in others to be seen associating with him and thought that this was indeed the case. This confirms assertions by Van Portfliet [102], who claims that whistleblowers are stigmatised and transition from 'employee' to 'whistleblower', and implicitly accept this label, together with all the negative connotations it carries. Kenny [47] also highlights the difficulties whistleblowers experience in gaining employment after they speak out.

Outcomes: The external whistleblowing failed. *Willingness to blow the whistle again:* When asked if he would blow the whistle again if he knew then what he knows now, he said he would – because he would consider it his duty to report wrongdoing. In saying this, he confirmed what Rost [90] found in his investigation, and what Gøtzsche [34] asserts, about what really motivates whistleblowers: a strong moral compass that compels them to act when they become aware of wrongdoing.

3.4 Interview 2: Internal Whistleblower (Bank)

All the previous discussions are related to external (public) whistleblowing. Here, we report on an internal whistleblowing incident related by a manager in a large South African bank in Pretoria, South Africa. We refer to him as W2. In the mid 1990s, W2 returned from holiday and asked his assistant for the ledger entries for all claimed expenses during his absence. Reviewing these was one of his responsibilities.

STAGE 1—Formulate Intention to Blow Whistle (Motivation). While reviewing expenses, W2 noticed a number of entries for exactly the same amounts, a week or two apart. He then asked to see the original receipts that had been submitted. The assistant only had half of the paperwork. The rest had been submitted via the Johannesburg-based head office. He then obtained the records from Johannesburg and discovered that a particular member of staff (we shall call him HJ) had been submitting claims for the same expense both in Pretoria at his local branch, and in Johannesburg, at the Head Office. This was the *trigger*. When W2 dug a bit deeper, it became clear that claims from HJ appeared to cover personal items that he had purchased, personal restaurant bills and petrol for personal use. The assistant then drew his attention to the fact that five very expensive African portraits had been removed from the board room, and replaced with discarded posters from the store room. HJ had decided, without consulting anyone, that these were worth ZAR100 each (when the more accurate price is estimated to be ZAR750 at that time). He then 'bought' them by depositing ZAR500 into the local branch expense account. W2 realised that HJ was engaged in widespread fraud and theft (*serious*), and that he, having uncovered it, had to report it (*personally responsible*).

STAGE 2—Blow Whistle Internally: HJ was W2's manager's manager, so W2 did not feel he could confront him directly. He was also concerned that HJ would destroy evidence if he realised W2 had become aware of his fraudulent activities. W2 made photocopies of all the paperwork covering the irregularities (*gathers evidence*). Using internal whistleblowing channels, W2 decided to blow the whistle internally. He contacted the central banking head office and attended a meeting with executives and a human resources representative where he presented all the evidence he had gathered related to HJ's activities. They asked W2 to return to work and not to say anything, while they immediately initiated an internal investigation.

STAGE 3—Evaluate Outcomes: Within a week, HJ was suspended. A hearing took place (where W2 testified) and HJ was fired. HJ appealed but lost the case. W2 was satisfied by the outcome. This, in his mind, concluded the incident.

STAGE 4—Blow Whistle Externally: N/A

STAGE 5—Reflection: *Consequences:* W2 was thanked for his actions, and felt that his employer had developed faith in his integrity. W2 was glad he had blown the whistle even though having to testify at the hearing was very stressful. He and HJ had been golf partners and on friendly terms before these events took place. This made it very difficult for him to report HJ, as argued by Hess [38] and included in Nicholls' [70] model. Some of W2's colleagues felt that he ought to have confronted HJ, but he explained that if this meant HJ was able to destroy evidence, that could have led to accusations of his helping HJ to hide his actions. W2 thus experienced what Trevino [98] found i.e., that ethical behaviours lead to peers disliking the person who blows the whistle on colleagues. The whistleblowing achieved its aims of curtailing the ongoing fraud. W2 remains convinced that he did the right thing in blowing the whistle internally and was satisfied that the bank acted correctly in response to his internal whistleblowing report.

3.5 Lessons Learned

A number of extra aspects emerged that needed to be reflected in the stage model, as follows:

Evidence gathering was a theme that ran through both case studies and interviews. This appeared to be carried out in order to substantiate claims. This is an activity that is eased and facilitated by the cyber era. Indeed W2's whistleblowing took place during the paper-based business era, and he mentioned making photocopies of everything. The others simply said they gathered electronic evidence. This is also highlighted by Martin [61].

Fear of Consequences/Retaliation: Snowden did not attempt to blow the whistle internally, having observed what happened to others who attempted this and the lack of action based on previous internal whistleblowing. This personal cost calculus ought to be included in Stage 2 because in both cases their calculations led to a decision to "go public".

Personal Action: W1 mentioned speaking to wrongdoers directly, a courageous action that not every whistleblower will have the capability to take. W2, for example, did not feel he was in a position to confront his manager's manager directly. As such, this optional action could precede internal whistleblowing in the stage model. Hess *et al.* [38] find that this is the preferred option when the employee and the wrongdoer have an amicable pre-existing relationship.

4 Whistleblowing Stage Model

We now combine insights from previous Sections, the case studies and interviews to develop a staged model of the whistleblower journey. This whistleblowing stage model delivers a nuanced understanding of the whistleblowing phenomenon and highlights key pressure points for averting the process explaining how and why events unfold over time [105]. We commenced with the stages depicted in Table 1 which were confirmed in the case studies and interviews. Finally, we incorporated the 'lessons learned' reported in Sect. 3.5. The integrated model is shown in Fig. 2. While stages 2–5 are clear from the figure, we briefly review Stage 1, which requires more explanation.

Trigger Event: Something untoward occurs which could be an unethical, immoral or illegal. (i) *Becoming Aware:* There might be a delay between the trigger event and awareness thereof. An employee's ethical sensitivity is influential after awareness occurs [3,35,96]. (ii) *Recognition of Seriousness:* This is an objective judgement related to how serious the consequences are [15,59,108].

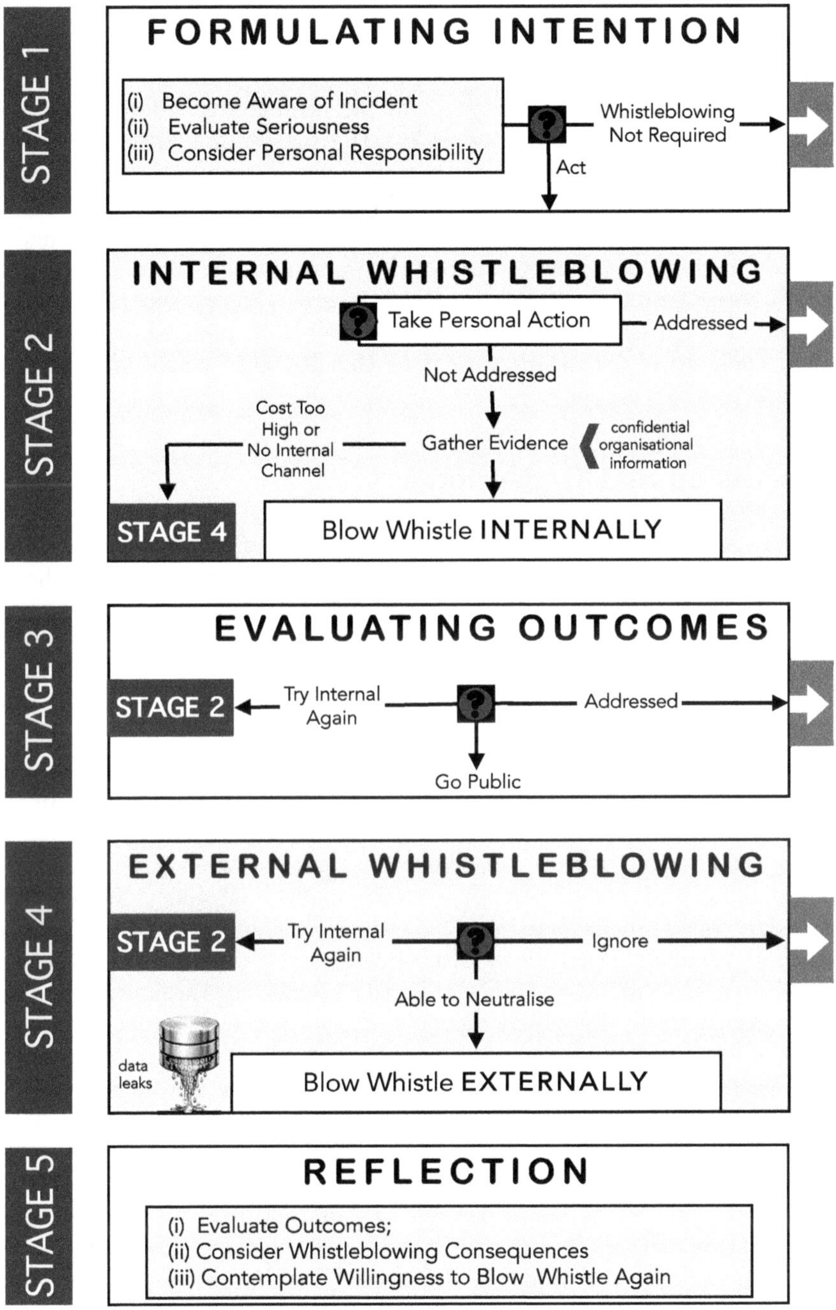

Fig. 2. Whistleblowing Model Stages aligned with Theory. Potential exits present opportunities for targeted interventions; Undesirable final exit allows unethical behaviours to continue. (?) denotes a choice to be made.

Fredin *et al.* [28] finds that whistleblowers will be relieved if they report on situations they consider serious. (iii) *Personal Responsibility:* Sometimes, employees see something, but do not take personal responsibility to act [89]. The whistleblower embraces the responsibility to do something, instead of remaining a bystander [59,89]. If the person considers the matter not to be serious [39], is not personally affected [91], or does not accept responsibility [89], the exit will be taken because there is no motivation to blow the whistle.

The whistleblowers we investigated gradually became aware of wrongdoing, considered it serious, and accepted personal responsibility to highlight the wrongdoing. W1 accepted responsibility because others who were aware of the problem did not have the capability to act. In the case of W2, he realised that the wrongdoer was likely to try to frame him if he did not act, which gave him extra motivation.

5 Discussion and Reflection

We propose a staged model of whistleblowing that reflects the journey in five distinct stages. This model provides a structured lens through which the complexities of whistleblowing can be understood and managed within organisations. While other researchers have proposed more static models of whistleblowing [63,103], we incorporate a temporal dimension into the model, reflecting the reality of a journey from awareness at the one end to blowing the whistle at the other. The model emphasises the evolving motivations, perceptions, and consequences of organisational and employee responses at each stage. It underscores whistleblowing as a process of ethical engagement, rather than an single disruptive act.

The staged model bridges the theoretical gap between individual ethical decision-making and organisational systems of accountability. Other models acknowledge the temporal nature of whistleblowing [65], but this model provides a more granular structure by distinguishing between internal and external whistleblowing. The inclusion of the reflective element of the journey is important because it reflects the whistleblower's emotional responses to having blown the whistle, and probably dealing with the negative consequences thereof [101].

By isolating temporal stages into the whistleblowing model we align with broader behavioural theories such as the theory of planned behaviour [1] and sense making theory [107]. This underscores the conflicts individuals have to wrestle with in navigating moral dilemmas in ethically-charged contexts. As such, the model provides a comprehensive explanatory framework of the whistleblower's journey.

For organisations, the value of this model lies in its capacity to support targeted, stage-appropriate interventions that: (1) create an ethical culture to minimise unethical behaviours, (2) prevent confidential organisational information from leaking, (3) protect whistleblowers [62], and (4) act upon reports and be seen to do so [77]. This means that managers should be trained to handle disclosures constructively.

This stage-based framework encourages organisations to retrospectively analyse whistleblowing failures—so as to prevent future whistleblowing events. Leaders can use specific ethically-tailored metrics to monitor the organisation's ethical climate and should pay attention to internal whistleblower outcomes, and system responsiveness.

In summary, the proposed staged model offers a theoretically-grounded framework for understanding and managing whistleblowing. Its emphasis on effective organisational intervention holds significant promise for both researchers and practitioners committed to fostering ethical transparency and accountability, and preserving organisational information security.

6 Conclusion and Future Work

In this paper, we carefully examined whistleblowing as a phenomenon. The outcome is a whistleblowing stage model (Fig. 2). This, then, is the main message to organisations, affirming the recommendations by [52,71]: '*if you don't want a full-blown external whistleblowing event, facilitate and encourage internal whistleblowing, act upon reports and protect whistleblowers.*' This research opens avenues for further empirical research. Longitudinal studies could examine the way individual whistleblowers traverse these stages in reality. Comparative studies might explore how cultural, sectoral, or regulatory differences influence stage transitions and outcomes.

Appendix: https://osf.io/63wg4/overview?view_only=2e1504ed3c564fcbacda2b279cd90227.

References

1. Ajzen, I.: The theory of planned behavior. Organ. Behav. Hum. Decis. Process. **50**(2), 179–211 (1991). https://doi.org/10.1016/0749-5978(91)90020-T
2. Alleyne, P., Hudaib, M., Pike, R.: Towards a conceptual model of whistle-blowing intentions among external auditors. Br. Account. Rev. **45**(1), 10–23 (2013). https://doi.org/10.1016/j.bar.2012.12.003
3. Antoh, A., Sholihin, M., Sugiri, S., Arifa, C.: A perspective on the whistleblowing intention of internal auditors: an integrated ethical decision-making model. Cogent Bus. Manag. **11**(1), 2292817 (2024). https://doi.org/10.1080/23311975.2023.2292817
4. Asp, D.: Espionage act of 1917 (2017). https://www.mtsu.edu/first-amendment/article/1045/espionage-act-of-1917 (Updated by Deborah Fisher in May 2019). Accessed 11 Apr 2021
5. Atkinson, K.: Book review: kate kenny, whistleblowing: toward a new theory. Work Employ. Soc. Br. Sociol. Assoc. **36**(1), 587–588 (2022). https://doi.org/10.1177/09500170211038523
6. Barnett, T., Cochran, D.S., Taylor, G.S.: The internal disclosure policies of private-sector employers: an initial look at their relationship to employee whistleblowing. J. Bus. Ethics **12**(2), 127–136 (1993). https://doi.org/10.1007/BF00871932

7. BBC: Enron witness: Lay was duped (2002). http://news.bbc.co.uk/2/hi/business/1820284.stm. Accessed 16 Dec 2021
8. Bergsma, T.: Fraud, whistleblowing, and sins of omission: a call to action. J. Biblical Integr. Bus. **25**(1), 19–28 (2022). https://doi.org/10.69492/jbib.v25i1.618
9. Berry, P.: Necessary Scars. CRC Press, London (2022)
10. Blenkinsopp, J., et al.: Whistleblowing over patient safety and care quality: a review of the literature. J. Health Organ. Manag. **33**(6), 737–756 (2019). https://doi.org/10.1108/JHOM-12-2018-0363
11. Bosupeng, M.: Whistle blowing: what do contemporary ethical theories say? Stud. Bus. Econ. **12**(1), 19–28 (2017). https://doi.org/10.1515/sbe-2017-0002
12. Carlson, M.: Embodying deep throat: mark felt and the collective memory of Watergate. Crit. Stud. Media Commun. **27**(3), 235–250 (2010). https://doi.org/10.1080/15295030903583564
13. Casal, J.C., Zalkind, S.S.: Consequences of whistle-blowing: a study of the experiences of management accountants. Psychol. Rep. **77**(3), 795–802 (1995). https://doi.org/10.2466/pr0.1995.77.3.79
14. Charmaz, K.: Constructing Grounded Theory. Sage, London (2014)
15. Cho, S., Yim, M.S.: Whistleblowing analysis for detection of insider threat in an multicultural environment. In: 56th Annual Meeting of the Institute for Nuclear Materials Management. Institute for Nuclear Materials Management (2015). http://hdl.handle.net/10203/199535
16. Contu, A.: Rationality and relationality in the process of whistleblowing: recasting whistleblowing through readings of Antigone. J. Manag. Inquiry **23**(4), 393–406 (2014). https://doi.org/10.1177/1056492613517512
17. Culiberg, B., Mihelič, K.K.: The evolution of whistleblowing studies: a critical review and research agenda. J. Bus. Ethics **146**(4), 787–803 (2017). https://doi.org/10.1007/s10551-016-3237-0
18. Dasgupta, S., Kesharwani, A.: Whistleblowing: a survey of literature. IUP J. Corp. Gov. **9**(4), 57–70 (2010)
19. davidhencke: Tribunal of the absurd part two: Dr chris day loses his whistleblowing appeal (2025). https://davidhencke.com/2025/08/20/tribunal-of-the-absurd-part-two-dr-chris-day-loses-his-whistleblowing-appeal/
20. Davies, D.: Journalist who helped break snowden's story reflects on his high-stakes reporting (2020). https://www.npr.org/2020/05/20/859376407/journalist-who-helped-break-snowdens-story-reflects-on-his-high-stakes-reporting
21. D'Cruz, P., Bjørkelo, B.: Sociocultural dynamics in whistleblowing: insights from India. Asia-Pacific J. Bus. Adm. **8**(2), 143–162 (2016). https://doi.org/10.1108/APJBA-07-2015-0061
22. Denzin, N.K.: The Research Act: A Theoretical Introduction to Sociological Methods. Routledge, New York (2009). https://doi.org/10.4324/9781315134543
23. Dozier, J., Miceli, M.: Potential predictors of whistle-blowing: a prosocial behavior perspective. Acad. Manag. Rev. **10**(823-836) (1985). https://doi.org/10.5465/amr.1985.4279105
24. Duska, R.: Whistle-blowing and employee loyalty. In: Duska, R. (ed.) Contemporary Reflections on Business Ethics, chap. 8, pp. 139–147. Springer (2007). https://doi.org/10.1007/978-1-4020-4984-2_8
25. Edgcumbe, D.: God bless the NHS roger taylor. Br. J. Gen. Pract. **63**(611), 320 (2013). https://doi.org/10.3399/bjgp13X668311

26. Editorial Board of The New York Times: Edward Snowden, Whistle-Blower, New York Times (2014). https://www.nytimes.com/2014/01/02/opinion/edward-snowden-whistle-blower.html
27. Francis, S.R.: Freedom to speak up, an independent review into creating an open and honest reporting culture in the NHS (2015). http://freedomtospeakup.org.uk/
28. Fredin, A., Venkatesh, R., Riley, J., Eldridge, S.W.: "The road not taken": a study of moral intensity, whistleblowing, and regret. Ethics Behav. **29**(4), 320–340 (2019). https://doi.org/10.1080/10508422.2018.1469088
29. Frøstrup, A.C., et al.: Varslingsutvalgets utredning om varsling i arbeidslivet (The whistleblowing committee's report on whistleblowing in working life). Technical Report NOU 2018:6, Norwegian National Human Rights Institution (2018). https://www.nhri.no/en/nou-2018-6-varsling-verdier-og-vern-varslingsutvalgets-utredning-om-varsling-i-arbeidslivet-2/
30. Glaser, B., Strauss, A.: Discovery of Grounded Theory: Strategies for Qualitative Research. Routledge, New York (2017). https://doi.org/10.4324/9780203793206
31. Gottschalk, P., Asting, C.: Crime signal detection theory: two case studies of the five-stage model from observer to whistleblower. Deviant Behav. **43**(4), 461–471 (2022). https://doi.org/10.1080/01639625.2020.1816147
32. Greenwald, G.: Edward Snowden: the whistleblower behind the NSA surveillance revelations, the Guardian (2013). https://www.theguardian.com/world/2013/jun/09/edward-snowden-nsa-whistleblower-surveillance
33. Gundlach, M., Martinko, D.: The decision to blow the whistle: a social information processing framework. Acad. Manag. Rev. **28**, 107–123 (2003). https://doi.org/10.5465/amr.2003.8925239
34. Gøtzsche, P.C.: Survival of a Whistleblower – Peter C. Gøtzsche at Summer Institute 2018 (2018). https://www.youtube.com/watch?v=G2mFHHWyTrc
35. Hanjani, A., Purwanto, A., Kusumadewi, R.K.A.: The impact of ethical judgment, locus of control, and organizational commitments to whistleblowing. J. Auditing Finance Forensic Account. **6**(2), 97–104 (2018). https://doi.org/10.21107/jaffa.v6i2.4937
36. Hedin, U.C., Månsson, S.A.: Whistleblowing processes in Swedish public organisations—complaints and consequences. Eur. J. Soc. Work **15**(2), 151–167 (2012). https://doi.org/10.1080/13691457.2010.543890
37. Hennequin, E.: Whistleblowing as a career crisis: recovering from retaliatory job loss through a process of bifurcation. Work Employ. Soc. **37**(2), 545–562 (2023)
38. Hess, M.F., Treviño, L.K., Chen, A., Cross, R.: Beyond silence or compliance: the complexities of reporting a friend for misconduct. Bus. Ethics Eur. Rev. **28**(4), 546–562 (2019). https://doi.org/10.1111/beer.12238
39. Hoffman, W.M., Schwartz, M.S.: The morality of whistleblowing: a commentary on Richard T. de George. J. Bus. Ethics **127**, 771–781 (2015). https://doi.org/10.1007/s10551-014-2186-8
40. Hole, Å.S., Sverdrup, T.E.: The influence of psychological contracts on decision-making in whistleblowing processes. In: by Peer Jacob Svenkerud, E., Sornes, J.O., Browning, L. (eds.) Whistleblowing, Communication and Consequences, pp. 185–200. Routledge (2021). https://doi.org/10.4324/9780367822033
41. Hopt, K.J.: Internal investigations, whistleblowing and external monitoring: comparative experiences, economic insights, findings from corporate practice. Eur. Company Financ. Law Rev. **18**(6), 863–898 (2022). https://doi.org/10.1515/ecfr-2021-0036

42. Hull, D.M., Schuetz, S.W., Lowry, P.B.: Tell me a story: the effects that narratives exert on meaningful-engagement outcomes in antiphishing training. Comput. Secur. **129**, 103252 (2023). https://doi.org/10.1016/j.cose.2023.103252
43. Hutchinson, C., Thorpe, N.: How should employers handle whistleblowing complaints? (2022). https://www.personneltoday.com/hr/legal-advice-on-how-hr-should-handle-whistleblowing-issues/
44. Jackson, D., et al.: Whistleblowing: an integrative literature review of data-based studies involving nurses. Contemp. Nurse **48**(2), 240–252 (2014). https://doi.org/10.1080/10376178.2014.11081946
45. Kaptein, M.: From inaction to external whistleblowing: the influence of the ethical culture of organizations on employee responses to observed wrongdoing. J. Bus. Ethics **98**(3), 513–530 (2011). https://doi.org/10.1007/s10551-010-0591-1
46. Keil, M., Park, C.: Bad news reporting on troubled it projects: reassessing the mediating role of responsibility in the basic whistleblowing model. J. Syst. Softw. **83**(11), 2305–2316 (2010). https://doi.org/10.1016/j.jss.2010.07.009
47. Kenny, K.: Whistleblowing: Toward a New Theory. Belknap Press, USA (2019)
48. Kenny, K., Fotaki, M.: The costs and labour of whistleblowing: bodily vulnerability and post-disclosure survival. J. Bus. Ethics **182**(2), 341–364 (2023)
49. Kenny, K., Fotaki, M., Vandekerckhove, W.: Whistleblower subjectivities: organization and passionate attachment. Organ. Stud. **41**(3), 323–343 (2020). https://doi.org/10.1177/0170840618814558
50. Khan, J., et al.: Examining whistleblowing intention: the influence of rationalization on wrongdoing and threat of retaliation. Int. J. Environ. Res. Public Health **19**(3), 1752 (2022). https://doi.org/10.3390/ijerph19031752
51. Kirkup, B.: Independent Investigation into East Kent Maternity and Neonatal Services (2022). https://www.ekhuft.nhs.uk/patients-and-visitors/news-centre/maternity-news/iiekms/
52. Kordestani, G., Rajabdorri, H.: The pattern of factors affecting the whistleblowing with emphasis on the bystander effect. Account. Auditing Rev. **28**(1), 135–160 (2021). https://doi.org/10.22059/acctgrev.2021.300584.1008372
53. Kostakopoulou, D.: 'j'accuse!': whistleblowing, critical citizenship and the eu directive on whistleblowers' protection. J. Eur. Integr. **46**(4), 477–497 (2024). https://doi.org/10.1080/07036337.2024.2307926
54. Lamb, T.M.: Whistleblowers: a study of employee decision-making. Ph.D. thesis, School of Business & Technology, Capella University (2009)
55. Lee, H., Kang, M.M., Kim, S.Y.: A psychological process of bureaucratic whistleblowing: applying the theory of planned behavior. Am. Rev. Public Adm. **51**(5), 374–392 (2021). https://doi.org/10.1177/0275074021100334
56. Leogue, J.: Air corps whistleblower was 'isolated, vilified' (2019). https://www.irishexaminer.com/opinion/commentanalysis/arid-30900464.html
57. Lincoln, Y.S., Guba, E.G.: Naturalistic Inquiry. Sage, California (1985)
58. Loeb, S.E., Cory, S.N.: Whistleblowing and management accounting: an approach. J. Bus. Ethics **8**, 903–916 (1989). https://doi.org/10.1007/BF00383425
59. Lowry, P.B., Moody, G.D., Galletta, D.F., Vance, A.: The drivers in the use of online whistle-blowing reporting systems. J. Manag. Inf. Syst. **30**(1), 153–190 (2013). https://doi.org/10.2753/MIS0742-1222300105
60. Lowry, P.B., Rouibah, K., Moody, G.D., Siponen, M.: Towards a cross-cultural model of online whistle-blowing systems use. In: Pan, S.L., Cao, T.H. (eds.) Proceedings of the 12th Pacific Asia Conference on Information Systems (PACIS 2012), Ho Chi Minh City, Vietnam, July, pp. 11–15 (2012)

61. Martin, B.: Research that whistleblowers want – and what they need. In: Brown, A., Lewis, D., Moberly, R., Vandekerckhove, W. (eds.) International Handbook on Whistleblowing Research, chap. 20, pp. 497–521. Edward Elgar, Cheltenham, UK (2014). https://doi.org/10.4337/9781781006795.00030
62. Mesmer-Magnus, J.R., Viswesvaran, C.: Whistleblowing in organizations: an examination of correlates of whistleblowing intentions, actions, and retaliation. J. Bus. Ethics **62**(3), 277–297 (2005). https://doi.org/10.1007/s10551-005-0849-1
63. Miceli, M.P., Near, J.P.: Blowing the Whistle: The Organizational and Legal Implications for Companies and Employees. Lexington, New York (1992)
64. Miceli, M.P., Near, J.P.: Individual and situational correlates of whistle-blowing. Pers. Psychol. **41**, 267–281 (1988). https://doi.org/10.1111/j.1744-6570.1988.tb02385.x
65. Miceli, M.P., Near, J.P., Dworkin, T.M.: Whistle-Blowing in Organizations. Routledge, New York (2008)
66. Munro, I.: An interview with Chelsea Manning's lawyer: Nancy Hollander on human rights and the protection of whistleblowers. Organization **26**(2), 276–290 (2019). https://doi.org/10.1177/1350508418779648
67. Near, J.P., Miceli, M.P.: Organizational dissidence: the case of whistle-blowing. J. Bus. Ethics **4**(1), 1–16 (1985). https://doi.org/10.1007/BF00382668
68. Near, J.P., Miceli, M.P.: Effective-whistle blowing. Acad. Manag. Rev. **20**(3), 679–708 (1995). https://doi.org/10.5465/amr.1995.9508080334
69. Near, J.P., Miceli, M.P.: After the wrongdoing: what managers should know about whistleblowing. Bus. Horiz. **59**(1), 105–114 (2016). https://doi.org/10.1016/j.bushor.2015.09.007
70. Nicholls, A.R., et al.: Snitches get stitches and end up in ditches: a systematic review of the factors associated with whistleblowing intentions. Front. Psychol. **12**, 631538 (2021). https://doi.org/10.3389/fpsyg.2021.631538
71. Norman, C.S., Rose, A.M., Rose, J.M., Suh, I., Xiao, X.: An investigation of the effects of allegation sources and consequences on corporate directors' responses to whistleblowing allegations. J. Inf. Syst. **35**(3), 223–232 (2021). https://doi.org/10.2308/ISYS-2020-049
72. Nurhidayat, I., Kusumasari, B.: Why would whistleblowers dare to reveal wrongdoings? An ethical challenge and dilemma for organisations. Int. J. Law Manag. **61**(3/4), 505–515 (2019). https://doi.org/10.1108/IJLMA-11-2018-0253
73. Oleson, P.C.: Assessing edward snowden whistleblower, traitor, or spy? Intelligencer J. U.S. Intell. Stud. **21**(2), 15–23 (2015)
74. O'Sullivan, P., Ngau, O.: Whistleblowing: a critical philosophical analysis of the component moral decisions of the act and some new perspectives on its moral significance. Bus. Ethics Eur. Rev. **23**(4), 401–415 (2014). https://doi.org/10.1111/beer.12058
75. Özeke, H.B.: Turkey: Whistleblowing Under Turkish Law (2010). https://www.mondaq.com/turkey/whistleblowing/105256/whistleblowing-under-turkish-law
76. Pacella, J.: Making whistleblowers whole. U.C. Irvine L. Rev. **12**, 1291–1330 (2022)
77. Park, H., Blenkinsopp, J.: Whistleblowing as planned behavior–a survey of South Korean police officers. J. Bus. Ethics **85**, 545–556 (2009). https://doi.org/10.1007/s10551-008-9788-y
78. Perrucci, R., Anderson, R.M., Schendel, D.E., Trachtman, L.E.: Whistle-blowing: professionals' resistance to organizational authority. Soc. Probl. **28**(2), 149–164 (1980). https://doi.org/10.2307/800148

79. Pohjanoksa, J., Stolt, M., Suhonen, R., Löyttyniemi, E., Leino-Kilpi, H.: Whistleblowing process in healthcare: from suspicion to action. Nurs. Ethics **26**(2), 526–540 (2019). https://doi.org/10.1177/0969733017705005
80. Powell, M.: The whistleblower stages model in British National Health Service Inquiries. J. Health Organ. Manag. **36**(3), 314–331 (2021). https://doi.org/10.1108/JHOM-10-2020-0388
81. Punch, K.F.: Survey Research: The Basics. Sage, University of Western Australia (2003)
82. Raval, A.: Corporate whistleblowing in the UK needs a shake-up (2025). https://www.ft.com/content/8dae5f0d-f481-4561-87b0-62066a4beaff
83. Renaud, K., Warkentin, M., Pogrebna, G., van der Schyff, K.: VISTA: an inclusive insider threat taxonomy, with mitigation strategies. Inf. Manag. **61**(1), 103877 (2023). https://doi.org/10.1016/j.im.2023.103877
84. Richardson, B.K., McGlynn, J.: Blowing the whistle off the field of play: an empirical model of whistle-blower experiences in the intercollegiate sport industry. Commun. Sport **3**(1), 57–80 (2015). https://doi.org/10.1177/216747951351749
85. Richardson, B.K., McGlynn, J.: Constructing the heroic whistleblower: a social scientific approach. Heroism Sci. **6**(2), Article 1 (2021). https://doi.org/10.26736/hs.2021.02.01
86. Richardson, B.K., Tran, J., James, E.: Using whistleblowers' metaphors to understand why organizational wrongdoing endures. Manag. Commun. Q. 08933189251322153 (2025)
87. Roberts, D.: NSA releases email in dispute over Snowden 'internal whistleblowing'. The Guardian (2014). https://www.theguardian.com/world/2014/may/29/nsa-email-snowden-surveillance-internal-whistleblowing
88. Roche, J.M.: The competitive system, to work, to preserve, and to protect. Vital Speeches Day **37**, 445 (1971)
89. Rosenthal, A.M.: Thirty-Eight Witnesses: The Kitty Genovese Case. Open Road Media, New York (2015)
90. Rost, P.: The Whistleblower: Confessions of a Healthcare Hitman. Soft Skull Press, Brooklyn, New York (2006)
91. Schmolke, K.U., Utikal, V.: Whistleblowing: incentives and situational determinants. J. Bus. Econ. 1–24 (2025). https://doi.org/10.1007/s11573-025-01223-0
92. Seaton, L.J.: Cognitive dissonance and rationalization's role in the whistle-blowing process. In: Institute for Global Business Research Conference Proceedings, vol. 5, no. 3, p. 48 (2021)
93. Stevens, E.: Chelsea Manning: Hero or Traitor? It's Complicated (2019). https://www.thegryphon.co.uk/2019/03/23/chelsea-manning-hero-or-traitor-its-complicated/. Accessed 11 Apr 2021
94. Stusinski, M.: NSA Whistleblower Edward Snowden's US Passport Revoked [Report] (2013). https://www.inquisitr.com/810915/nsa-whistleblower-edward-snowdens-us-passport-revoked-report/. Accessed 22 July 2021
95. de Swardt, C.: Silencing a Whistleblower. A Story of Hypocrisy. Springer (2021)
96. Taylor, E.Z., Curtis, M.B.: An examination of the layers of workplace influences in ethical judgments: whistleblowing likelihood and perseverance in public accounting. J. Bus. Ethics **93**(1), 21–37 (2010). https://doi.org/10.1007/s10551-009-0179-9
97. Thomas, D.R.: A general inductive approach for analyzing qualitative evaluation data. Am. J. Eval. **27**(2), 237–246 (2006). https://doi.org/10.1111/beer.12290

98. Trevino, L.K., Victor, B.: Peer reporting of unethical behavior: a social context perspective. Acad. Manag. J. **35**(1), 38–64 (1992). https://doi.org/10.2307/256472
99. Trevino, L., Weaver, G., Reynolds, S.: Behavioral ethics in organizations: a review. J. Manag. **32**, 951–990 (2006). https://doi.org/10.1177/0149206306294258
100. United Nations High Commissioner for Refugees: Convention on the reduction of statelessness, uNHCR (1961). https://www.unhcr.org/ibelong/wp-content/uploads/1961-Convention-on-the-reduction-of-Statelessness_ENG.pdf
101. Uys, T., Smit, R.: Resilience and whistleblowers: coping with the consequences. South African Rev. Sociol. **47**(4), 60–79 (2016). https://doi.org/10.1080/21528586.2016.1204242
102. Van Portfliet, M.: Resistance will be futile? The stigmatization (or not) of whistleblowers. J. Bus. Ethics **175**, 451–464 (2020). https://doi.org/10.1007/s10551-020-04673-4
103. Vandekerckhove, W., Lewis, D.: The content of whistleblowing procedures: a critical review of recent official guidelines. J. Bus. Ethics **108**(2), 253–264 (2012). https://doi.org/10.1007/s10551-011-1089-1
104. Vandekerckhove, W., Phillips, A.: Whistleblowing as a protracted process: a study of UK whistleblower journeys. J. Bus. Ethics **159**(1), 201–219 (2019). https://doi.org/10.1007/s10551-017-3727-8
105. Van de Ven, A.H., Poole, M.S.: Alternative approaches for studying organizational change. Organ. Stud. **26**(9), 1377–1404 (2005). https://doi.org/10.1177/0170840605056907
106. Watts, L.L., Buckley, M.R.: A dual-processing model of moral whistleblowing in organizations. J. Bus. Ethics **146**(3), 669–683 (2017). https://doi.org/10.1007/s10551-015-2913-9
107. Weick, K.E.: Sensemaking in Organizations. Sage, USA (1995)
108. Zakaria, M.: Antecedent factors of whistleblowing in organizations. Procedia Econ. Finance **28**, 230–234 (2015). https://doi.org/10.1016/S2212-5671(15)01104-1
109. Zamzami, A.H., Mahmudah, H.: Whistleblowing intention: a study of situation factors. SSRG Int. J. Econ. Manag. Stud. (SSRG-IJEMS) **6**(12), 132–140 (2019). https://doi.org/10.2139/ssrn.3535021
110. Zhang, J., Chiu, R., Wei, L.Q.: On whistleblowing judgment and intention: the roles of positive mood and organizational ethical culture. J. Manag. Psychol. **24**(7), 627–649 (2009). https://doi.org/10.1108/02683940910989020

Security and AI

Cyber Resilience Through LLM-Enhanced Vulnerability Reports

Mario Hoffmann[1(✉)] and Erik Buchmann[2,3]

[1] IT Department, Leipzig University, Leipzig, Germany
mario.hoffmann@uni-leipzig.de
[2] Department of Computer Science, Leipzig University, Leipzig, Germany
erik.buchmann@uni-leipzig.de
[3] ScaDS.AI Dresden/Leipzig, Leipzig, Germany

Abstract. Decentralized, heterogeneous, and self-administered IT infrastructures pose major challenges to cyber resilience. For example, in academia and research institutions, each research group requires its own set of IT systems. Frequently, those systems are managed part-time by local system administrators (LSAs) without formal IT security education. A common option to increase cyber resilience in such scenarios is to perform vulnerability scans on the institution's IP address range. However, LSAs are often overwhelmed by the technical depth of the scan reports. A promising solution is to have a large language model (LLM) translate the reports into a less demanding, more actionable form. In this paper, we present a mixed-method study ($n = 21$) with LSAs from Leipzig University. Our study investigates how LLM-enhanced vulnerability scan reports affect the LSAs' remediation capabilities. Our findings reveal that LSAs' backgrounds are even more diverse than one might think, with varying levels of technical understanding. LLM-translated reports allow them to fix common vulnerabilities much faster, without requiring central IT support. However, complex issues still require expert support and knowledge. Our results help design vulnerability communication approaches that strengthen cyber resilience in heterogeneous environments.

Keywords: AI · LLM · Vulnerability Reports · Cybersecurity · Survey

1 Introduction

In business and industry, a central IT department controls installed IT systems, enforces security policies, and manages vulnerability remediation and incident response. In universities and similar research institutions with self-managed departments, cyber resilience is even more demanding because the departments are individually staffed and have unique requirements for their IT infrastructure. Each research group operates local IT systems of various ages with different operating systems and applications. In many cases, the local system administrators (LSAs) are PhD students who manage IT systems part-time alongside their research.

© The Author(s), under exclusive license to Springer Nature Switzerland AG 2027
K. Renaud et al. (Eds.): EICC 2026, CCIS 3012, pp. 109–128, 2027.
https://doi.org/10.1007/978-3-032-28957-5_7

A well-known measure for improving cyber resilience in such environments is that the IT department periodically executes vulnerability scans on all IP addresses, then automatically generates vulnerability reports like the one shown in Fig. 1, and shares them with the responsible LSAs.

We already observed [16] that LSAs often lack formal IT training and are overwhelmed by the technical depth of vulnerability reports. This led to repetitive support tickets to the IT department and unresolved vulnerabilities, degrading the university's security posture. LSAs also postponed mitigating vulnerabilities in favor of research deadlines or other lecturing responsibilities. A promising approach to reduce time-to-remediation and to build the LSAs' capabilities is to have the vulnerability reports translated [16] by a Large Language Model (LLM). The LLM can generate less technical representations and augment them with supplementary information, e.g., threat intelligence explaining why a vulnerability must be fixed quickly.

In this paper, we present a mixed-method study with 21 LSAs to investigate how such LLM-enhanced vulnerability reports help LSAs to maintain cyber resilience. In particular, we used ChatSEC to generate intuitive reports that accommodate heterogeneous LSA backgrounds by combining technical information with contextual guidance on severity and remediation steps. We sent these reports to LSAs with an online questionnaire containing quantitative and qualitative questions.

We pursue three research goals: We want to learn the LSAs' professional and demographic background, including potential gaps between self-assessed and actual cybersecurity knowledge. We want to know whether the rewritten LLM reports enable independent vulnerability remediation. Finally, we are interested in finding out how the rewritten vulnerability reports can be improved.

Thus, we make three contributions:

- We describe a concrete use case for enhancing organizational cyber resilience by using LLMs to simplify vulnerability reports for non-expert administrators at a university.
- We analyze the results of a mixed-method study, based on an online questionnaire with 7 quantitative and 8 qualitative questions and 21 participants from Leipzig University.
- We derive recommendations for vulnerability representation approaches in heterogeneous institutions that strengthen proactive security capabilities.

Our results indicate that the background knowledge of our LSAs is even more heterogeneous than expected. Enhanced reports enabled independent remediation of less complex vulnerabilities and improved the institution's security posture. Incidentally, we also observed that ChatSEC did not generate hallucinated vulnerability reports that would pose a serious risk to security-critical application domains. We think that the cybersecurity domain is exceptionally well represented in the training data of the LLM used by the tool.

Paper structure: Sect. 2 contains related work. Section 3 describes our LLM approach. In Sect. 4 we present our study design, which is evaluated in Sect. 5 and discussed in Sect. 6. Section 7 concludes the paper.

Vulnerability Scan Report

Scan Time: Thu, Jun 27, 2024 1:00 AM - Fri, Jun 28, 2024 1:09 AM, Hosts scanned: 1

Results (1/1)

Vulnerability	*Severity*	*Host*	*Location*
Microsoft Remote Desktop Protocol RCE Vulnerabilities (2671387) - Active Check	9.3 (High)	2.3.4.5	package

Summary

This host is missing a critical security update according to Microsoft Bulletin MS12-020.

Vulnerability Detection Result

Vulnerability was detected according to the Detection Method.

Details: Microsoft Remote Desktop Protocol RCE Vulnerabilities (2671387) - Active Check
OID: 1.3.6.1.4.1.25623.1.0.902818
Version used: 2022-04-27T12:01:52Z

Solution

Solution Type: Vendorfix. The vendor has released updates.

Affected Software/OS

Microsoft Windows 7 Service Pack 1 and prior
Microsoft Windows XP Service Pack 3 and prior
Microsoft Windows 2K3 Service Pack 2 and prior
Microsoft Windows Vista Service Pack 2 and prior
Microsoft Windows Server 2008 Service Pack 2 and prior

CVSS Vector String

AV:N/AC:M/Au:N/C:C/I:C/A:C

Vulnerability Insight

The flaws are due to the way Remote Desktop Protocol accesses an object in memory that has been improperly initialized or has been deleted and the way RDP service processes the packets.

References

CVE: CVE-2012-0002 CVE-2012-0152 DFN-CERT-2012-0477
Other: http://blog.binaryninjas.org/?p=58
http://www.securityfocus.com/bid/52353 ...

Hosts 1 of 1

IP-Address	*Hostname*	*OS*	*High*	*Medium*	*Low*	*Total*	*Severity*
2.3.4.5	host.domain.tld	Microsoft Windows	1	0	0	1	9.3 (High)

Fig. 1. Typical vulnerability report, as generated by Greenbone

2 Related Work

Organizational cyber resilience depends on technical controls and human capability [24,26]. Thus, we review related work on vulnerability scans, cybersecurity in academia, LLMs in cybersecurity, and how users react to security warnings.

2.1 Understanding and Using Vulnerability Results

Vulnerability scans and penetration testing represent fundamental security practices for identifying system weaknesses. Studies revealed significant differences between the results of various vulnerability scanners, concluding that "the output of the vulnerability scanners should be interpreted with care" [18,32]. This shows that vulnerability scan reports are not designed for non-cybersecurity experts, requiring specialized knowledge for reasonable interpretation.

A study investigated usability gaps in penetration testing reports and found substantial challenges in understanding, prioritizing, and acting on the provided insights [13]. Research also identified the absence of a standardized format for penetration and vulnerability testing reports as a significant issue, leading to potential misunderstandings between technical staff and management [3,36]. Thus, the organizational security posture is directly impacted.

Although ubiquitous in vulnerability assessment, the Common Vulnerability Scoring System (CVSS) has shown significant limitations in real-world application [19,25]. Expert-based CVSS investigations revealed substantial agreement issues among security professionals when scoring identical vulnerabilities [17]. This inconsistency, combined with the technical complexity of CVSS metrics, creates barriers for administrators without specialized cybersecurity backgrounds. Research on vulnerability assessment frameworks consistently emphasizes that present tools require domain-specific knowledge [33].

In heterogeneous academic environments where LSAs are often non-experts, communication gaps and technical misunderstandings delay remediation and weaken organizational cyber resilience. Reports must therefore be transformed into actionable guidance that enables an independent response.

2.2 Cybersecurity in Academia

Universities face unique cybersecurity challenges that distinguish them from traditional enterprise environments due to heterogeneity and decentralized structures [1,8]. Individual faculties and departments operate under separate IT structures with varying technological needs. As shown in [6], academic institutions are among the least secure organizational types, documenting vulnerabilities resulting from cultural, structural, and resource-based factors. Resource constraints, short-staffed security teams, a decentralized academic culture, and prevalent shadow IT create fundamental tensions in protecting sensitive information and led to inherent vulnerabilities. This describes our observed university setting.

2.3 Users Awareness and Behavior

User Awareness: Research on warning perception examined how users react to warnings and established a model to understand human factors in handling warnings [7]. Among several key factors, comprehension is critical. The receivers must understand the issue, not just perceive the warnings.

Studies evaluated intrinsic and external motivational factors for IT security-related behavior, finding that most approaches focus only on external motivation [29]. Building an intrinsic understanding of security policies strengthens administrators' independent response capabilities, reducing dependence on external enforcement, and improving organizational resilience.

Examination of mental models of LSAs found that most LSAs do not understand 23% of their supervised systems [20], indicating a possibly higher cybersecurity risk within heterogeneous LSA groups. Based on this, we assume that these groups need support.

User Behavior: Observations of several LSA work environments concluded that communicational and technical difficulties between stakeholders can complicate problem-solving [4]. Improving LSAs' understanding of vulnerabilities and threats can address these difficulties.

As system updates are a crucial part of IT system security, research examined how LSAs behave during update procedures, finding that even experienced LSAs are skeptical about possible failures and service consequences [31]. Studies showed that previous updates often change familiar features or break functionalities [34], resulting in "if it ain't broke, don't fix it.", or "if I don't understand it, I don't update" attitudes. Although LLM-enhanced reports cannot eliminate update risks, they reduce uncertainty and enable LSAs to make informed decisions faster, thus shortening remediation cycles and improving security posture.

Furthermore, employees tend to ignore security rules when the effort required outweighs the resulting advantages [5]. Research showed that users evaluate effort against benefit when handling cybersecurity demands [15]. Reducing the cognitive effort required for remediation through simplified guidance can accelerate response times and strengthen organizational cyber resilience.

2.4 Large Language Models in Cybersecurity

Large Language Models (LLMs) have emerged as transformative tools in cybersecurity, demonstrating capabilities in threat detection, vulnerability analysis, automated security operations, and threat intelligence processing [21,37].

Studies documented their effectiveness in enriching vulnerability descriptions and generating contextualized explanations, detecting anomalies in system logs, and mapping threat behaviors to frameworks like MITRE ATT&CK [9,22,35]. Surveys found that LLMs combined with transformer-based architectures such as BERT achieve effective extraction and classification of adversary behaviors from cyber threat intelligence [2].

Although LLMs excel at pattern recognition and contextual reasoning, challenges remain in ensuring accuracy, preventing hallucinations, and maintaining performance across diverse domains [10]. Therefore, research emphasizes the importance of validation mechanisms to ensure that LLM outputs provide reliable guidance. These findings suggest that LLMs can effectively support vulnerability communication, but domain-specific validation and knowledge is still required.

3 Rewriting Vulnerability Scan Reports

In this section, we briefly describe ChatSEC [16], the tool we used to generate the vulnerability scan reports for our study.

Concept. ChatSEC was designed based on two observations: First, many LSAs administer their systems besides their research and teaching activities, and have limited time available for this task. Second, many LSAs have little experience in IT administration or cybersecurity. To take this into account, ChatSEC uses two steps:

Step 1: Generate a brief, self-contained, intuitive report from the original vulnerability scan.

Step 2: Add threat intelligence information necessary to understand and prioritize the vulnerability.

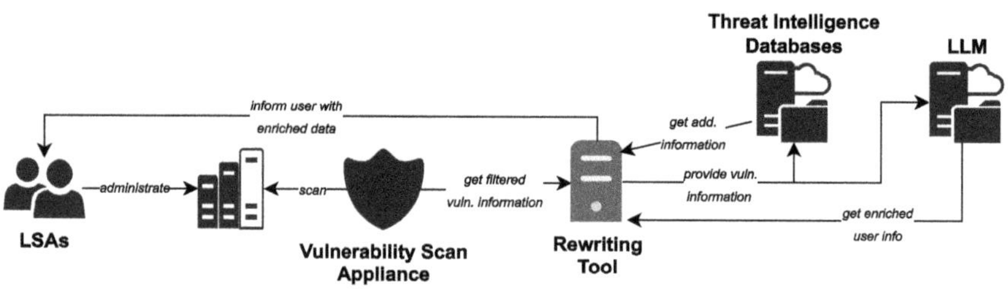

Fig. 2. ChatSEC's architecture and data flow.

Figure 2 illustrates ChatSEC's data flow. ChatSEC extracts the *IP address* of the vulnerable system, the *vulnerability name*, the *vulnerability description/details/insights*, and the *CVSS* vector string [12] from the original vulnerability report. The extracted IP address is not part of the provided information to the LLM. For Step (1), ChatSEC then uses an LLM to generate an intuitive explanation from the vulnerability name and the vulnerability description. For Step (2), the tool uses the CVSS vector string from the original report. The CVSS vector string describes different dimensions of a certain threat, such as "Attack Vector", "Attack Complexity", or "Impact on Confidentiality". Those dimensions are the basis for the CVSS base score [11]. As the CVSS base score indicates the severity of a vulnerability, the tool uses the CVSS vector string to let an LLM explain the potential impact of the vulnerability on the LSA's systems.

Implementation. ChatSEC is implemented in Python and uses regular expressions to extract information from the original vulnerability scan report. It reads reports from OpenVAS scan engine (Greenbone appliance) and connects to OpenAI's ChatGPT using "gpt-4o-mini" (Created: 1721172741) with a "temperature" setting of 0.2 via the API. ChatSEC utilizes ChatGPT, because it is a state-of-the-art general-purpose LLM well-suited for experimenting with generated explanations. Note that the results of this study will serve as a basis for fine-tuning a locally hosted model that will translate vulnerability reports without involving external parties. Figures 3 and 4 show the role-based prompts for Steps 1 and 2, with "###" as a placeholder for input data.

```
You are a skillful IT security expert with expertise in explaining security issues.
Summarize the following vulnerability description and explain it without using
technical terms.

Vulnerability Description:
###Vulnerability description input here ###
```

Fig. 3. System prompt translate vulnerability description and details

```
You are a skillful IT security expert with expertise in explaining security issues. You
will be provided with a comma separated list of properties of an IT security severity
score. The list is formed as a key-value list, where the values are on the right side of
the colon. Assume that you answer to a non-technician. Help in the following order:
Explain each list item and assume that 'low', 'None' or 'Network' is highly critical.
Answer in one sentence per property.

List:
### List input here ###
```

Fig. 4. System prompt to translate the CVSS vector string

To provide an intuitive example, reconsider Fig. 1. From this report, ChatSEC extracts the vulnerability name "*Microsoft Remote Desktop Protocol RCE Vulnerabilities (2671387) - Active Check*". The extracted vulnerability description is "*This host is missing a critical security update according to Microsoft Bulletin MS12-020.*", with the vulnerability details "*The flaws are due to the way Remote Desktop Protocol accesses an object in memory that has been improperly initialized or has been deleted and the way RDP service processes the packets.*" Finally, ChatSEC extracts from the vulnerability report the CVSS string "*AV:N/AC:M/Au:N/C:C/I:C/A:C*". After that, it invokes "gpt-4o-mini" to generate the report with the prompts in Figs. 3 and 4. Figure 5 shows the resulting report, which can be sent to the LSA.

Microsoft Remote Desktop Protocol RCE Vulnerabilities (2671387) - Active Check

Host: *2.3.4.5*

There is a serious security issue with a feature called Remote Desktop Protocol (RDP) used by Microsoft. This problem arises because RDP doesn't handle certain data correctly, which can lead to unauthorized access to a computer. Essentially, if someone exploits this flaw, they could potentially take control of the affected system. Microsoft has released an important update to fix this issue, and it's crucial for users to install it to protect their computers from potential attacks.

Attack Complexity: Medium *This means that the attack requires some level of skill or effort to execute, making it more likely that a determined attacker could succeed.*

Confidentiality Impact: Complete *This indicates that if the attack is successful, all sensitive information could be fully exposed to unauthorized individuals.*

Integrity Impact: Complete *This suggests that the attacker could completely alter or destroy the data, compromising its accuracy and trustworthiness.*

Availability Impact: Complete *This means that the attack could render the system or service completely unavailable to users, disrupting normal operations entirely.*

Fig. 5. Transformed scan report from Fig. 1

4 Study Methodology

In this section, we describe our study concept and procedure. We start by deriving three study objectives from our research goals from Sect. 1:

O1: What is the professional and demographic profile of our LSAs? This provides the foundation for assessing the gap between self-perceived expertise and actual knowledge while revealing resource limitations.

O2: To what extent are the LLM-generated reports sufficient for our LSAs to close vulnerabilities? This shows whether the provided reports enable independent remediation and validates knowledge gaps.

O3: What can we learn about rewriting vulnerability reports for LSAs? This captures the LSAs' perception of useful components and necessary improvements.

4.1 Study Concept

We decided on a mixed-method study based on [14,23,30]. This allows quantitative (multiple-choice) and qualitative (free text) questions [14] on various scales [30] to be combined with answers that complement each other. In line with [23], we will evaluate the answers to both types of questions separately and subsequently correlate them. We do not want to make assumptions about the profiles of our participants. Therefore, we decided to conduct an inductive evaluation [27] for both types of questions, i.e., we derive clusters of responses from the answers. To cluster the qualitative answers, we repeatedly iterate over them to obtain a coding scheme. Quantitative answers can be clustered on their own.

4.2 Study Procedure

To obtain real-world data, we aligned our study with regular vulnerability scans of our IT department, which uses OpenVAS (Greenbone appliance) for scanning. Our pool of participants consists of all LSAs responsible for an IT system in which a vulnerability with a CVSS v3.x rating of "critical" or "high" was detected [11,28]. This is in line with our IT department, which filters out scan results with a CVSS v3.x rating < 7.0 to avoid overloading the LSAs with less urgent vulnerabilities. Note that the scanner might find multiple vulnerabilities on the same host and one LSA can be responsible for multiple hosts.

We then used ChatSEC to rewrite the vulnerability reports and distributed them among our LSAs together with an online questionnaire. After 3 weeks, we sent a reminder to complete the questionnaire, if not already done. After 6 weeks, we filtered, encoded, and analyzed the completed questionnaires.

Questionnaire. Our questionnaire consists of 7 quantitative and 8 qualitative questions. Table 1 shows the quantitative questions, asking for employment status (Q1, Q2), time spent on administration (Q4) and IT knowledge (Q3 and Q7). We provided examples on how to interpret Q3: Answer A2 means using a computer if needed, A3 allows one to answer general questions about using a computer, A4 refers to the capability to solve basic IT problems independently, and A5 requires deep, domain-specific knowledge in IT. To assess how reliable this self-assessed IT knowledge actually is, we also asked to evaluate two different attack scenarios (Q5, Q6) as control questions. We omit questions about sex or age as these are not related to our research objectives.

Table 2 shows the qualitative questions of our questionnaire. To relate the answers given to a vulnerability, we asked our participants to name the highest-ranked vulnerability in the received report (Q8). The next questions examined which additional information was required, beyond the generated report, to understand the vulnerability description (Q9) and the vulnerability assessment (Q10). In particular, we asked the LSAs whether they required additional information to determine the relevance of the vulnerability for their IT systems (A1), as well as to identify the system commands needed to locate (A2) and remediate (A3) the vulnerability. Question Q11 captures additional information needs not covered by the preceding items. Question Q12 assessed whether the LSA was able to remediate the vulnerability and, if not, the reasons for this outcome. Finally, Questions Q13 and Q14 ask for a general evaluation of the generated report, and Q15 invites further suggestions.

Table 1. Quantitative Questions

Questions	Answer Options
Q1: Are you employed in an IT related field?	(A1) Yes, (A2) No, (A3) No answer
Q2: What is your employment status?	(A1) Student assistant, (A2) Temporary/part time, (A3) Temporary/full-time, (A4) Permanent/part-time, (A5) Permanent/full-time, (A6) Other, (A7) No answer.
Q3: Which statement best describes your IT expertise?	(A1) None, (A2) Novice user, (A3) Proficient user, (A4) Basic administrator, (A5) Proficient administrator, (A6) No answer.
Q4: How much time do you spent administrating your systems per week?	(A1) I do not know, (A2) Less than one hour, (A3) 1–2 h, (A4) 2–4 h, (A5) 4–7 h, (A6) More than eight hours, (A7) No answer
Q5: How severe is an attack where the attacker can execute arbitrary programs with root privileges?	(A1) Not relevant, (A2) Less relevant, (A3) Neutral, (A4) Relevant, (A5) Catastrophic, (A6) I do not know, (A7) No answer
Q6: How severe is an attack where the attacker with access to your system can read user data?	(A1) Not relevant, (A2) Less relevant, (A3) Neutral, (A4) Relevant, (A5) Catastrophic, (A6) I do not know, (A7) No answer
Q7: Which statement describes your ability to mitigate the vulnerability without our provided report?	(A1) No answer, (A2) Completely unable, (A3) Partially, (A4) Completely able

4.3 Encoding and Pre-processing

To analyze the completed questionnaires, we **summarized the quantitative answers** (Q1–Q7). Consider Q5 and Q6, where we asked the participants to assess a given threat scenario against their systems. Q5 and Q6 can be answered from "not relevant" (A1) to "catastrophic" (A5), in addition to an "I do not know" answer. In both scenarios, the threat actor has gained access to the respective IT system. From a security perspective, both threat scenarios should be rated at least "relevant", regardless of the respective IT system. However, the threat in Q5 is more critical than the threat in Q6. Q5 and Q6 are meant to be control questions, to classify the self-assessed IT knowledge in Q3, and to put subsequent answers into context. Given the heterogeneous cybersecurity knowledge levels among participants, we cannot ensure reliable differentiation between the fine-grained and overlapping rating categories 'relevant'/'catastrophic' and 'less relevant'/'neutral'. To eliminate this uncertainty and obtain more generalizable

Table 2. Qualitative Questions

Questions	Answer Options
Q8: Name the highest ranked vulnerability from the provided reports.	(A1) Free text, (A2) No answer
Q9: Which information did you researched to understand the provided vulnerability description?	(A1) Relevance, (A2) System commands, (A3) Update commands, (A4) Other
Q10: Which information did you researched to understand the provided vulnerability assessment?	(A1) Relevance, (A2) System commands, (A3) Update commands, (A4) Other
Q11: Which information would have helped to mitigate the vulnerability?	(A1) Free text, (A2) No answer
Q12: Did you close the vulnerability? If not, why?	(A1) Yes, (A2) No, (A3) Free text, (A4) No answer
Q13: What did you like the most on the provided report?	(A1) Free text, (A2) No answer
Q14: What did you like the least on the provided report?	(A1) Free text, (A2) No answer
Q15: What would you like to see within the next vulnerability report?	(A1) Free text, (A2) No answer

results, these category pairs were aggregated into three clearly distinguishable categories: 'not relevant', 'neutral', and 'relevant'. The summarization is shown in Table 3. We summarized the answers from Q3 and Q4 similarly (see Tables 4a and 4b).

Table 3. Summarization for Questions Q5 and Q6

Answer (Freq. Q5 |Freq. Q6)	Coding
A1 - not relevant (0 | 1)	Not relevant
A2 - less relevant (1 | 3) A3 - neutral (1 | 3)	Neutral
A4 - relevant (13 | 10) A5 - catastrophic (5 | 4)	Relevant
A6 - I do not know. (1 | 0)	Unknown

We **encoded the qualitative answers** for further analysis. Questions Q8, Q11, and Q13–Q15 accepted (non-compulsory) free text answers. The other qualitative questions used structured text fields to enter answers, plus a free text field for cases where participants felt that the answer options were not appropriate. For example, Table 5 shows the answer options and the coding for Q9. The answer option A1 asked the participants to indicate how relevant the vulnerable

Table 4. Summarization of the answers from Q3 and Q4.

(a) Summarization for Question Q3

Answer (Freq.)	**Coding**
A1 - None (1)	None
A2 - Novice User (0)	
A3 - Proficient User (1)	
A4 - Basic Administrator (6)	Basic
A5 - Proficient Admin. (13)	Expert

(b) Summarization for Question Q4

Answer (Freq.)	**Coding**
A1 - I do not know. (0)	Unknown
A2 - Less than one hour. (10)	<2 h
A3 - 1–2 h (6)	
A4 - 2–4 h (1)	2–7 h
A5 - 4–7 h (0)	
A6 - More than eight hours. (4)	$\geq$ 8 h

Table 5. Coding for Question Q9

Answer (Freq.)	Coding
A1 - System relevancy. (3) A2 - System commands. (3) A3 - Update commands. (4)	System related instructions and relevance.
A4 - Others. (5) A4.1 - Enough informative value. A4.2 - None.	No additional information needed.
A4.3 - Consultation with the IT department.	Consultation with the IT department.
A4.4 - See above. A4.5 - Uncertain, maybe no action required.	Wrong or misleading answers.

application is to the system. A2 asked for the system commands needed to fully understand which application the vulnerability referred to. A3 demands the system commands needed to fix the vulnerable application. Because we observed that some participants confused A1–A3, we aggregated and coded these answer options together, and we added the codes for A4 (Other). Individual answers from the participants are marked with the option code plus a number, e.g. A4.5 for the fifth individual answer within option A4. Finally, we added a code for incorrect or misleading answers. We applied the same coding to Q10.

5 Study Results

This section presents the findings of our study, starting with general statistics in Fig. 6. Our IT department's vulnerability scan detected vulnerabilities with CVSS v3.x rating $\geq$ 7.0 on 207 IT systems, i.e., ChatSEC had to rewrite 207 different LLM vulnerability reports. 57 LSAs were registered as responsible for these 207 systems. The LSAs are spread over different institutions, faculties, research groups, and the IT department itself. Due to duplicates, we had 47 participants in our study, most of them receiving more than one vulnerability report. We obtained 21 completed questionnaires ($\approx$ 45% response rate).

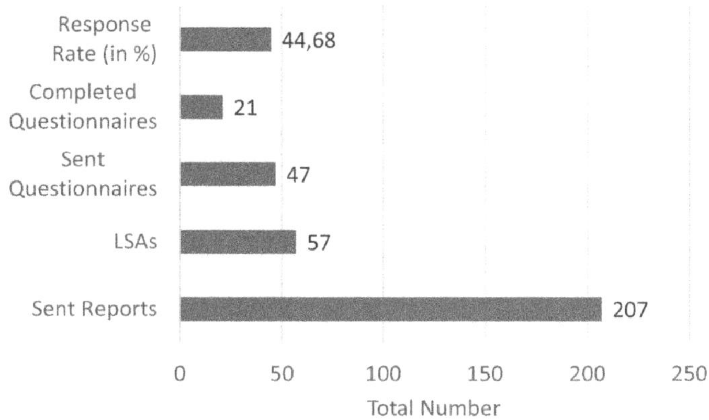

Fig. 6. Study overview

What is the professional and demographic profile of our LSAs? (O1) The answers from Q1 and Q2 tell us that 15 of 21 participants work full-time in an IT-related field, five of them with a permanent contract, seven temporarily employed, and three with a part-time employment status. The remaining six participants do not work full-time in an IT-related field.

Their available time to administer IT systems (Q4) varies significantly (see Fig. 7a). Participants who do not work primarily in IT (orange) tend to spend less than two hours a week. LSAs who work primarily in IT (blue) spend less than two hours or more than eight hours per week maintaining their systems. We observe that a high share of LSAs spend only a small part of their working time on IT administration.

Figure 7b correlates administration time (Q4) with self-assessed knowledge on cybersecurity (Q3). LSAs with "none" (green) or "basic" (blue) knowledge spend less than two hours administering their IT systems per week, while the "expert" status (orange) is associated with up to eight hours per week. Q5 and Q6 are control questions to validate the self-assessed level of cybersecurity knowledge (Q3). Executing arbitrary code with root privileges on a machine harms the data stored on the machine. Thus, the correct answer to Q5 must be coded as "relevant". Q6 should be "relevant" – we also accepted "neutral", because in science an IT system could store only non-sensitive experiment data. Now reconsider Table 3. The numbers in brackets show how many LSAs have chosen which answer. The answers from our LSAs vary considerably.

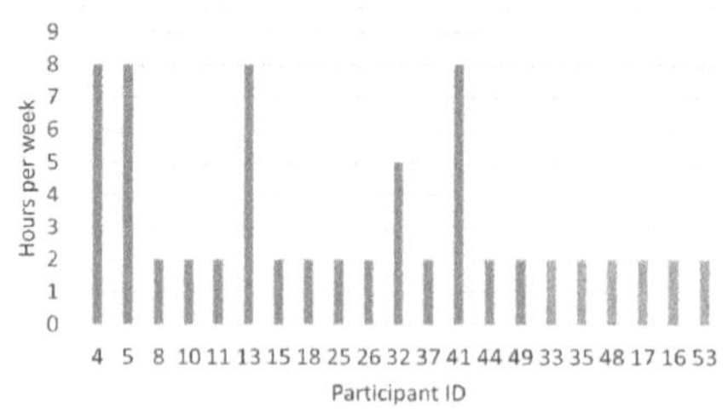

(a) Administration time vs. employment status (full-time: blue, part-time: orange)

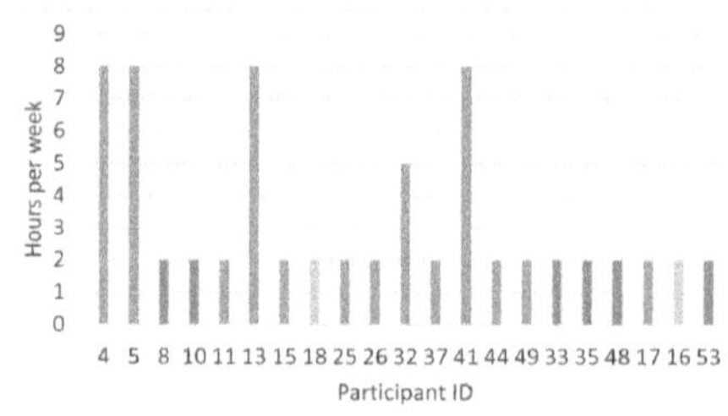

(b) Administration time vs. self-assessed cybersecurity knowledge (none: green, basic: blue, expert: orange)

Fig. 7. Time budget for administration vs. employment status and knowledge (Color figure online)

Table 6 shows the relationship between the self-assessed IT knowledge of Q3 and the assessed attack scenarios of Q5 and Q6. Two LSAs who rated themselves as having "no knowledge" correctly classified both scenarios. Four of six LSAs with "basic" knowledge correctly ranked both scenarios. Three LSAs with self-assessed "expert" knowledge did not correctly rank the attack scenarios. Thus, the relationship between Q3, Q5, and Q6 shows that the self-assessed IT knowledge does not automatically result in a correct assessment of threats. However, it also shows that an awareness of cybersecurity does not necessarily correlate with a self-assessed IT knowledge. Thus, the proportion of LSAs who overestimate their expertise is not negligible, and the distribution of cybersecurity knowledge among the LSAs is more heterogeneous than we expected.

Table 6. Self-assessed IT knowledge Q3 vs. control questions Q5, Q6

Q3 (IT-knowledge)	Q5 (wrong)	Q5 (correct)	Q6 (wrong)	Q6 (correct)
none	0	2	0	2
basic	2	4	2	4
expert	1	12	2	11

Are the generated reports sufficient for our LSAs to close vulnerabilities? (O2) With Q9-Q11, we aimed to find out whether the generated report provided sufficient information to understand, identify, and close vulnerabilities. The answers to Q9 revealed that seven participants needed additional information to understand the vulnerability description. Two were unable to answer the question properly and one insisted on asking the IT department. Eleven participants either provided no answer or indicated that the provided information was sufficient. Q10 and Q11 had a similar distribution of answers.

We now correlate this data with the answers to the other questions. In nine returned questionnaires, the LSAs checked "I did not close the vulnerability" (Q12). Six of 21 LSAs did not correctly name the highest-ranked vulnerability from their reports (Q8). Six participants were unable to solve their problems themselves, either partially or completely, without further assistance (Q7).

When asking for missing advice (Q11), a "No answer" can be interpreted both as "I do not need additional information" and as "I do not know which information would have helped me". In summary,

- 12 LSAs were able to close vulnerabilities using the enhanced reports, and
- 9 LSAs required additional information (Q9–Q11) or direct IT assistance beyond our enhanced reports.

What can we learn for rewriting vulnerability reports for LSAs? (O3) To reveal the challenges our LSAs face in closing vulnerabilities, we combined the answers to questions Q1-Q12. As Column 2 of Table 7 shows, we have identified four major challenges. Column 3 contains the combinations of answers for Q1–Q12, which support these challenges. Finally, Column 4 shows how many of our LSAs face each challenge. Note that an LSA might struggle with multiple challenges.

To derive specific and differentiated design recommendations for generating enhanced vulnerability reports that actually help LSAs to close vulnerabilities, we aligned the identified challenges with the answers from Q13-Q15. Question Q13 was answered by nine LSAs. Six of them appreciated the enhanced reports with the simplified and concise vulnerability explanations. In Q14, six of twelve LSAs asked for more details within the report. The last question (Q15) was answered by twelve LSAs. Six LSAs asked for any kind of communication channel with the IT department. Column 5 of Table 7 summarizes the recommendations we have developed to address these challenges.

Table 7. Challenges and Design Recommendations

ID	Challenge	Rationale	Freq.	Recommendation
C1	Limited time available	(Q1.A1/A2 AND Q2.A1/A2/A3/A4/A5) OR (Q4.A1/A2/A3/A4 AND Q12.A2)	16	Concise and summarize fixes
C2	Heterogeneous IT and cybersecurity knowledge	(Q1.A1/A2 AND Q3.A1/A2/A3) OR Q7.A1 OR Q9.A1/A2/A3/A4	21	Adaptive information depth within report
C3	Inconsistent self-perception	(Q3.A1/A2/A3 AND ((Q5.A1/A2/A3/A4 AND Q6.A1/A2/A3/A4) OR Q8.A1/A2/A3/A4) OR Q13.A1 OR Q15.A1	6	Add detailed mitigation strategies
C4	Broad complexity of vulnerabilities	Q7.A1 OR Q9.A1/A2/A3/A4 OR Q12.A1/A2	7	Add further support mechanisms (e.g. contact IT department).

The identified challenges and recommendations reveal that effective LLM-enhanced vulnerability reports require an adaptive design rather than a "one-size-fits-all" approach.

6 Discussion

In this section, we discuss the relevance and generalizability of our study, together with the methodical limitations.

6.1 Relevance of Our Study

Our study results shed light on a practical issue in vulnerability management: It is not sufficient to detect vulnerabilities. Their remediation depends on LSAs, who often have limited time and uneven cybersecurity expertise. In our study, many LSAs spent only a small portion of their working time on administration, and those not primarily working in IT tended to have at most two hours per week available for IT administration. This has an impact on their ability to interpret a report, locate the vulnerable component, and eventually fix the vulnerability.

As expected, our control questions on cybersecurity knowledge revealed mismatches between self-rated knowledge and correct threat assessment: some participants who rated themselves as having no knowledge assessed scenarios correctly, while several self-rated experts did not. This is important because LSAs who overestimate their expertise may refrain from actively seeking help in interpreting vulnerability reports. LLM-enhanced reports that simplify such reports can overcome this barrier, but simplification alone is insufficient when recipients misjudge severity or do not know which steps are most relevant for their systems.

Finally, the outcomes speak to the operational value of report rewriting. In our study, 12 of 21 LSAs reported being able to close vulnerabilities using the enhanced reports, while 9 required additional information or direct IT assistance. This indicates that enhanced reports can reduce support burden for a portion of cases, but also that a non-trivial subset will still require escalation paths and tailored guidance.

6.2 Generalizability of Our Study

In self-managed organizations, IT systems cannot always be operated by the central IT department. Our findings are applicable to all organizations with (i) centralized vulnerability scanning, (ii) distributed responsibility for remediation across many LSAs, and (iii) heterogeneous administrator roles where IT administration is often not the primary job function. In such contexts, our main contribution is not that LLM-enhanced reports universally solve remediation challenges, but that they can meaningfully improve remediation for some recipients while identifying needs for adaptive content and communication paths.

6.3 Methodical Limitations

Our study results are based on 21 completed questionnaires from 47 LSAs, yielding a response rate of $\approx 45\%$. Although the questionnaire was not mandatory, due to the wide spectrum of answers shown in Sect. 5 we think that we can rule out selection bias towards participants with exclusively negative or positive perceptions. Thus, we assume that we obtained a sufficient set of LSAs from Leipzig University. However, a different organizational culture, other support structures, and local policies may shape how LSAs from other organizations might respond to reports and whether they seek help, limiting the external validity of our study. Furthermore, we were unable to verify that the vulnerabilities had actually been fixed, but relied on self-reported success. For questions about missing advice, a missing answer is difficult to interpret. "No answer" may indicate sufficiency of information, but may also signal uncertainty about what was missing. We also observed that other LLMs have a different style of generating text, which might have an impact on the generated reports.

In summary, our findings support that LLM-enhanced vulnerability reports are a useful component of vulnerability remediation pipelines, but their value depends on adaptive design and explicit escalation mechanisms. Future work should address several key areas:

First, targeted interviews with LSAs to refine guidance and follow-up mechanisms. Second, comparative evaluations with multiple LLM models would provide insights into model-specific capabilities. Third, expert evaluation of report quality improvements, potentially using "LLM-as-a-judge" methodologies. Fourth, systematic evaluation of LLM reliability and robustness for productive deployment. Fifth, prompt engineering refinement through comparative studies to optimize report quality. Finally, investigating scalability across different vulnerability types would enhance practical applicability.

6.4 Use of ChatGPT

We used ChatSEC to translate technical vulnerability reports into intuitive reports. For the time being, ChatSEC uses ChatGPT gpt-4o-mini for this purpose, which means that information about vulnerabilities is sent to an external host. This is problematic from a security perspective. However, the purpose of this study was to evaluate the extent to which the reports generated by LLM help LSAs address vulnerabilities and identify opportunities for further improvement. Thus, our study results are an important step towards fine-tuning a compact LLM, which can be operated locally, without disclosing information to externally hosted services.

7 Conclusion

In this paper, we studied whether LLM-generated vulnerability reports can improve vulnerability management for LSAs in self-managed organizations, e.g., from academia and research. We conducted a mixed-method study, examining the capabilities of the local administrators at Leipzig University and to what extent the LLM-generated reports helped them to close the identified vulnerabilities without having to search for further information.

Our study addressed three research goals: First, we identified substantial diversity in the LSAs' professional backgrounds and cybersecurity knowledge based on our survey. Thus, we revealed gaps between self-assessed and actual capabilities. Second, we found that enhanced reports enabled LSAs to independently remediate less complex vulnerabilities, such as installing patches, without requiring IT department support. Third, based on these findings, we derived design recommendations (Table 7) to improve the enhanced reports. Those recommendations particularly emphasize the need for more detailed mitigation strategies and context-specific guidance for sophisticated vulnerabilities involving complex configuration errors.

While our results demonstrate the potential of LLM-enhanced vulnerability reports for low-complexity issues, further research is needed. This research includes testing different LLM models, validating report quality improvements, and assessing the approach's scalability across different vulnerability types.

References

1. van Acken, J.P., Gadellaa, J.F., Jansen, S., Labunets, K.: Poster: the unknown unknown: cybersecurity threats of shadow it in higher education. In: Proceedings of the SIGSAC Conference on Computer and Communications Security, CCS 2023, pp. 3633–3635. ACM, New York, NY, USA (2023)
2. Al-Sada, B., Sadighian, A., Oligeri, G.: Mitre ATT&CK: state of the art and way forward. ACM Comput. Surv. **57**(1) (2024)
3. Barik, K., et al.: Penetration testing analysis with standardized report generation. In: Proceedings of the Conference on Integrated Intelligent Computing Communication & Security (ICIIC 2021) (2021)
4. Barrett, R., et al.: Field studies of computer system administrators: analysis of system management tools and practices. In: Proceedings of the Conference on Computer Supported Cooperative Work. CSCW 2004. ACM (2004)
5. Beautement, A., Sasse, M.A., Wonham, M.: The compliance budget: managing security behaviour in organisations. In: Proceedings of the New Security Paradigms Workshop, NSPW 2008, pp. 47–58. ACM, New York, NY, USA (2008)
6. Bongiovanni, I.: The least secure places in the universe? A systematic literature review on information security management in higher education. Comput. Secur. **86**, 350–357 (2019)
7. Börger, W., Iacono, L.L.: User perception and response to computer security warnings. In: Mensch und Computer 2015 Workshopband, pp. 621–646. Oldenbourg Wissenschaftsverlag, Stuttgart (2015)
8. Cheng, E.C.K., Wang, T.: Institutional strategies for cybersecurity in higher education institutions. Information **13**(4) (2022)
9. Cuong Nguyen, H., et al.: Towards effective identification of attack techniques in cyber threat intelligence reports using large language models. In: Companion Proceedings of the ACM on Web Conference 2025, WWW 2025, pp. 942–946. ACM, New York, NY, USA (2025)

10. Ferrag, M.A., et al.: Generative ai in cybersecurity: a comprehensive review of LLM applications and vulnerabilities (2025). https://arxiv.org/abs/2405.12750
11. Forum of Incident Response and Security Teams: Common Vulnerability Scoring System (CVSS-SIG) (2000). https://www.first.org/cvss. Accessed Aug 2024
12. Forum of Incident Response and Security Teams (First): Common Vulnerability Scoring System: Specification Document (2000). https://www.first.org/cvss/specification-document. Accessed Jan 2025
13. Galanska, K., et al.: From reports to actions: bridging the customer usability gap in penetration testing. IEEE Access **13**, 73975–73986 (2025)
14. Hammersley, M.: The relationship between qualitative and quantitative research: paradigm loyalty versus methodological eclecticism. In: Handbook of Qualitative Research Methods for Psychology and the Social Sciences, vol. 1, no. 1, pp. 159–174 (2002)
15. Herley, C.: So long, and no thanks for the externalities: the rational rejection of security advice by users. In: Proceedings of the Workshop on New Security Paradigms Workshop, NSPW 2009, pp. 133–144. ACM, New York, NY, USA (2009)
16. Hoffmann, M., Buchmann, E.: Chatsec: spicing up vulnerability scans with AI for heterogeneous university it. In: Buchmann, E., Hoppe, D. (eds.) AISyS 2024 The First International Conference on AI-based Systems and Services, vol. 1, p. 70. IARIA, IARIA Press, Venice, Italy (2024)
17. Holm, H., Afridi, K.K.: An expert-based investigation of the common vulnerability scoring system. Comput. Secur. **53**, 18–30 (2015)
18. Holm, H., Sommestad, T., Almroth, J., Persson, M.: A quantitative evaluation of vulnerability scanning. Inf. Manag. Comput. Secur. **19**(4), 231–247 (2011)
19. Howland, H.: CVSS: ubiquitous and broken. Digit. Threats **4**(1) (2022)
20. Hrebec, D.G., Stiber, M.: A survey of system administrator mental models and situation awareness. In: Proceedings of the ACM SIGCPR Conference on Computer Personnel Research, SIGCPR 2001, pp. 166–172. ACM, New York, NY, USA (2001)
21. Jaffal, N.O., Alkhanafseh, M., Mohaisen, D.: Large language models in cybersecurity: a survey of applications, vulnerabilities, and defense techniques. AI **6**(9) (2025)
22. Jiang, Y., et al.: Mitre ATT&CK applications in cybersecurity and the way forward (2025). https://arxiv.org/abs/2502.10825
23. Kelle, U.: Mixed methods. In: Handbuch Methoden der empirischen Sozialforschung, pp. 163–177. Springer, Wiesbaden (2022)
24. van der Kleij, R., Leukfeldt, R.: Cyber resilient behavior: integrating human behavioral models and resilience engineering capabilities into cyber security. In: Ahram, T., Karwowski, W. (eds.) AHFE 2019. AISC, vol. 960, pp. 16–27. Springer, Cham (2020). https://doi.org/10.1007/978-3-030-20488-4_2
25. Koscinski, V., Nelson, M., Okutan, A., Falso, R., Mirakhorli, M.: Conflicting scores, confusing signals: an empirical study of vulnerability scoring systems (2025). https://arxiv.org/abs/2508.13644
26. Malatji, M., Von Solms, S., Marnewick, A.: Socio-technical systems cybersecurity framework. Inf. Comput. Secur. **27**(2), 233–272 (2019)
27. Mayring, P., Fenzl, T.: Qualitative inhaltsanalyse. In: Handbuch Methoden der empirischen Sozialforschung, pp. 633–648. Springer Fachmedien Wiesbaden, Wiesbaden (2019)
28. National Institute of Standards and Technology (NIST): Vulnerability Metrics (CVSS) (2000). https://nvd.nist.gov/vuln-metrics/cvss. Accessed Aug 2024
29. Son, J.Y.: Out of fear or desire? Toward a better understanding of employees' motivation to follow is security policies. Inf. Manag. **48**(7), 296–302 (2011)

30. Tashakkori, A.: Mixed Methodology: Combining Qualitative and Quantitative Approaches, vol. 46. Sage, Thousand Oaks, London and New Dehli (1998)
31. Tiefenau, C., Häring, M., Krombholz, K., von Zezschwitz, E.: Security, availability, and multiple information sources: exploring update behavior of system administrators. In: Sixteenth Symposium on Usable Privacy and Security (SOUPS 2020), pp. 239–258. USENIX Association, Virtual Conference (2020)
32. Tundis, A., Mazurczyk, W., Mühlhäuser, M.: A review of network vulnerabilities scanning tools: types, capabilities and functioning. In: Proceedings of the Conference on Availability, Reliability and Security. ARES 2018. ACM, New York, NY, USA (2018)
33. Tyagi, Y., Bhardwaj, S., Shekhar, S., P, A.: Efficient vulnerability assessment and penetration testing: a framework for automation. In: 2023 International Conference on Computational Intelligence and Sustainable Engineering Solutions (CISES), pp. 553–557 (2023)
34. Vaniea, K.E., Rader, E., Wash, R.: Betrayed by updates: how negative experiences affect future security. In: Proceedings of the Conference on Human Factors in Computing Systems, CHI 2014, pp. 2671–2674. ACM, New York, NY, USA (2014)
35. Xu, H., et al.: Large language models for cyber security: a systematic literature review. ACM Trans. Softw. Eng. Methodol. (2025, just accepted)
36. Zakaria, M.N., et al.: A review of standardization for penetration testing reports and documents. In: International Conference on Research and Innovation in Information Systems (ICRIIS), pp. 1–5. IEEE, Johor Bahru, Malaysia (2019)
37. Zhang, J., et al.: When LLMs meet cybersecurity: a systematic literature review. Cybersecurity **8**(1), 55 (2025)

SecGrad-XAI: Secure and Accurate Explainable AI with Secure Multi-party Image Classification

Daphnee Chabal[1(✉)], Dolly Sapra[1], and Zoltán Ádám Mann[2]

[1] Parallel Computing Systems Group, Informatics Institute, University of Amsterdam, Amsterdam, The Netherlands
d.n.m.s.chabal@uva.nl

[2] Institut für Informatik, Fachbereich Mathematik und Informatik der Universität Münster, Münster, Germany

Abstract. Explainable AI (XAI) is critical for interpreting Neural Network behavior. Existing XAI methods require full access to model weights and/or inputs and outputs, which may be sensitive data. This transparency requirement creates a fundamental conflict in Secure Neural Network Inference (SNNI) environments, where model weights are protected intellectual property and client inputs/outputs are confidential. We present SecGrad-XAI, a novel framework that provides a fully secure input-to-explanation pipeline using Secure Multi-Party Computing (SMPC). SecGrad-XAI allows a client to receive both an inference result and a corresponding XAI explanation without revealing the input to the server or exposing the server's model parameters to the client. Using SecGrad-XAI, we implement and evaluate three prominent XAI methods—Vanilla Gradient, Gradient × Input, and Integrated Gradients—across three convolutional architectures and three image classification datasets. Our results demonstrate that despite the constraints of SMPC, SecGrad-XAI maintains high utility; XAI outputs provided by SecGrad-XAI are nearly identical to the ones created in plaintext.

1 Introduction

The rapid adoption of Artificial Intelligence (AI) and Machine Learning (ML) across diverse domains has made AI automation an indispensable driver of modern technology. As ML models grow in complexity, particularly deep Neural Networks, their "black-box" nature presents significant challenges to adoption, especially in regulated or sensitive environments. This deficiency has fuelled the demand for Explainable AI (XAI), i.e., techniques that render model predictions comprehensible [1]. Researchers developed various XAI methods, focusing on generating local explanations, often visualized as *saliency maps*, highlighting the importance of different input features for a specific prediction. These methods include *gradient-based* [18] and other *back-propagation methods* [11,12]. A parallel and equally crucial requirement in ML pertains to confidentiality. Many high-stakes applications demand that both the user's input data and the proprietary model parameters remain private during the inference process [9]. Secure

© The Author(s), under exclusive license to Springer Nature Switzerland AG 2027
K. Renaud et al. (Eds.): EICC 2026, CCIS 3012, pp. 129–146, 2027.
https://doi.org/10.1007/978-3-032-28957-5_8

Multi-Party Computing (SMPC) is the cryptographic primitive often employed to achieve this. SMPC allows multiple parties to jointly compute a function over their private inputs without revealing the inputs to each other.

To date, research at the intersection of XAI and secure inference suffers from two major gaps:

1. To the best of our knowledge, no prior work has successfully implemented gradient-based or back-propagation-based XAI techniques entirely in an SMPC context. Existing secure inference solutions typically only provide the prediction securely, but do not provide functions to calculate gradients for example, to implement XAI methods.
2. There is no comprehensive analysis on how the constraints of secure computation impact the utility of securely generated explanations. When implementing XAI methods securely, it is critical to determine if the loss in precision inherent to practical SMPC protocols (e.g., approximation of functions or bitlength limitations) significantly degrade the quality of the explanation.

Our Contribution. We address these gaps by presenting the first comprehensive implementation and evaluation of a suite of XAI methods—including Vanilla Gradient, Gradient × Input, and Integrated Gradients—specifically designed to operate in a secure mode. Our core innovation is the capability to extend secure inference using SMPC by seamlessly feeding the resulting secret-shared intermediate values (activations and final class scores) directly into the XAI calculation, which also uses SMPC to guarantee confidentiality. Crucially, when an XAI method requires the input image or data for its computation, we provide this data in a secret-shared state, ensuring that the entire explanation process remains cryptographically secure from start to finish. The prediction and the saliency map are the only information reconstructed in plaintext.

We conduct a rigorous evaluation across three Neural Network architectures and three datasets. We measure inference and XAI execution times, for both plaintext and secure modes as well as the amount of communication between parties during both secure inference and secure XAI. We assess resulting XAI saliency maps using two metrics, and then perform a detailed quantified comparison between the plaintext/secure XAI outputs. Our results demonstrate that secure XAI is not only feasible but also highly accurate, successfully preserving the utility of the original XAI output, thus paving the way for trustworthy, privacy-preserving AI systems. Our implementation of SecGrad-XAI is available at https://github.com/edgesnni/SecGrad-XAI.

2 Preliminaries

2.1 XAI: Explainable AI

We focus on *feature attribution*, an important family of XAI methods [3,10]. These methods assign relevance scores to input features, such as pixels, to indicate their impact on a specific prediction [18]. Gradient-based methods, a prominent sub-category of feature attribution, determine this importance by calculating the partial derivative of the model's output with respect to the input features.

Explanations are often visualized in the form of saliency maps, which highlight the features of the input that contributed the most to the model's output [18]. Specifically, we use the following XAI methods in this paper.

(1) Vanilla Gradients. The saliency map $\mathbf{S}$ is computed by backpropagating the gradient of a target class score through the network to the input pixels [14]. $\mathbf{S}$ highlights the location of the values in the input data to which the output is most sensitive.

Before the method is applied, we hold:

- the input image $\mathbf{x} \in \mathbb{R}^{C \times H \times W}$ (C = Channels, H = Height, W = Width).
- the model $f(\cdot)$ with parameters $\mathbf{P}$. The model has L layers. The lth layer computes $\sigma_l(\mathbf{x}_{l-1}) = \mathbf{x}_l$, where σ_l is the specific function for layer l; $\mathbf{x}_0 = \mathbf{x}$.

The saliency map $\mathbf{S}$ is computed as follows:

1. **Forward Pass:** We compute the forward pass $\mathbf{y} = f(\mathbf{x}, \mathbf{P})$ where $\mathbf{y}$ is a vector of K elements, with K being the number of classes in a specific classification task. Let t be the index of the highest value in $\mathbf{y}$ so that y_t is the target score, that is, the score attributed by the model to the class that the model considers the input data to belong to.
2. **Backpropagation:** The gradient of y_t is propagated backward from the output layer to the input layer. Using the chain rule, the raw saliency map $\mathbf{A}$ is defined as:
$$\mathbf{A} = \frac{\partial y_t}{\partial \mathbf{x}} = \frac{\partial y_t}{\partial \mathbf{x}_L} \cdot \prod_{l=L}^{1} \frac{\partial \mathbf{x}_l}{\partial \mathbf{x}_{l-1}}$$
where $\mathbf{x}_l$ is the output of layer l. Each local derivative $\frac{\partial \mathbf{x}_l}{\partial \mathbf{x}_{l-1}}$ is evaluated at the activation values $\mathbf{x}_{l-1}$ recorded during the forward pass.
3. **Aggregation:** To transform the multi-channel gradient map $\mathbf{A} \in \mathbb{R}^{C \times H \times W}$ into a single-channel saliency map $\mathbf{S} \in \mathbb{R}^{H \times W}$, we extract the maximum absolute value across the channel dimension: $\mathbf{S}_{h,w} = \max_{1 \leq c \leq C} |(\mathbf{A})_{c,h,w}|$.

(2) Gradient × Input. This method, introduced by [13], is a variation of the Vanilla Gradient approach that aims to improve the sharpness of the saliency map by accounting for the magnitude of the input features. It adds an element-wise multiplication after Step 2 of the Vanilla Gradient procedure: $\mathbf{A} := \mathbf{A} \odot \mathbf{x}$. This operation scales each pixel of the raw saliency map with the intensity of the corresponding input pixel, attenuating the contribution of gradients in regions of the input where $x \approx 0$, focusing the final saliency map on pixels where the gradient aligns with actual feature content [13,18]. The single-channel saliency map $\mathbf{S}$ is then produced from $\mathbf{A}$ in Step 3 as in the Vanilla Gradient method.

(3) Integrated Gradients. This method was developed to improve XAI clarity by calculating an integral of gradients along a path, rather than a local derivative from the scalar y_t [15]. The method follows the steps detailed in Algorithm 1:

Algorithm 1: Integrated Gradients Computation

Input : Input image $\mathbf{x}$, Target class t, Baseline $\mathbf{x}'$, Number of steps M
Output: Saliency map $\mathbf{S}$

1 $\mathbf{\Delta x} \leftarrow \mathbf{x} - \mathbf{x}'$;
2 $\mathbf{AccumulatedGrad} \leftarrow \mathbf{0}$;
3 **for** $m = 1$ *to* M **do**
4 $\quad \alpha_m \leftarrow m/M$;
5 $\quad \hat{\mathbf{x}}_m \leftarrow \mathbf{x}' + \alpha_m \cdot \mathbf{\Delta x}$;
6 $\quad \mathbf{Grad}_m \leftarrow \frac{\partial f(\hat{\mathbf{x}}_m)_t}{\partial \hat{\mathbf{x}}_m}$;
7 $\quad \mathbf{AccumulatedGrad} \leftarrow \mathbf{AccumulatedGrad} + \mathbf{Grad}_m$;
8 **end**
9 $\overline{\mathbf{Grad}} \leftarrow \mathbf{AccumulatedGrad}/M$;
10 $\mathbf{A} \leftarrow \mathbf{\Delta x} \odot \overline{\mathbf{Grad}}$;
11 $\mathbf{S} \leftarrow \max_{1 \leq c \leq C} |(\mathbf{A})|$;
12 **return S**

1–5 **Path Construction:** The difference vector $\mathbf{\Delta x} = \mathbf{x} - \mathbf{x}'$ from a given baseline vector x' is calculated and a sequence of M interpolated inputs is generated. M determines the precision of the approximation; a higher M reduces the discretization error. For each step $m \in \{1, \ldots, M\}$, the interpolated input $\hat{\mathbf{x}}_m$ is defined as: $\hat{\mathbf{x}}_m = x' + \alpha_m \Delta x$, where $\alpha_m = m/M$, leading to a linear trajectory in the feature space from the baseline to the input image.

6 **Forward Pass and Backpropagation:** For each of the M steps, a forward pass is performed using $\hat{\mathbf{x}}_m$ as the input to compute the target class score y_t while caching intermediate activations $\mathbf{x}_{l,m}$. Backpropagation is also executed to find the gradient: $\mathbf{Grad}_m = \frac{\partial f(\hat{\mathbf{x}}_m)_t}{\partial \hat{\mathbf{x}}_m}$, representing the local saliency at point $\hat{\mathbf{x}}_m$ (i.e., how much y_t changes with changes in $\hat{\mathbf{x}}_m$). This is computationally intensive, as it requires M complete forward-backward cycles through the neural network.

7–9 **Riemann Sum:** To approximate the integral, the computed gradients are added and averaged as: $\overline{\mathbf{Grad}} = \frac{1}{M} \sum_{m=1}^{M} \mathbf{Grad}_m$.

10 **Saliency Scaling:** The raw saliency map $\mathbf{A} \in \mathbb{R}^{C \times H \times W}$ is calculated as the element-wise multiplication $\mathbf{\Delta x} \odot \overline{\mathbf{Grad}}$, which ensures that the saliency (i.e. the total contribution of a feature to the change in output) of a feature from the original input $\mathbf{x}$ is proportional to the feature's displacement from the baseline; if the baseline is a zero-filled vector, the saliency of a feature is then proportional to its intensity (i.e., dark pixels are less likely to be considered important).

11 **Aggregation:** $\mathbf{S}$ is then calculated similarly to the previous XAI methods.

2.2 Secure Neural Network Inference (SNNI)

SNNI protocols enable privacy-preserving collaboration between a client and a model holder during AI inference. These protocols allow the client to provide input data without revealing it in plaintext, while the model holder keeps the model weights confidential, though the model architecture may be public. The parties jointly execute the inference protocol without accessing each other's private data; only the client learns the output.

SNNI protocols are built upon cryptographic primitives such as Secure Multi-Party Computing (SMPC) and Homomorphic Encryption (HE), which ensure strong privacy guarantees [9]. SMPC-based approaches, including secret sharing and garbled circuits [9,22], implement neural network computations using collaborative protocols, while HE-based protocols allow computations directly on encrypted data [9,22]. Each approach involves trade-offs in terms of performance, supported operations, and security assumptions.

The fundamental SMPC mechanism for ensuring privacy is *Additive Secret Sharing*. Let $\mathbf{x}$ be some sensitive data held by only one party in plaintext. To be able to perform algebraic operations on $\mathbf{x}$ in a multi-party system without revealing it to other parties, $\mathbf{x}$ is decomposed into n shares: $\mathbf{x} = \sum_{i=1}^{n} [\mathbf{x}]_i$, where $n-1$ shares are random. For a two-party SNNI setting, where $n = 2$, Party 1 is the client and $\mathbf{x}$ is the input data the client holds. The client creates the shares such that $\mathbf{x} = [\mathbf{x}]_1 + [\mathbf{x}]_2$, and sends only one share to the server. The server does the same with $\mathbf{P}$, the trained weights of the neural network it holds, which represent sensitive intellectual property. $\mathbf{P}$ is decomposed into $[\mathbf{P}]_1$ and $[\mathbf{P}]_2$. The model architecture is known to both parties. Following the agreed-upon SNNI protocol, both parties execute collaboratively a series of steps to compute the SMPC-equivalent of the functions that the neural network comprises.

For efficient inference, SNNI protocols typically follow the layer-wise decomposition of neural network computations, applying cryptographic techniques best suited for each layer type. Linear layers (e.g., fully connected or convolutional) are typically more efficient to compute securely than non-linear ones (e.g., ReLU or SoftMax), which pose greater challenges due to their non-polynomial nature [17]. Modern frameworks adopt specialized, layer-optimized cryptographic routines to maintain both performance and accuracy. Typically, the protocol for each layer outputs secret-shared numbers, which are then processed by the protocol for the next layer. Only the final output is restored from its shares.

In this work, we use CrypTen [6], an SMPC-based SNNI library. CrypTen uses a combination of arithmetic and Boolean secret sharing, with the former for linear and the latter for non-linear layers. CrypTen implements SMPC protocols for the conversion between arithmetic and Boolean secret sharing. CrypTen assumes a semi-honest threat model, in which parties follow the protocol but try to infer additional information. While this is a common assumption, some SNNI protocols also protect against malicious adversaries that may deviate from the protocol. Our approach could also be extended to such SNNI protocols.

3 Related Work

The fields of Explainable AI (XAI) and Secure Neural Network Inference (SNNI) have historically developed separately, driven by distinct concerns. XAI research focuses on rendering the decisions of deep neural networks comprehensible, typically assuming full access to model parameters and input data [3]. Conversely, SNNI research prioritizes the acceleration of protocols without compromising the confidentiality of proprietary model weights and sensitive client inputs [23].

Our work aims to bridge the gap between these fields to mend the problematic scenario where users in high-stakes SNNI environments cannot currently, prior to our work, verify the reasoning behind an inference result without compromising the underlying security guarantees.

[4] is the only paper in our knowledge that investigates this intersection by implementing secure occlusion-based XAI—a perturbation method that identifies feature importance by masking parts of the input. Unlike our use of gradient-based XAI, which relies on backpropagating through the network, their approach requires numerous forward-pass inferences on perturbed versions of the input image. Thus, [4]'s methods are significantly more expensive than ours, both in terms of execution times ($4-20\times$ slower than ours for same architecture SmallNet for MNIST) and amount of data exchanged (they use approximately twice more communication rounds and GB of data exchange than SecGrad-XAI), despite the SNNI protocol used being identical in both works.

4 The SecGrad-XAI Approach

4.1 Challenges

The integration of XAI methods in SMPC frameworks introduces several non-trivial technical hurdles that do not exist in traditional plaintext environments:

1. Transitioning from plaintext to secret shares requires a conversion from floating-point numbers to fixed-point arithmetic. In SMPC, values are typically represented as integers in a large prime field $\mathbb{Z}_p$. To represent decimals, a scaling factor 2^k is used (fixed-point). Repeated multiplications and truncations required by deep learning and XAI algorithms lead to a cumulative loss of precision, which may degrade the fidelity of the saliency maps, potentially leading to misleading explanations and a reduction in model utility.
2. Most state-of-the-art XAI methods (e.g., Integrated Gradients) rely on gradients. In plaintext libraries like PyTorch or TensorFlow, these are computed efficiently via *autograd* engines. SMPC libraries like CrypTen often lack fully-featured supported differentiation. The resulting manual implementations to achieve secure backward passes leave room for error, and are often slower than optimized libraries developed for the purpose.
3. Unlike plaintext XAI, where data remains local, SMPC requires communication between parties for every multiplication or comparison. Gradient-based XAI methods often require traversing the entire neural network multiple times. In a multi-party setting, this may result in a communication bottleneck, where the total time is dominated by network latency, making high-fidelity explanations difficult to achieve over standard internet connections.

4.2 Overall System

SecGrad-XAI provides explanations in a multi-party setting (Fig. 1). Model provider (server) and client maintain the confidentiality of their respective assets. The client obtains both the inference result and the corresponding saliency map for a private input image.

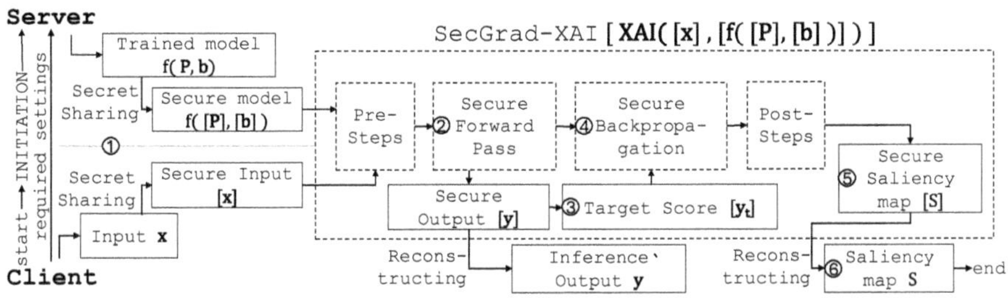

Fig. 1. Overview of SecGrad-XAI. [·]: Secret Shared Data; →: Flow of Data/Computations; -----: Collaborative SMPC between Client & Server.

We assume two privacy constraints: 1. The server possesses a trained model $f(\cdot)$ and its associated weights $\mathbf{P}$. These must remain private to protect intellectual property. 2. The client possesses a private input image $\mathbf{x}$. To ensure query privacy, the server must not learn the contents of $\mathbf{x}$, the classification result $\mathbf{y}$, the target class $\mathbf{t}$, nor the resulting saliency map $\mathbf{S}$.

Across all supported XAI methods (Vanilla Gradient, Gradient × Input, and Integrated Gradients), the execution follows a standardized workflow: 1. The client secret-shares the input as $[\mathbf{x}]$ and the server secret-shares the model as $[f(\cdot)]$. Some additional steps are required by some methods to produce additional inputs to the next step. 2. They co-execute the SNNI protocol to produce secret-shared output $[\mathbf{y}]$. 3. Upon obtaining y, the client identifies the target class index t. To preserve query privacy during the XAI phase, the client converts this index into a one-hot vector and secret-shares it as $[\mathbf{e}_t]$. This ensures the server does not learn which target class which feature importance is being calculated. This method is faster than the secure `ArgMax` SMPC protocol which could also be used for this purpose. 4. The secret-shared inference results $[\mathbf{y}]$, input $[\mathbf{x}]$, the target selector $[\mathbf{e}_t]$, and model $[f(\cdot)]$ serve as inputs for the backpropagation part of the secure XAI algorithm. The protocol leverages the cached activations $[\mathbf{x}_l]$ from the secure forward pass to compute the raw saliency map in a secret-shared form $[\mathbf{A}]$. 5. The final single-channel saliency map is generated as a set of secret shares $[S]$. 6. The server sends its share of $[S]$ to the client, so that the client can reconstruct the saliency map in plaintext.

4.3 SecGrad-XAI Methods

In the following, we describe the secure implementation of the XAI methods from Sect. 2.1 in SecGrad-XAI.

(1) Secure Vanilla Gradient. While the mathematical chain rule remains identical to the plaintext version, the algorithm shifts to an interactive protocol:

1. **Forward Pass with Caching:** Similarly to the plaintext version, a forward pass through the model computes $[y]$.
2. **Secure Target Selection:** Unlike the plaintext version where y_t is obtained directly, we must compute $[y_t] = \sum [y_i] \cdot [\mathbf{e}_t]_i$ using an interactive dot product protocol. This hides which class the client is investigating from the server to preserve query privacy.
3. **Secure Backpropagation:** Securely computing $\prod \frac{\partial [\mathbf{x}_l]}{\partial [\mathbf{x}_{l-1}]}$ now involves a chain of secret-shared multiplications and securely determining the partial derivatives of each layer.
4. **Secure Aggregation:** To compute $[\max |\mathbf{A}|]$ from secret-shared $[\mathbf{A}]$, one could theoretically reconstruct $\mathbf{A}$ in plaintext. However, this is prohibited, as $\mathbf{A}$ contains high-resolution gradient information that could reveal sensitive internal weights of the server's model and specific feature responses to the client's input. To maintain privacy, we perform the aggregation directly on the shares using a secure comparison protocol [6].

(2) Secure Gradient $\times$ Input. The additional element-wise multiplication $[\mathbf{A}] := [\mathbf{A}] \odot [\mathbf{x}]$ is computed on secret-shared vectors, adding one synchronization round to the Secure Vanilla Gradient pipeline.
(3) Secure Integrated Gradients. This method introduces the highest overhead due to the iterative nature of the Riemann sum approximation.

1. **Baseline & Path Construction:** The client creates in plaintext and then secret-shares the baseline $[\mathbf{x}']$. The interpolation $[\hat{\mathbf{x}}_m] = [\mathbf{x}'] + \alpha_m([\mathbf{x}] - [\mathbf{x}'])$ is performed by the parties locally on their respective shares because α_m is a public constant, and addition and multiplication by public scalars are linear operations that can be computed independently on each share.
2. **Iterative Forward Passes and Backpropagations:** For each step m, the forward pass and the backpropagation are performed.
3. **Secure Riemann Approximation:** The summation $\sum [\mathbf{Grad}_m]$ is local, performed on each party's share of $\mathbf{Grad}_m$ respectively. Similarly, the scaling by $1/M$ is performed locally. The multiplication by $[\Delta \mathbf{x}]$ requires the use of SMPC. The final map $[\mathbf{A}]$ is aggregated in secret-share to produce $[\mathbf{S}]$.

5 Experiments

5.1 Datasets and Neural Networks Training

We trained benchmark Convolutional Neural Networks (CNNs) AlexNet, VGG11, and VGG19, named here in increasing order of number of parameters and number of layers [21]. We also used a small 3-fully connected layer model, coined SmallNet here, provided with CrypTen. We used three benchmark datasets for image classification tasks: MNIST (classifying handwritten digits [8]), Tiny-ImageNet (classifying images into one of 200 animal and object classes [7]), and ImageNet (classifying higher resolution images into one of 1000 animal and object classes [2]). Images of all 3 datasets were resized to $3 \times 224 \times 224$ to be compatible with the used CNNs, except for MNIST for SmallNet, which uses the size $1 \times 28 \times 28$. Following standard community guidelines, we performed additional preprocessing steps to promote task accuracy, during both training and inference.

We manually implemented these models in both plaintext (using PyTorch) and secure (using CrypTen) formats, because CrypTen necessitated statically-defined networks. We eliminated the use of dynamic PyTorch constructs – such as `nn.Sequential`, `nn.ModuleList`, and complex Python logic like list comprehensions – to ensure compatibility with CrypTen's secure computation primitives.

We conducted the training on a Linux (v4.18.0) cluster managed by SLURM, using x86_64 nodes (64 logical cores) equipped with 1–2 × NVIDIA A40/A10 GPUs. We used standard open-source training paradigms specific to each of the 10 model-dataset combinations, to achieve state-of-the-art accuracy. We trained the PyTorch models and, when using the CrypTen models, transferred the trained parameters accordingly.

Recursive Models. To explore the limits of the chosen secure computation framework, we also implemented custom-made GoogLeNet, ResNet18, ResNet50, and ResNet152 twice, using either PyTorch or CrypTen. While forward passes were successful, we were unable to use these models when performing secure backpropagation, whether using ONNX-converted PyTorch models or CrypTen models. This was because the `crypten.grad()` function, calculating gradients using secret shares, proved incompatible with the recursive (or non-linear logic) structure inherent in these highly-nested architectures. However, to enable further experimentation, we include these models in the code we provide.

5.2 Inference Setup

The multi-party inference setup used two identical machines, acting as a server and a client. The server held the trained models and waited for incoming connections; the client held the inference images and initiated connection to start inference using the server's IP address. Each machine operated on Debian GNU/Linux 11 (bullseye) and was powered by a single Intel Xeon E-2378 CPU @ 2.60GHz featuring 8 double-thread cores and 62 GB of memory. Network communication between nodes was controlled using `tc` (traffic control) to enforce a maximum bandwidth of 10 GBit/s (measured on average as 9.56 Gbit/s by `iperf3`) and to introduce a network emulation latency of 1 millisecond. Our implementation uses Python 3.9.2 and CrypTen 0.4.1.

5.3 XAI Implementation Details

We investigated the effect of two parameters in plaintext and secure **Integrated Gradient** XAI on the utility of the generated saliency map and the comparative plaintext/secure fidelity. The first one was the number of steps M, determining how the path integral is approximated during the Riemann sum calculation. Higher M values should lead to a more accurate approximation. However, we found both qualitatively upon visualizing resulting saliency maps (see Fig. 2 for an example) and quantitatively using the Sparsity and Alignment metrics, that higher M values did not increase XAI utility. For these reasons, we set $M = 2$ for both plaintext and secure Integrated Gradient XAI to decrease the number of iterations and the associated computational overhead.

The second one was the type of baseline $\mathbf{x}'$ used as starting point for the integration. We implemented 5 types of baselines:

- **BLACK**: $\mathbf{x}'$ is a zero-filled vector, which leads to dark pixels being less likely to be rendered salient in the saliency map.

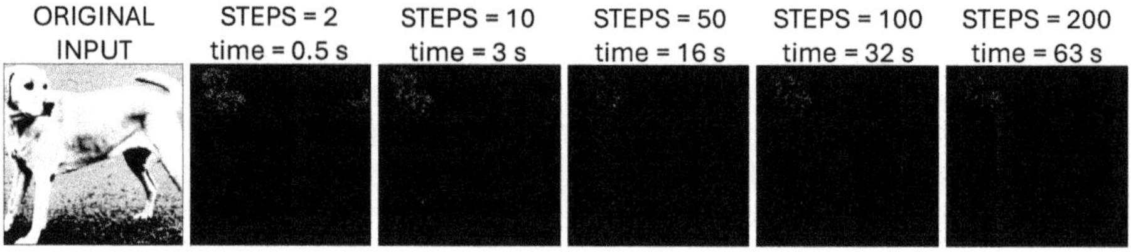

Fig. 2. Comparison of Integrated Gradient XAI, varying the number of steps M, for one image of ImageNet using VGG19. Increasing M resulted in a linear increase in execution time ($\mathcal{O}(M)$) but did not improve explanation quality.

- **WHITE**: $\mathbf{x}'$ is a one-filled vector (the highest value for a pixel), which similarly penalizes light pixels.
- **GREY**: $\mathbf{x}'$ is filled uniformly with value 0.5. This penalizes regions of the input that lack discriminative features or significant visual information, as the baseline provides a neutral, non-informative reference.
- **RANDOM**: $\mathbf{x}'$ is a Gaussian random noise vector with range $[0, 1]$. This baseline serves to marginalize the influence of fine-grained local variations and high-frequency noise in the input image, highlighting only the structural components that consistently drive the model's output.
- **BLUR**: $\mathbf{x}'$ is a heavily blurred version of the input, generated by applying a Gaussian filter with a 11×11 kernel and $\sigma = 5$, which only preserves the global spatial structure and luminance of the input while removing any detail; this favors high-frequency characteristics, such as textures and sharp edges, in contributing to the model's output.

The last two methods ensure that pixels are not under-represented in the saliency map due to their intrinsic intensity or color, which is especially useful for multi-color datasets such as Tiny ImageNet and ImageNet.

CAM & LRP XAI. We also implemented both plaintext (in PyTorch) and secure (in CrypTen) versions of two additional backpropagation-based XAI methods: Class Activation Mapping (CAM) [12] and Layer-wise Relevance Propagation (LRP) [11]. Despite high Sparsity of their resulting saliency maps, we found low fidelity when comparing the plaintext/secure XAI saliency maps generated by these methods. We attribute this failure to specific functions missing in the CrypTen framework, which led to a significant difference in the steps and logic between the plaintext and secure methods. Therefore, we did not include these methods in our results but we provide their implementation in the code to facilitate future research.

5.4 Quantitative Assessment

XAI Utility Evaluation. We use the Sparsity, Alignment, and Consistency metrics to quantitatively evaluate how successful the XAI methods [16], whether plaintext or secure, were individually. Each metric is calculated once per saliency map, whether from a plaintext or secure XAI method, and is therefore subscripted with $k \in \{plt, sec\}$, representing values obtained using plaintext (denoted as *plt*) or secure (denoted as *sec*) XAI.

Sparsity quantifies the concentration of the explanation, reflecting how compact the saliency map is [5]. A higher sparsity value indicates that the most relevant pixels are confined to a smaller, more focused region. This is measured using a variation of the Gini index:

$$\mathrm{SP}_k(\mathbf{S_k}(\mathbf{x})) = 1 - 2\sum_{i=1}^{N} \frac{\mathbf{S_k}(\mathbf{x})_i}{\|\mathbf{S_k}(\mathbf{x})\|_1} \cdot \frac{N - i + 0.5}{N}$$

In this equation, $\mathbf{S_k}(\mathbf{x})_i$ represents the components of the flattened saliency map $\mathbf{S_k}(\mathbf{x}) \in \mathbb{R}^{H \cdot W}$ sorted in ascending order, for input x. The term $\|\mathbf{S_k}(\mathbf{x})\|_1 = \sum_{j=1}^{N} |\mathbf{S_k}(\mathbf{x})_j|$ is the L^1 norm of the saliency map vector, and N is the total number of pixels in the saliency map. The term $\frac{N-i+0.5}{N}$ serves as a weighting factor. The resulting metric ranges from 0 to 1, where a value of 1 signifies a perfectly sparse and concentrated explanation.

Alignment measures the explanation's structural consistency relative to the original input [20]. An ideal explanation should produce a saliency map that closely aligns with the features of the input image x. We employ the normalized root mean square as:

$$\mathrm{AL}_k = \frac{\|\mathbf{x} - \mathbf{S_k}(\mathbf{x})\|_2}{\|\mathbf{x}\|_2}$$

The minimum value of the result is 0, indicating identical images. A value near 1 suggests that the saliency map is as distant from the input as a zero-filled array would be.

Consistency assesses the visual similarity between two arrays. We use the Structural Similarity Index Measure [19] for this purpose:

$$C_k(\mathbf{x}, \mathbf{S}_k) = \frac{(2\mu_1\mu_2 + E)(2\sigma_{12} + D)}{(\mu_1^2 + \mu_2^2 + E)(\sigma_1^2 + \sigma_2^2 + D)}$$

where $\mathbf{x}, \mathbf{S}_k \in \mathbb{R}^{1\times H\times W}$. μ_1 and μ_2 are the mean values of the inputs, σ_1^2 and σ_2^2 are their respective variances, and σ_{12} is the covariance between $\mathbf{x}$ and $\mathbf{S_k}$. The constants E and D are included to prevent instability when the denominators are close to zero, often calculated as $E = (KL)^2$ and $D = (K'L)^2$. Here, L represents the dynamic range of the pixel values, with $K = 0.01$ and $K' = 0.03$. The resulting value is bounded between -1 and 1, where values approaching 1 denote a higher degree of visual similarity between the two arrays.

Plaintext/Secure XAI Fidelity Evaluation. We evaluate fidelity, that is, the difference/dissimilarity between the plaintext and secure XAI saliency maps. This quantitative assessment is key to determine whether SecGrad-XAI affects the utility of XAI methods through the use of SMPC.

Correlation: We used Pearson's Correlation to measure the linear element-wise correlation between a plaintext and secure saliency map pair, by flattening each saliency map into a 1-dimensional vector; each saliency map pair's coefficient is then averaged with the others'.

Differences: We re-used the Consistency metric to compute $C_{comp}(\mathbf{S}_{Sec}, \mathbf{S}_{Plt})$, to capture how alike the two explanations are. Moreover, we compute $C_{diff} = C_{plt} - C_{sec}$, to further assess plaintext/secure similarity in relation to the same input. Similarly, we compute $AL_{diff} = AL_{plt} - AL_{sec}$. A value close to 0 in all three difference metrics indicates high plaintext to secure XAI fidelity.

Other: L^1 norm and Mean Squared Error (MSE) are also available in our implementation but not discussed in Sect. 6 as they yield redundant insights.

Implementation of Assessment. The metrics were assessed by the client after reconstructing the inference and XAI outputs. For consistency and robustness, where format $1\times H\times W$ is required, the cross-channel average of 3-channel inputs (e.g., for ImageNet) is used. Accuracy metrics were calculated using functions of the `sklearn.metrics` module. For XAI evaluation, we implemented the Sparsity, Alignment, Consistency, and Mean Square Error metrics using custom functions in Python, while Pearson Correlation was calculated using the `scipy.stats` module. The amount of data exchanged during secure inference and secure XAI was measured in bytes using CrypTen's `comm_bytes` function and converted to Gigabytes as GB = bytes/1000^3. The number of communication rounds between parties was measured using CrypTen's `comm_rounds` function.

6 Results and Discussion

While every evaluation metric was calculated for all experiments, we only report highlight findings here. All reported significant findings used statistical tests with significance threshold $\alpha_{stats} = 0.001$. The Pearson correlation and C_{comp} metrics were almost perfectly correlated, with a correlation coefficient of $\rho > 0.99$.

6.1 Accuracy of SecGrad-XAI

The core objective was to show that secure XAI can be achieved with minimal disruption to the explanation itself. We found that in most cases, the plaintext and secure saliency maps were nearly identical for a given image and neural network (Table 1). A Pearson correlation test confirmed that plaintext and secure saliency maps had a strong positive correlation ($\rho > 0.99$). Further analyses showed that the top 10% highest (i.e. most important) values of plaintext and secure saliency maps are the same pixels 70–100% of the time.

In some cases (AlexNet on Tiny ImageNet and VGG11 on MNIST for all the Vanilla and Gradient × Input methods; see Fig. 3 for Integrated Gradients results), C_{comp} was lower, indicating lower plaintext-secure XAI fidelity.

In all these cases, however, C_{diff} and AL_{diff} remained $\approx 0 \pm 0.001$.

Table 1. XAI assessment results. *: no plaintext vs. secure differences. **Bold**: secure XAI scored significantly higher than plaintext XAI saliency maps.

Model	Dataset	Vanilla Gradient (VG)				Gradient × Input (GI)			
		$SP_{(plt/sec)}$	C_{comp}	C_{diff}	AL_{diff}	$SP_{(plt/sec)}$	C_{comp}	C_{diff}	AL_{diff}
SmallNet	MNIST	0.49*	1.00	0.00	0.00	0.63*	1.00	0.00	0.00
AlexNet	MNIST	0.39*	0.81	0.00	0.00	0.40*	1.00	0.00	0.00
	T-ImNet	**0.40/0.97**	0.03	0.00	0.02	**0.50/0.98**	0.05	0.01	0.02
	ImNet	0.45*	0.63	-0.01	0.02	0.57/0.61	0.69	0.01	0.01
VGG11	MNIST	0.31*	0.54	0.00	0.00	0.34/0.31	0.51	0.00	-0.01
	T-ImNet	0.41*	0.99	0.00	0.00	0.53*	1.00	0.00	0.00
	ImNet	0.48*	1.00	0.00	0.00	0.59*	1.00	0.00	0.00
VGG19	MNIST	0.36*	0.78	0.00	0.00	0.36*	0.74	0.00	0.00
	T-ImNet	0.48*	1.00	0.00	0.00	0.57*	1.00	0.00	0.00
	ImNet	0.49*	1.00	0.00	0.00	0.60*	1.00	0.00	0.00

		Integrated Gradients ($M = 2$; $x' =$ BLACK)					
Model	Dataset	$SP_{(plt/sec)}$	$C_{(plt/sec)}$	$Se_{(plt/sec)}$	C_{comp}	C_{diff}	AL_{diff}
SmallNet	MNIST	0.63*	-0.02*	0.93*	1.00	0.00	0.00
AlexNet	MNIST	0.40*	-0.01*	1.03*	0.80	0.00	0.00
	T-ImNet	**0.50 / 0.98**	0.01 / 0.00	1.00*	0.06	0.01	0.00
	ImNet	0.57*	0.02*	1.01*	0.66	0.01	0.01
VGG11	MNIST	0.32*	-0.01*	1.05*	0.50	0.00	0.01
	T-ImNet	0.52*	0.01*	1.01*	1.00	0.00	0.00
	ImNet	0.59*	0.03*	1.00*	1.00	0.00	0.00
VGG19	MNIST	0.36*	-0.01*	1.03*	0.73	0.00	0.00
	T-ImNet	0.55*	0.00*	1.00*	1.00	0.00	0.00
	ImNet	0.61*	0.03*	1.00*	1.00	0.00	0.00

These lower C_{comp} values were consistently associated with lower SP_{diff} values, where SP_{sec} is significantly higher compared to SP_{plt} values. Qualitative inspections (See Fig. 4) confirm this finding. Because this difference did not affect C_{diff} nor AL_{diff}, we conclude that a single metric alone, especially without visual inspection, does not assess the utility of explanations.

The problematic C_{comp} drop specifically for AlexNet on Tiny ImageNet was thoroughly investigated. We found the cause to be the numerical instability within the model's Batch Normalization (BN) layers, a phenomenon that occurred only for this specific model-dataset combination. Through training, the running variance—the parameter used to rescale activations based on their statistical spread—reached a minimum of $\approx 3 \times 10^{-5}$ in the final BN layer. In secure inference, this value was constrained by its 16-bit fixed-point secret share representation $\approx 1 \times 10^{-5}$. This caused the backpropagating signal to accumulate significant rounding errors at each layer, resulting in a gradient flow that represented numerical artifacts rather than the learned features of the model.

Individual XAI Utility: For all experiments using the Vanilla Gradient, the Gradient $\times$ Input, and Integrated Gradients (BLACK/BLUR baselines), for both plaintext and secure saliency maps individually, we observe $C_k \approx 0.1$ and $AL_k \approx 1.01$ with $k \in \{plt, sec\}$, scores attributable to the high-frequency noise and pixel-wise sparsity inherent to gradient-based XAI [18]. Qualitative assessment (Fig. 5) confirms that the saliency maps successfully localize the salient semantic structures, providing functional utility and meaningful interpretability of the XAI results.

Classification Accuracy: Although not central to this work, classification accuracy is reported in Table 2, confirming standard task performance.

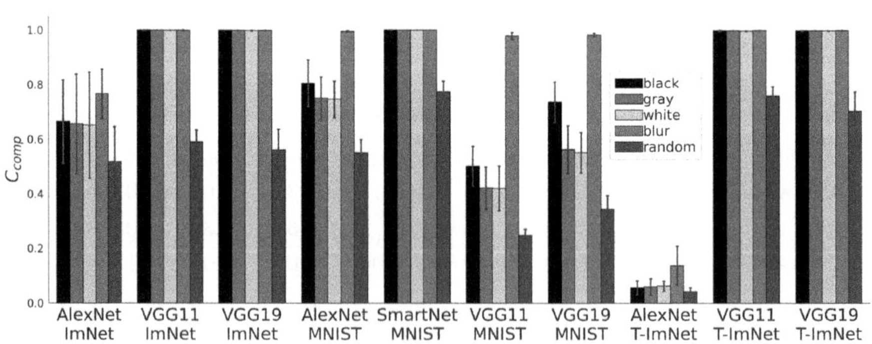

Fig. 3. Average C_{comp} values of the secure Integrated Gradients XAI method for each dataset and model combination.

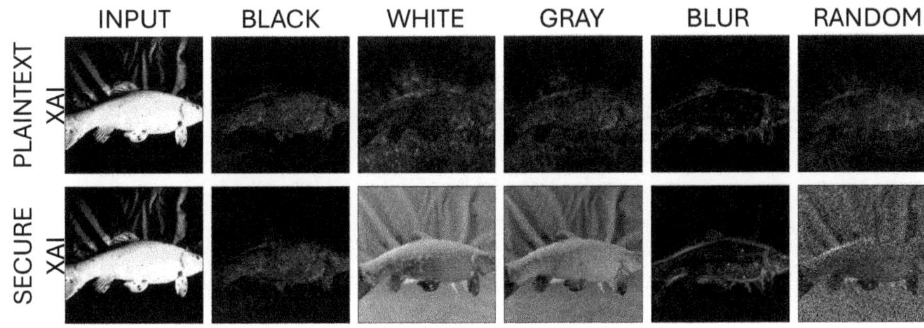

Fig. 4. Example ImageNet original input $\mathbf{x}$ (top) & reconstructed from secret shares $[\mathbf{x}]$ (bottom), next to saliency maps S generated by plaintext and secure Integrated Gradients using each baseline on AlexNet.

6.2 Cost of SecGrad-XAI

As expected, the transition from plaintext to secure XAI introduced substantial overhead in XAI execution time for all XAI methods (see Table 3). The communication cost of secure XAI is equally significant and scales with model depth, confirming that the SNNI protocols become increasingly expensive as the Neural Network architecture grows and the XAI method increases in algorithmic complexity. There was no difference in execution time or communication cost between baseline methods for plaintext nor secure Integrated Gradients; this is because the baselines are created in PyTorch before being secret-shared, which is inexpensive compared to the costly SNNI protocols, and all baselines were of identical sizes, which render them identical when computed in secret shares.

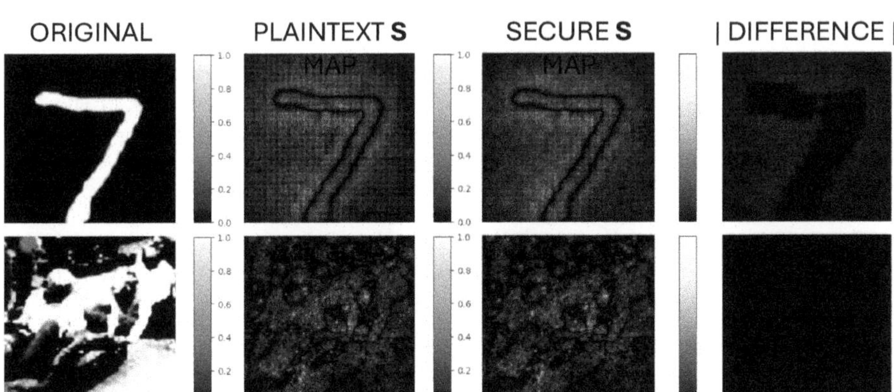

Fig. 5. Comparison of saliency maps for a test image produced by plaintext vs. secure Gradient × Input for AlexNet on MNIST (top) and plaintext vs. secure Vanilla Gradient for VGG11 on Tiny ImageNet (bottom).

Table 2. Classification Performance Metrics per Dataset.

Model	Dataset	Acc	F1	Rec	Prec
SmallNet	MNIST	98.17	98.17	98.17	98.17
AlexNet	MNIST	99.65	99.65	99.65	99.65
	T-ImNet	52.98	52.4	52.98	52.54
	ImNet	58.17	58.17	58.17	58.17
VGG11	MNIST	99.32	99.32	99.31	99.32
	T-ImNet	59.51	59.18	59.51	59.4
	ImNet	72.63	72.63	72.63	72.63
VGG19	MNIST	99.33	99.33	99.32	99.33
	T-ImNet	62.93	62.88	62.93	63.3
	ImNet	76.69	76.69	76.69	76.69

Table 3. Plaintext and secure execution time (mean ± std over 15 saliency maps) and secure data exchange. Secure GB/Rds: data exchanged in gigabytes/communication rounds in SecGrad-XAI.

XAI method	Model	Dataset	Plaintext time [s]	Secure time [s]	Secure GB/Rds
Vanilla Gradient	SmallNet	MNIST	0.001 ± 0.000	0.396 ± 0.049	$5.81 \cdot 10^{-3}/78$
	AlexNet	ALL	0.049 ± 0.005	19.673 ± 0.226	4.28/371
	VGG11	ALL	0.113 ± 0.040	138.88 ± 0.67	29.25/427
	VGG19	ALL	0.159 ± 0.085	241.01 ± 0.61	56.64/531
Gradient × Input	SmallNet	MNIST	0.001 ± 0.000	0.264 ± 0.048	$4.06 \cdot 10^{-3}/43$
	AlexNet	ALL	0.049 ± 0.005	17.661 ± 2.545	3.84/325
	VGG11	ALL	0.113 ± 0.033	138.13 ± 0.49	28.54/382
	VGG19	ALL	0.159 ± 0.071	240.53 ± 2.2	55.11/486
Integrated Gradients ($M = 2$; any baseline)	SmallNet	MNIST	0.001 ± 0.000	0.612 ± 0.040	$9.96 \cdot 10^{-3}/112$
	AlexNet	ALL	0.101 ± 0.005	38.104 ± 0.458	8.58/688
	VGG11	ALL	0.309 ± 0.070	276.20 ± 0.695	61.45/800
	VGG19	ALL	0.507 ± 0.161	480.596 ± 1.36	112.97/1008

High communication cost is a characteristic of the cryptographic primitives underlying the specific SNNI protocol used in our implementation, rather than a limitation to our framework itself. SecGrad-XAI is SNNI protocol-agnostic. For instance, substituting with Fully Homomorphic Encryption (FHE) would significantly reduce communication costs by eliminating the need for frequent interactive rounds, though it would likely result in a substantial increase in execution times. This trade-off is well-documented [9]. Our framework remains functionally identical regardless of the underlying protocol, allowing practitioners to choose the specific protocol that best aligns with their deployment constraints.

For all methods, the forward and backward passes had nearly identical execution durations, and together dominated the execution profile, representing $96 - 99\%$ of total runtime. The cost of secure Integrated Gradients is proportional to its number of steps; $M = 2$ roughly doubled secure execution time compared to Vanilla. For all experiments with Integrated Gradients, the time spent secret-sharing model parameters, input, and baseline (including plaintext creation of the baseline and its secure transmission), and reconstructing inference output and saliency map took $< 2\%$ of the time, except for SmallNet where aggregation made up a higher portion of the < 1 second total execution time.

7 Conclusion and Future Work

While our framework establishes a high-fidelity baseline for secure XAI, the technical constraints of current SMPC libraries suggest several research avenues.

The primary hurdle remains the lack of native `autograd` and common `torch` utilities (e.g., `max_unpool`, `interpolation`) in CrypTen. Manual backward pass implementation and architectural obfuscation during ONNX conversion currently increase complexity and risk numerical instability. Consequently, expanding open-source SMPC frameworks to support automated gradient calculation for non-linear architectures is essential.

To facilitate this, we provide our full repository—including plaintext and secure implementations for recursive models (ResNets, GoogLeNet) and preliminary modules for CAM and LRP. These resources open several research directions. First, analyzing the bandwidth trade-offs between element-wise multiplication protocols and high-throughput matrix-based operations for XAI. Second, investigating the fidelity drop observed for the CAM and LRP methods. Third, enhancing CrypTen to natively support the structural mapping required for secure model conversion and gradient-calculating functionalities.

References

1. Aysel, H.I., Cai, X., Prugel-Bennett, A.: Explainable artificial intelligence: advancements and limitations. Appl. Sci. **15**(13), 7261 (2025)
2. Deng, J., Dong, W., Socher, R., Li, L.J., Li, K., Fei-Fei, L.: ImageNet: a large-scale hierarchical image database. In: IEEE Conference on Computer Vision and Pattern Recognition, pp. 248–255 (2009)
3. Dwivedi, R., Dave, D., Naik, H., et al.: Explainable AI (XAI): core ideas, techniques, and solutions. ACM Comput. Surv. **55**(9) (2023)
4. Fiedler, V.M., Mann, Z.Á.: Secure and explainable inference with neural networks (2025, unpublished manuscript)
5. Hurley, N., Rickard, S.: Comparing measures of sparsity. IEEE Trans. Inf. Theory **55**(10), 4723–4741 (2009)
6. Knott, B., Venkataraman, S., Hannun, A., Sengupta, S., Ibrahim, M., van der Maaten, L.: CrypTen: secure multi-party computation meets machine learning. In: Advances in Neural Information Processing Systems (NeurIPS), vol. 34, pp. 4961–4973 (2021)

7. Le, Y., Yang, X.: Tiny ImageNet visual recognition challenge (2015)
8. LeCun, Y.: The MNIST database of handwritten digits (1998). http://yann.lecun.com/exdb/mnist/
9. Mann, Z.Á., Weinert, C., Chabal, D., Bos, J.W.: Towards practical secure neural network inference: the journey so far and the road ahead. ACM Comput. Surv. **56**(5), art. 117 (2023)
10. Minh, D., Wang, H.X., Li, Y.F., Nguyen, T.N.: Explainable artificial intelligence: a comprehensive review. Artif. Intell. Rev. **55**(5), 3503–3568 (2022)
11. Montavon, G., Binder, A., Lapuschkin, S., Samek, W., Müller, K.-R.: Layer-wise relevance propagation: an overview. In: Samek, W., Montavon, G., Vedaldi, A., Hansen, L.K., Müller, K.-R. (eds.) Explainable AI: Interpreting, Explaining and Visualizing Deep Learning. LNCS (LNAI), vol. 11700, pp. 193–209. Springer, Cham (2019). https://doi.org/10.1007/978-3-030-28954-6_10
12. Selvaraju, R.R., Cogswell, M., Das, A., Vedantam, R., Parikh, D., Batra, D.: Grad-CAM: visual explanations from deep networks via gradient-based localization. In: IEEE International Conference on Computer Vision, pp. 618–626 (2017)
13. Shrikumar, A., Greenside, P., Kundaje, A.: Learning important features through propagating activation differences. In: International Conference on Machine Learning, pp. 3145–3153. PMLR (2017)
14. Simonyan, K., Vedaldi, A., Zisserman, A.: Deep inside convolutional networks: visualising image classification models and saliency maps. arXiv preprint arXiv:1312.6034 (2013)
15. Sundararajan, M., Taly, A., Yan, Q.: Axiomatic attribution for deep networks. In: International Conference on Machine Learning, pp. 3319–3328. PMLR (2017)
16. Van Der Waa, J., Nieuwburg, E., Cremers, A., Neerincx, M.: Evaluating XAI: a comparison of rule-based and example-based explanations. Artif. Intell. **291**, 103404 (2021)
17. de Vries, R., Mann, Z.Á.: Secure neural network inference as a service with resource-constrained clients. In: Proceedings of the IEEE/ACM 16th International Conference on Utility and Cloud Computing (UCC) (2023)
18. Wang, Y., Zhang, T., Guo, X., Shen, Z.: Gradient based feature attribution in explainable AI: a technical review. arXiv preprint arXiv:2403.10415 (2024)
19. Wang, Z.: Image quality assessment: form error visibility to structural similarity. IEEE Trans. Image Process. **13**(4), 604–606 (2004)
20. Yeh, C.K., Hsieh, C.Y., Suggala, A., Inouye, D.I., Ravikumar, P.K.: On the (in)fidelity and sensitivity of explanations. NeurIPS **32** (2019)
21. Yu, W., Yang, K., Bai, Y., Xiao, T., Yao, H., Rui, Y.: Visualizing and comparing AlexNet and VGG using deconvolutional layers. In: 33rd International Conference on Machine Learning, vol. 2 (2016)
22. Zhao, C., et al.: Secure multi-party computation: theory, practice and applications. Inf. Sci. **476**, 357–372 (2019)
23. Zhou, I., Tofigh, F., Piccardi, M., Abolhasan, M., Franklin, D., Lipman, J.: Secure multi-party computation for machine learning: a survey. IEEE Access **12**, 53881–53899 (2024)

Cybersecurity Specialists' Perspectives on the Use of Generative AI in Professional Practice

Alexander Lindström[1(✉)], Leonardo A. Martucci[2], Simone Fischer-Hübner[2], and Matthias Beckerle[2]

[1] Dizparc Secured, Karlstad, Sweden
alexander.lindstrom@dizparc.se
[2] Karlstad University, Karlstad, Sweden
{leonardo.martucci,simone.fischer-huebner,matthias.beckerle}@kau.se

Abstract. The use of *Generative Artificial Intelligence* (Gen AI) in cybersecurity is receiving increasing attention, with existing research largely focusing on technical applications such as penetration testing, malware detection, and threat hunting. However, comparatively little is known about how cybersecurity specialists perceive Gen AI, how they currently use it in practice, and what challenges and opportunities they associate with its adoption. To address this gap, we conducted a qualitative study using semi-structured interviews with 11 cybersecurity specialists. The interviews explored participants' current work practices, their experiences and attitudes toward Gen AI, and their views on how Gen AI could support or improve their work. The interviews were transcribed and analysed using a thematic analysis approach. In addition, a user experiment was included to capture insights into commonly used cybersecurity tools and to inform discussions about desired features and requirements for future Gen AI-based solutions. Our findings provide insights into practitioners' perceived benefits and barriers of Gen AI, as well as key requirements they consider essential for the effective and safe integration of Gen AI into cybersecurity workflows.

Keywords: Cybersecurity · Gen AI · User study · Penetration testing

1 Introduction

Cybersecurity refers to the methods, techniques, and policies employed to mitigate the risk of cyberattacks and their potential impact. In recent years, threat actors have become increasingly efficient, often adopting more business-like and professionalized approaches to illicit activities. This has resulted in substantial financial losses and data breaches affecting organizations worldwide. At the same time, the growing availability of Gen AI has lowered the barrier to entry for conducting cyber attacks, enabling individuals with limited technical expertise to engage in increasingly sophisticated forms of malicious activity.

© The Author(s), under exclusive license to Springer Nature Switzerland AG 2027
K. Renaud et al. (Eds.): EICC 2026, CCIS 3012, pp. 147–166, 2027.
https://doi.org/10.1007/978-3-032-28957-5_9

Although extensive research has been conducted on preventing and mitigating malicious cyber activity, much of this work has focused on technical solutions rather than the needs and experiences of cybersecurity practitioners. Consequently, there is limited understanding of what cybersecurity specialists themselves expect from Gen AI, how they perceive its risks and benefits, and how such technologies could meaningfully support their daily work. Addressing this gap is essential for ensuring that emerging Gen AI-based solutions align with real-world security practices and constraints.

Research on the application of Gen AI in cybersecurity has expanded steadily over the past few years. For example, several studies have investigated the effectiveness of Gen AI in penetration testing. Shao et al. [12] evaluate widely used models by benchmarking their performance on *Capture The Flag* (CTF) challenges. Other work explores the use of local models in combination with techniques such *Retrieval-Augmented Generation* (RAG) and fine-tuning to improve performance in CTF challenges and penetration testing [5]. Additional areas of applied Gen AI research in cybersecurity include malware detection [1], vulnerability assessment, and threat hunting [8], among others.

However, the direct use of these models, either locally or through *Application Programming Interface* (API)s and web-based applications, may introduce new security risks. Opinions regarding the capabilities and trustworthiness of Gen AI systems remain divided. Many widely used models are developed by large technology companies, often with limited transparency regarding how training data are collected, curated, and cleaned. Furthermore, a persistent trust challenge arises from the phenomenon of hallucinations in generated content [15]. When applied to code generation, hallucinations may, at best, result in code that fails to compile, and at worst, produce seemingly functional programs that contain subtle logical errors or security vulnerabilities [9].

Despite the growing body of research on novel techniques and applications of Gen AI in cybersecurity, the perspectives of cybersecurity specialists remain underexplored. In particular, it is not well understood how practitioners currently integrate Gen AI into their workflows, what limitations they encounter, and how they perceive the associated risks and opportunities. This paper aims to address this gap by examining the work practices and perspectives of cybersecurity specialists on the use of Gen AI. Through an interview-based study, we explore perceived barriers and limitations, identify desired features for future Gen AI-based tools, and elicit requirements for the effective and safe adoption of Gen AI in professional cybersecurity workflows. In the longer term, the insights gained from this study are intended to inform the design of Gen AI-based tools that better align with practitioners' needs, thereby increasing the likelihood of successful adoption and practical usefulness.

The research questions we aim to answer in this study are as follows:

- RQ1. What characterizes the current work practices of cybersecurity specialists, including workflows and tools of trade?
- RQ2. How do cybersecurity specialists perceive the use of generative AI in cybersecurity contexts?

RQ3. What requirements and desired features do cybersecurity specialists identify for future Gen AI cybersecurity tools?

To answer these questions, we conducted semi-structured interviews with cybersecurity specialists to understand how they perform their work, focusing on their day-to-day practices and the tools they use. We asked about their experience with Gen AI, including how they use it, and whether they see any challenges or drawbacks with using it within their workflow. Finally, we explore their views on the potential opportunities and benefits that Gen AI could bring to their work. Our results show The most relevant aspect for cybersecurity specialists to use a cybersecurity tool built upon Gen AI is that no data is sent to third parties due to security risks. Other requirements included mitigating hallucinations and ensuring good explainability. Some desired features included automated reporting, automatic generation of *Proof of Concept* (POC), and analysis and guidance on paths during penetration testing. The paper continues as follows: Sect. 2 reviews related work. Section 3 describes the research methods and ethical considerations. Section 4 presents the pilot study and its results, which were later used to refine our user study. Section 5 describes the recruitment process, data collection procedures, and interview analysis. Section 6 summarizes the results from our user experiment. Section 7 presents the results of the thematic analysis. Section 8 presents a discussion on the results and limitations of the study. In Sect. 9 we present the conclusions and future work.

2 Related Work

The application of *Artificial Intelligence* (AI) in cybersecurity has received increasing research attention. Existing studies predominantly emphasize defensive security practices, leaving offensive cybersecurity contexts comparatively underexplored. Earlier work includes an explorative study interviewing participants in managerial positions about *Machine Learning* (ML)-based solutions in defensive cybersecurity, another survey study examining issues of trustworthiness in defensive cybersecurity solutions due to the lack of explainability, and a literature review assessing the future potential of AI in cybersecurity, primarily focused on defensive security [2–4].

Alshaikh et al. [3] conducted an exploratory study examining the perceptions of specialists and decision-makers regarding ML-based solutions in defensive cybersecurity. Their study involved 25 participants in managerial positions and revealed a mixed understanding of ML capabilities and limitations. They concluded that a more standardized approach is needed to bridge these perception gaps. While their findings offer valuable insights into managerial views on defensive applications, they excluded offensive cybersecurity contexts. Moreover, their participant group and study objective differ from ours, as we target hands-on cybersecurity specialists and their experiences and opinions on Gen AI [3].

Al Siam et al. [2] presented a comprehensive literature review on the future potential of AI in cybersecurity, focusing mainly on defensive areas such as threat detection, endpoint protection, phishing and fraud detection, network security, and adaptive authentication. Their work aimed to compare various AI models, highlighting their use cases, strengths, and limitations to identify future research directions. While their study shares our interest in the potential of AI for cybersecurity, it takes a technical rather than human-centered approach, concentrating on model performance and taxonomy rather than practitioner perspectives [2].

Capuano et al. [4] highlight that Explainable AI remains insufficiently addressed by design in most cybersecurity solutions. This challenge was also echoed by participants in our study, who emphasized the lack of transparency as a key barrier to adopting Gen AI in their work. Capuano et al. conducted a large-scale survey and reviewed over 300 papers in the cybersecurity domain, focusing primarily on defensive applications, including intrusion detection systems, malware detection, phishing and spam detection, and botnet identification. Their motivation stemmed from the limited interpretability of existing ML models and the opacity of classification decisions. Although their review primarily focused on traditional ML approaches, the underlying issue of explainability remains equally relevant for contemporary Gen AI systems [4].

Collectively, these studies demonstrate the growing academic interest in AI applications for cybersecurity. However, existing research primarily emphasizes defensive use cases, technical evaluations, and managerial perceptions, leaving a notable gap in understanding how cybersecurity specialists themselves perceive, evaluate, and envision the use of Gen AI in their day-to-day work.

3 Methodology

Due to the limited existing research capturing cybersecurity specialists' own perspectives on how Gen AI can be applied effectively and safely within their work environments, this study was designed to address this gap. The aim of the study was to develop a foundational understanding of (1) how cybersecurity specialists currently conduct their work, (2) how they perceive and make use of Gen AI in their professional practice, and (3) how they envision Gen AI being applied to support their work. The latter includes desired functionalities, requirements, and the conditions necessary for such applications to be feasible in practice.

To address these questions, the study employed *semi-structured interviews.* This method was selected to enable an open-ended exploration of participants' experiences and perspectives, while still ensuring a degree of consistency across interviews. Semi-structured interviews are particularly well-suited for exploratory research, as they strike a balance between flexibility and structure, allowing for in-depth probing of topics that may not be fully anticipated in advance. Through this approach, we sought to gain an understanding of cybersecurity specialists' work practices, as well as their experiences and envisions towards Gen AI.

All interviews were audio-recorded with participants' consent and later transcribed. The transcripts were analyzed using a two-phased *qualitative coding.* In the first phase, descriptive coding and sub-coding were applied to capture key concepts and themes in the data. In the second and final phase, pattern coding was used to identify broader themes and relationships across participants.

In addition, our user study includes a *user experiment*: a ranking exercise designed to provide insights into commonly used cybersecurity tools. This exercise served two purposes: (1) to identify which tools are most widely used and how participants evaluate their strengths and weaknesses; (2) to provide a foundation for later parts of the interview, where participants reflected on desired features and requirements for potential AI-based solutions in cybersecurity.

Ethical Considerations. The research has been conducted in accordance with the rules of the Swedish Ethical Research Review Act involving humans [14] and the EU *General Data Protection Regulation* (GDPR).

An ethical review was submitted to our *Institutional Review Board* (IRB), which decided that we could proceed with our research.

Before each interview, participants were informed about the project's purpose, and all were asked to provide consent before the interviews began. During the interviews, voice recordings were used. Recordings were transcribed using a GDPR-compliant transcription service. All data collected is safely archived locally, in accordance with agreements on archiving research to prevent fraud.

4 Pilot Study

We adjusted and refined our user studies from the results of a pilot study. The goals of the pilot study are: to identify weaknesses and logical flaws in the semi-structured interview guide, to assess the clarity of the questions, to estimate an average time, and to evaluate the user experiment. The pilot study was conducted from late December 2024 to early February 2025.

The participants were three university employees recruited on-site using our personal contacts, who had no prior involvement or knowledge about our study. One has extensive cybersecurity experience, while the others have some cybersecurity experience but considerable expertise in computer networks and software development.

The pilot study employed an iterative process. After the first participant concluded their participation, we reflected on the results, identified shortcomings, and refined the interview guide and the user experiment. We repeated the process for the second and third participants. The outcomes of the pilot study are: (1) a revised interview guide and a modified procedure (we decided not to give the list of questions to the participants ahead of the interview to allow for spontaneity in their answers and to avoid participants from tailoring their responses in expectation of upcoming questions), and (2) a revised ranking exercise (at first we used sticky notes on a two-dimensional matrix but that proved cumbersome, so we decided to use colour-coded sticky notes and written down marks on a 1–10 scale, which simplified scoring and reduced confusion).

The pilot study improved clarity and reduced bias of our user studies, and ensured that each interview could be completed in 30 to 60 min.

Table 1. Participants' demographics ($N = 11$).

Gender	Man	11 (100%)
Age (years)	18–30	1 (9.0%)
	31–40	5 (45.5%)
	41–50	5 (45.5%)
Educational background	High school	4 (36.4%)
	Bachelor	3 (27.2%)
	Master or higher	4 (36.4%)
Experience in cybersecurity (years)	2–5	2 (18.2%)
	5–10	2 (18.2%)
	10+	7 (63.6%)
Sector	Industry	10 (90%)
	Academia	1 (10%)

5 Semi-structured Interviews

The study was conducted in Sweden with eleven cybersecurity specialists over a two-month period, from late February 2025 to late April 2025. The remainder of this section describes each step of the interviews in detail.

5.1 Participants and Recruitment

We recruited through the authors' professional networks and peer-to-peer, rather than contacting companies directly, which we expected would yield a higher acceptance rate. Five were recruited directly through the author's professional network, and six were recruited through peer-to-peer networks. Most were contacted directly by email, while a few were first contacted through LinkedIn and then by email. All but one were working in the private sector at the time of their interview. The only participant from the public sector works for a cybersecurity institute. Three participants are from a cybersecurity *Small and Medium-sized Enterprises* (SME), while the others are from different companies and organizations. Participation was voluntary, and no compensation was offered. Tables 1 and 2 summarize the participants' demographics. To distinguish participants, we use pseudonyms P1 to P11 throughout the study.

5.2 Interviews and Data Collection

Eleven interviews were conducted, four in person and seven via Zoom. Eight interviews lasted 40–60 min, and the three others lasted 30–40 min, 60–70 min, and 80–90 min. All interviews were conducted in Swedish, except for P2 and P8, which were in English. All quotes were translated into English.

Table 2. Participants' demographics (P1–11) breakdown.

Part.	P1	P2	P3	P4	P5	P6	P7	P8	P9	P10	P11
Age	41–50	31–40	41–50	41–50	41–50	31–40	31–40	31–40	31–40	18–30	41–50
Edu.	Bach.	MSc+	H.S.	MSc+	Bach.	H.S.	H.S.	MSc+	MSc+	Bach.	H.S.
Exp.	2–5	10+	10+	10+	10+	10+	5–10	10+	2–5	5–10	10+
Sec.	Ind.	Acd.	Ind.	Ind.	Ind.	Ind.	Ind.	Ind.	Ind.	Ind.	Ind.

The interview guide contains 18 questions, divided into five parts, and was carried out in the following order: (1) general information, (2) the participant's background, (3) cybersecurity tools and their limitations, (4) their current workflow and potential improvements, and (5) AI integration in cybersecurity.

The interview guide, its questions, the consent form, and the results from the user experiment are available in our `git` repository.[1]

5.3 Thematic Data Analysis

The thematic analysis was conducted using NVivo 14. We followed the coding procedure described in [11]. In the first coding cycle, *descriptive* and *sub-coding* were used. Sub-coding was applied to pre-existing categories and questionnaire questions. For the descriptive part, we used nouns or short phrases to capture the topics from participants' answers. The codes were revised and relabelled with more suitable description phrases to gain a more comprehensive understanding of the codes before starting the second cycle.

In the second coding cycle, *pattern coding* was used to find themes related to the codes. Three main themes, related to the three interview guide topics, and 18 subthemes were identified in this stage.

6 User Experiment: A Ranking Exercise

Around the start of the second quarter of the interview, we introduced a user experiment. We asked the participant to write down the cybersecurity tools they use on sticky notes.[2] They were then asked to score their selected tools according to their configurability, usability, and versatility on a [1–10] scale. The ISO 25010 definitions were used to explain and clarify those terms to participants [7].

Configurability. How much configuration is needed to use a tool effectively. It is derived from ISO 25010's "Modifiability" and "Installability." Together, they describe the extent to which a system or product can be effectively and efficiently modified without compromising its quality, introducing defects, or being installed and/or uninstalled.

[1] https://github.com/EEIC-2026/Appendix.

[2] We used a web-based application with virtual sticky notes in the online interviews.

Usability. The perception of how easy or difficult it is to use a tool. It is derived from ISO 25010's "Operability", "Self-descriptiveness", and "Learnability." These aspects assess how intuitive a system is to operate and control, and whether adequate documentation and guidance are available.

Versatility. The usefulness of a tool across multiple contexts or purposes. It is derived from ISO 25010's "Reusability" and "Functional completeness." The goal is to measure the extent to which a tool can be reused across different systems and how comprehensively it supports users' tasks.

We set an undisclosed soft limit of 10 tools per participant, i.e., the participants were not informed of it. The reasons were: (1) we didn't want them to set a target number of tools for them to list, and (2) we wanted them to concentrate on the most important cybersecurity tools they use in their work. To keep the upper threshold, when a participant wrote down his eighth tool, we would intervene with a question: "*If you could only write two more tools, which would those be?*" Nevertheless, we included all tools listed by the participants, and none were discarded.

A limitation of the experiment was pointed out by one of our participants. He mentioned that the use of certain tools is a company secret, meaning they could not be disclosed publicly. Therefore, it is reasonable to assume that some participants refrained from listing all of the tools they regularly use for similar confidentiality reasons, without necessarily mentioning it.

The key results from the experiment are: (1) there is a high divergence for configurability, usability, and versatility scores of most of the listed tools, and (2) web browsers and chatbots got favourable ratings but were mentioned by less than half of the participants. The results are summarized in Fig. 1, which lists the most mentioned tools (all experiment results are available on our `git`) (see footnote 1).

7 Thematic Analysis

In the thematic analysis, we identified three themes: 'source of information', 'workflow', and 'challenges and possibilities'.

Source of Information. We grouped topics in which participants discussed how they search for and find information relevant to their work, along with ideas for improving these information-gathering tasks. This includes ideas for enhancing efficiency and streamlining workflows, e.g., using local Gen AI setups and RAG solutions integrated with the OWASP Guide to better search for relevant information. This theme is visualised in Fig. 2.

Worflow. We identified topics related to their daily tasks and duties. This includes their point of view on what parts of their work they enjoy, dislike, or find very time-consuming. This theme is visualised in Fig. 3.

Challenges and Possibilities. This theme primarily covers content related to Gen AI. We identified topics around the participants' experiences with Gen AI and

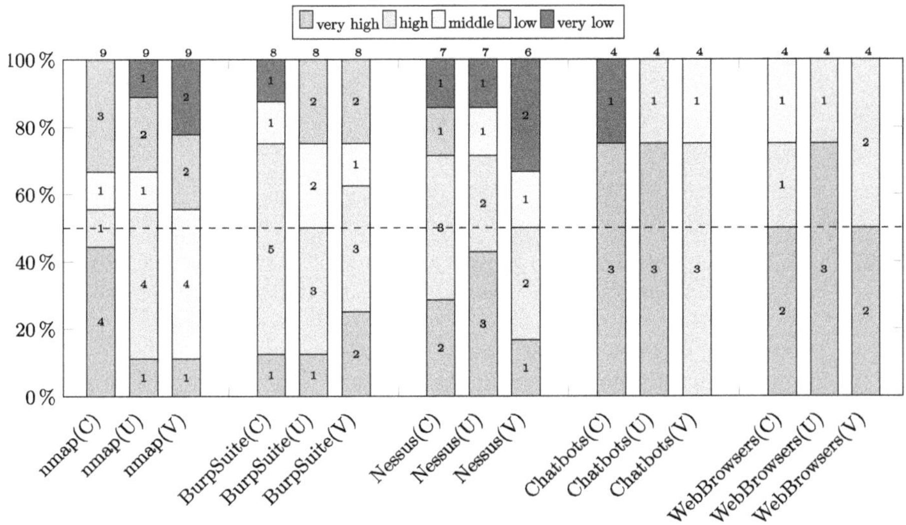

Fig. 1. Perceived ease of configurability (C), usability (U), and versatility (V) for tools mentioned by participants ($n \geq 4$). Bar segments show the number of participants who rated each tool in a given category. Numbers above bars indicate the total number of distinct participants who rated the tool.

their ideas on Gen AI drawbacks, challenges, and possibilities of further integrating it into their workflow. The sub-themes and their corresponding codes are visualized in Fig. 4.

In the remainder of this section, we look into each of these themes in detail.

7.1 Theme: Source of Information

This theme contains topics related to finding and searching for information, both AI and non-AI related. The theme and its sub-themes are illustrated in Fig. 2.

When asked about what resources or databases they usually use to obtain information, we got a variety of responses. Many mentioned relying on internal or contextual information sources: for example, P3 stated that he primarily uses internal documentation, P7 prefers to consult colleagues or review customer-provided documents, and P6 emphasized that the documentation for the target application is often the most valuable source during a penetration test.

A majority reported using Gen AI tools to some extent for information gathering. Some described how they have increasingly incorporated chatbots into their workflows for searching and summarizing information. P3 and P4 noted that they increasingly use ChatGPT in the same way they once relied on traditional web searches or printed references, indicating an ongoing shift in how cybersecurity specialists access and interact with technical information. P2 describes how he uses chatbots to gather information through hacking challenges, to learn how to proceed on a challenge, or to generate commands, e.g.,

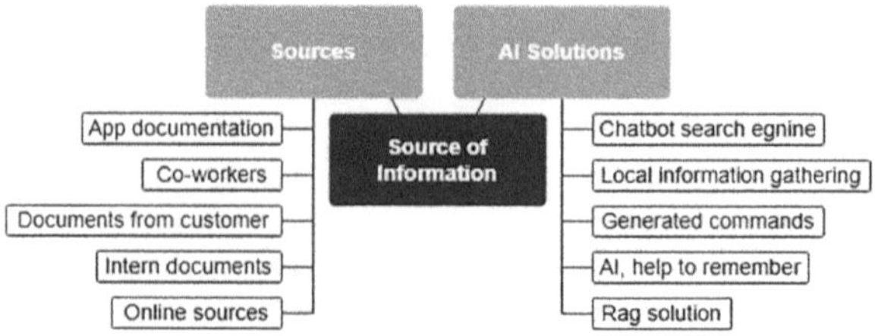

Fig. 2. Thematic map of the "Sources of Information" theme and its sub-themes, "Sources" and "AI Solutions," and their associated codes derived from the interviews.

SQL injection commands. P5 describes the problem of keeping track of all commands and configurations across different tools and environments, such as `bash`, `curl`, and `nmap`. P5 says that chatbots are very useful for providing answers about specific setups and their commands.

P7 says he does not use Gen AI tools, but they could be beneficial in conjunction with resources such as the OWASP Guide, provided they retrieve factual information without altering it. This corresponds to what is commonly known as RAG solution. P7 noted that such a tool could reduce the time spent manually scanning lengthy documents, allowing for a greater focus on the penetration test. P4 describes a similar solution where it would be beneficial to integrate an AI solution with local tools for gathering information. However, both P1 and P4 have mentioned the problems and risks associated with data leakage. For these solutions to be attractive to use, users need to be completely reassured that no data is being transmitted from the local device or local network.

As an example of valuable non-AI tools, P1 highlighted that RSS readers are a good solution for gathering information (an RSS reader is a software application that allows users to view updates from multiple websites in one place). However, he prefers in-house solutions, citing the risks of tags stored online. The AI part of the solution would be to learn how we work and highlight the information we commonly find interesting. He highlights that a good feature would be for it to keep track over a long time span and check for information that was deemed irrelevant earlier but could become of interest given the news feed.

7.2 Theme: Workflow

During the study, we asked the participants about their workflow. We wanted to understand which parts of their work they find time-consuming and tedious, and which they enjoy. By asking these questions, we expected participants to later reflect on how AI could support and improve their workflow. For this theme, we identified 7 sub-themes, depicted in Fig. 3, each with its respective code.

Threat Hunting. P1, a blue team member who works partly on threat hunting, describes the determination and construction of effective queries, as well as reviewing all hits, as very time-consuming and less enjoyable when the query results in generating a lot of noise to go through. He describes the trade-off:

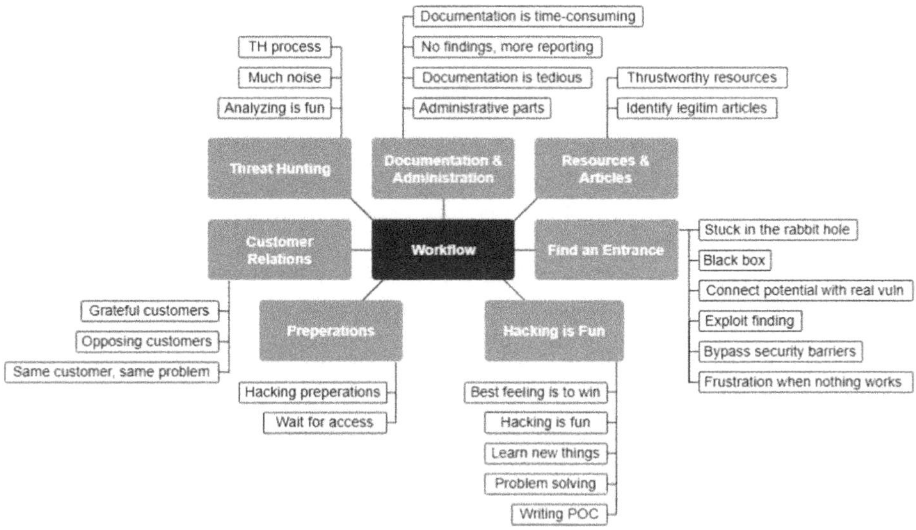

Fig. 3. Thematic map of the "Workflow" theme and its sub-themes, shown with their corresponding codes identified during the qualitative analysis of the interviews.

"*spending time formulating good queries may reduce the amount of data to review, but carries the risk of filtering out too much and potentially missing threats.*" P1 also described the process of determining whether a finding represents a real threat or a false positive as time-consuming, especially when something looks suspicious. P4, who also primarily works on blue-team assignments, describes the threat-hunting process as very time-consuming overall, largely due to the broad scope that must be covered. For the more enjoyable parts. Both P1 and P4 explained that the most enjoyable aspect is analysing and piecing together the puzzle to identify signs of suspicious activity. P1 also mentioned that blue-team versus red-team competitions are very enjoyable. In these exercises, the red team attempts to hack and escalate their activities as far as possible, while the blue team tries to detect their actions and prevent them.

Resources and Articles. P5, who generally works on red teaming assignments, mentioned that searching for reliable, reasonably priced services can be a very time-consuming and not-so-enjoyable part of his work. Here, he provided an example of searching for a supplier of useful password lists. He has to contact providers who turn out to be criminals or enterprises that do not advertise prices and end up being too expensive. Finding a good supplier for these purposes is, therefore, often a very time-consuming process. A similar issue was described by P1, who also works with threat intelligence. Identifying trustworthy sources of information is a time-consuming process, as it requires sorting through numerous articles. P1 also mentioned this as one of the less enjoyable tasks, especially when articles turn out to be advertisements or contain little of real value. The skimming process itself was described as tedious.

Hacking is Fun! Most participants work with penetration testing to some degree. When it comes to pentesting, P9–11 emphasized that discovering a vulnerability and exploiting it is a very enjoyable part of their work. This gives them a sense of victory. P9 and P10 added that presenting such findings to customers is also a rewarding experience. Being able to demonstrate both the vulnerability and its potential exploitation. P6 pointed out that he generally enjoys hacking in all its aspects, especially reviewing source code, understanding the domain, and building a mental model of how everything connects. P7 expressed similar views and described a pentest having three phases: (1) a free hacking/exploration phase to understand the domain, and experiment with different inputs; (2) a more structured phase with systematic testing; and (3) report writing. He describes the first two phases as more fun, especially the first, the hacking/exploration phase, which he finds most enjoyable.

P8 explained that learning something new, working with unfamiliar applications or domains, and piecing together breadcrumbs to identify possible exploits is always fun. P11 describes the problem-solving aspect as enjoyable, since solving these problems often means discovering ways to do things that should not be possible. P2, P5–6 mentioned that writing an POC is fun, as it demonstrates the existence of the vulnerability and shows how it can be exploited.

Finding Your Way In. According to P2, the initial stages of a pentest, i.e., information gathering, reconnaissance, and enumeration, are fairly straightforward. The most time-consuming part is constructing an exploit for a suspected vulnerability. This can, in many cases, lead you down the wrong path and cause you to become stuck in a rabbit hole, as you spend time constructing an exploit for something that is not exploitable, which can be frustrating. P11 expressed a similar experience, describing the effort of connecting potential vulnerabilities to real ones as one of the most time-consuming aspects of the process. He explained that during a pentest, one may spend hours trying different attacks with no success, which can lead to frustration and a loss of creativity. At this early stage, when nothing seems to work, it is not an enjoyable process, but the moment the exploit finally succeeds is highly rewarding. P10 notes that black-box pentesting is very time-consuming and tedious, especially when the domain is large and requires extensive enumeration to identify entry points. P7 described bypassing security barriers as one of the most time-consuming tasks. He also noted that when starting a web application pentest, he must map every page, button, and input field using tools such as BurpSuite. After mapping, all these parts must then be tested for unexpected behaviours, which is time-consuming, but he thinks it is enjoyable work.

Preparations. P3 states that one of the most frustrating aspects of his work is waiting for access before a pentest due to unfinished preparations, which he finds to be a waste of time. P2 instead highlighted that the preparations themselves can be less enjoyable. For example, setting up a device using an open-source project or a private testing platform can be painful and time-consuming.

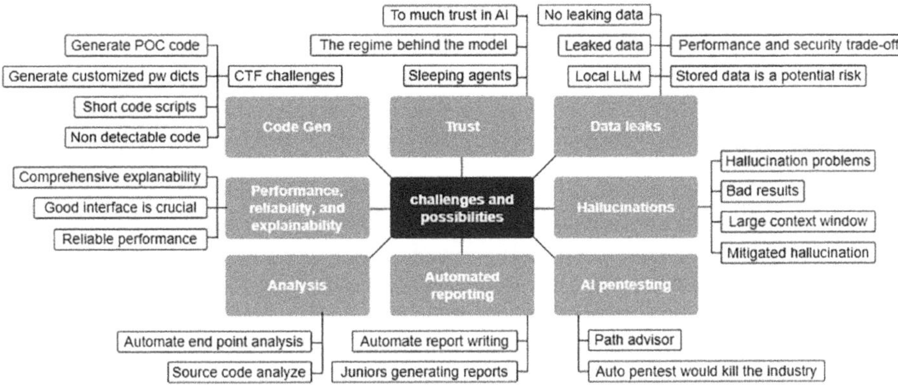

Fig. 4. Thematic map of the "Challenges and possibilities" theme and its nine sub-themes, shown with their corresponding codes derived from the interviews.

Customer Relations. P9 described customers' gratitude when reporting vulnerabilities as one of the most enjoyable parts of his work, especially when customers are friendly and genuinely interested in the findings. However, he also mentioned that dealing with customers who are dismissive or question his competence is one of the least enjoyable aspects of his job. He suspects that it happens because either the person he reports to is not the same one who ordered the pentest, or when the customer does not have a full understanding or awareness of the contract's 'rule of agreement'. P8 mentions another frustration: customers requesting new pentests without first addressing previously identified vulnerabilities from the previous tests. He considered this a waste of time and explained that they usually try to prevent this by asking the customers beforehand what they plan to do with the findings once the report is delivered.

Documentation and Administrations. The most frequently mentioned and agreed-upon time-consuming aspect of their work is documentation and reporting. Six participants (P3–4, P6–9) said that documentation and reporting are very time-consuming. P8 noted that while it might not be the most time-consuming task, it feels that way because nobody really enjoys doing it. P3, P6–10 also specifically mentioned reporting as one of the most boring tasks. P4, on the other hand, did not find the documentation too bad, but considered administrative tasks, such as creating new cases when identifying security issues, to be more boring. P10 emphasized that when there are fewer findings to report, documentation becomes even more important. More detailed reporting is then required to show customers how time was spent and what tests were performed, making this task even more time-consuming.

7.3 Theme: Challenges and Possibilities

In this theme, we identified subjects related to risks, benefits, possibilities, and requirements of Gen AI from the participants' perspective. The theme comprises nine sub-themes, which are visualized in Fig. 4.

Code Gen. Several participants reported experience using Gen AI to generate code and scripts. P10 noted that while he finds Gen AI impressive, he approaches it with care and critical thinking. He mentioned a few use cases, such as script generation for pentests. P11 described a similar use case. He finds Gen AI effective for creating small, highly specific, and undetectable code snippets to circumvent the *Intrusion Detection System* (IDS), which can usually detect well-known tools with recognizable patterns. P5 mention his main reason using Gen AIs is for code correction. P6 also describes that Gen AI can be useful for building skeleton code for CTF challenges. P6 and P8 highlighted the potential of using Gen AI to generate POC code that can be shared with their customers, including scripts to automate pentest demonstrations. P7 mentioned using chatbots occasionally to generate code, without providing any specific details about the type of code. P9 points out that he does not use Gen AI frequently, but finds it useful for producing reverse shell scripts.

Trust. Trust in Gen AI was generally low among participants. P3 emphasized the importance of being cautious about which models are used, even when working with local ones. He pointed out the potential risk of backdoors being implemented, sometimes referred to as 'sleeping agents' [6]. He also emphasized the importance of considering which regime or organization is behind the model, as some may be more trustworthy than others. Another recurring concern was the risk of placing excessive trust in Gen AI models' response. This issue was raised by P3–8. P8 provided the example of someone trusting a chatbot's claim that everything is fine, even if problems exist. P8 compared this to running a Nessus scan: "*just because no vulnerabilities are found, that does not mean they do not exist, ... and manual testing is still necessary.*" P9 expressed doubts about trusting Gen AI tools. He pointed out the risk that it might perform unintended actions, such as locking accounts by launching a password-spraying attack on a domain. Nonetheless, he acknowledged potential use cases, such as collecting company information (e.g., email addresses) and using that to generate possible passwords. P7 warned that relying on outputs without thoroughly verifying their accuracy could constitute professional misconduct. Similarly, P4–6 highlighted the broader risk of people assuming the tool can perform entire tasks without errors. For instance, P5 described a case in which a chatbot might correctly follow parts of a guide but complete only half the required steps. Even if the test runs without issues, "*it remains the pentester's responsibility to ensure the task is fully and correctly carried out.*" P3 added that over-reliance on Gen AI may widen the gap between developers' understanding of code and the code they adopt from chatbot outputs. He cautioned that this lack of understanding could introduce significant security risks.

Data Leaks. Four participants (P5–7, P10) highlighted data privacy as a key requirement for adopting Gen AI tools in their work. They all stressed that preventing data leakage was essential. They emphasized that, as a first step to use Gen AI in their work, no information should leave their local machine. P8 expressed a similar requirement, stating that any pentesting tool using Gen AI would need to run locally or on a server under his full control.

The most frequently mentioned risk of using Gen AI solutions was data leakage. Eight participants (P2–8, P10) identified it as the greatest challenge in adopting Gen AI for their work. P10 explained that not knowing where information ends up is a major concern. P2 concern is that web-based Gen AI tools may use all their data input for training future models. P4 emphasized the importance of understanding how Gen AI providers use the uploaded data before relying on such web applications. P7 and P8 mentioned similar concerns about using web-based Gen AI tools. P6 stated that he would never upload customer code to a model without having full control over the data.

P2 noted that setting up a local model could mitigate such risks, but acknowledged that doing so is not simple. P3 noted that one of the biggest challenges is currently providing employees with an internal solution to prevent the leakage of company secrets. P7 further added that even with a local model, there should be some assurance that no data is stored on hard drives, so that sensitive information from completed assignments remains securely managed. P5 raised similar concerns about data exposure to third parties, but added that existing local models generally do not match the quality of web-based ones. This trade-off between performance and security was also discussed by P3, who nevertheless believes that smaller local models will soon reach a performance level comparable to that of web-based models.

Hallucinations. Concerns about hallucinations were raised by several participants as a major drawback of using Gen AI tools, built for pentesting. P5 and P11 specifically stated that mitigating hallucinations would be necessary in order to adopt such tools. P1, P4–5, P7 identified hallucinations as a key obstacle to trusting Gen AI tools. Although this relates to the broader issue of trust, it differs from the limitation of *Large Language Models* (LLMs) being restricted to their training data and training time, which can lead to outdated or unavailable information about a topic, often resulting in hallucinations.

P1 explained that he often uses chatbots to summarize multiple articles, but always double-checks the output to ensure it has not hallucinated. P4 reported a recurring problem of receiving false information. P5 mentioned that during programming, hallucinations typically occur once the context window becomes too large, causing the model to hallucinate. P5 described how he has tried experimenting with dividing a problem into smaller sub-problems to mitigate hallucinations. P7 mentioned that this kind of problem makes it risky to trust responses, if using it in his work, as "*hallucinations may result in missing steps or introducing steps that do not exist.*" P7 and P11 also described scenarios where Gen AI failed to generate usable code. P7 described a scenario where he specifically asked for examples using the Linux `pandas` module, but none of the generated code was useful. P11 reported that the model often hallucinated by inventing libraries or suggesting solutions that later failed to compile.

AI Pentesting. P2, P6, P8, and P10 mentioned that Gen AI could be beneficial if used as a path advisor. P2 suggested that Gen AI could help map large web applications containing many different pages, providing guidance on which pages to examine first and identifying where vulnerabilities or sensitive data might be found. P6 described a similar use case, giving the example of working with many different endpoints. He noted that it would be useful if Gen AI could suggest which endpoints to prioritize. However, P6 emphasized the difficulty of providing such sensitive data to an Gen AI system, which currently prevents him from using it in this way. P8 expanded on this idea, explaining how Gen AI could support students learning how to perform penetration testing. Since "*beginners often struggle to know where to start, and Gen AI integrated into the exploit chain could guide them on how to proceed and what to focus on.*" P10 mentioned the benefit of using Gen AI as a potential path advisor, suggesting that it could analyse different parts of a system and provide guidance on next steps, as well as its possible usefulness for brainstorming ideas. P5 cautioned that while automated penetration testing might sound appealing, he fears that fully achieving it could risk undermining the industry and the jobs it provides.

Automated Reporting. As discussed in Sect. 7.2, P3–4, P6–10 agreed that documentation and reporting are tedious and time-consuming tasks. As a result, P8–10 view report writing and documentation as areas where Gen AI could be particularly helpful. They suggested that feeding all relevant information into an Gen AI system and having it generate a summary or report would save considerable time. However, P5 disagrees with that. He expressed concern that if junior colleagues relied on Gen AI to produce reports, customers might reasonably expect to pay significantly less, since the report would not have been written by the consultant themselves.

Analysis. P6 described source code analysis as a highly promising use case for Gen AI. He sees potential in using it to identify patterns, particularly in large codebases developed by multiple teams. In such contexts, where time is limited and the amount of material to review is extensive, Gen AI could serve as a valuable starting point to indicate where to focus attention first. A similar perspective was shared by P7, who notes that projects often involve large amounts of code to review. He suggested that in these situations, Gen AI could provide useful guidance on which parts of the code deserve closer inspection. P6 further added that the same approach would be very beneficial for endpoint analysis. When faced with numerous endpoints to examine, it would be valuable if Gen AI could recommend which endpoints to prioritize.

Performance, Reliability, and Explainability. P3 emphasized that a good interface is always crucial. While the model itself is the most important component, it is ultimately the interface that determines whether users will want to work with it. P9 raised another requirement: comprehensive explainability. He stressed the importance of an Gen AI tool being able to clearly describe what it has tested, with the primary purpose to inform customers which tests have been conducted. According to P9, one of the most difficult challenges with automated Gen AI-based penetration testing is ensuring that the tool behaves exactly as expected. Not only must it remain within the agreed scope of a penetration test, but it must also provide sufficient explainability. P9 also raised the requirement: "*there has to be some reassurance it will behave as expected and won't break anything.*" To illustrate this, P9 compared the situation to running a script: you know precisely what it does and what it does not. He emphasized that the same level of certainty should apply to Gen AI tools. He stressed that reliable performance is critical and added that for him to use such a tool, there would need to be guarantees that nothing unexpected would happen or assurances that the company behind the tool would take responsibility for any malfunctions.

8 Discussion

Throughout the study, we obtained insights from the participants on their current workflow, addressing the first part of RQ1 (Sect. 1), which aimed to understand their current work practices. This was presented in Sect. 7.2, where we identified different aspects of their tasks and gained insights into which parts they enjoy, dislike, and find time-consuming.

The user experiment in (Sect. 6) provided a qualitative answer to the second part of RQ1, and yielded interesting results. While unsurprising that `nmap`, a widely used tool for network scanning and reconnaissance in pentesting [10], is the most mentioned tool, it is highly interesting that the

`nmap` scores show significant divergence across configurability, usability, and versatility. A possible explanation is that `nmap` is flexible and complex. For some participants, it serves as a straightforward utility for mapping ports, while for others it is a powerful, highly extensible framework that supports automation and advanced scanning with the Nmap Scripting Engine. This range of use cases, from basic scans to customized script-based probing, may explain the variability in perceived configurability and usability among participants [10]. It indicates a high potential for AI-based tools to help pentesters extract the most from `nmap` and similar tools.

From the interviews, we also gained interesting insights into how participants perceive Gen AI, the barriers they face to higher adoption, and their requirements and desired features for the future evolution of Gen AI in the cybersecurity domain. From these insights, we were able to answer RQ2 and RQ3. In how cybersecurity specialists perceive the use of Gen AI, the feedback we got included both positive and negative perspectives. They were mostly positive about including Gen AI more in their work going forward, though some expressed skepticism about security and privacy concerns. The most mentioned barriers to higher adoption of Gen AI are data leaks to third parties and the risk of hallucination, including fabricated data and misleading responses. Some requirements pointed out included comprehensive explainability, a good interface, full control over the data, and assurances that nothing unexpected would happen. The most mentioned desired feature is automated reporting.

Limitations. This study has a few limitations that should be considered when interpreting its findings. First, all participants are based in Sweden. While this enabled a focused exploration of local practices, it limits the transferability of the findings to cybersecurity specialists working in other geographical locations. Second, the study used a mix of purposeful and snowball sampling, recruiting participants through professional networks and peer-to-peer recruitment. Although this may introduce motivation and selection bias, this approach is commonly used in qualitative studies; the high level of expertise and trust it offers provides rich, nuanced data, even if certain perspectives may not be fully captured [13]. Third, the study sample did not include any female participants. This partly reflects the male-dominated nature of the cybersecurity pentester profession. However, it also means that women's perspectives on the use of Gen AI in cybersecurity are underrepresented in this study. Future research should include a more diverse range of participants to capture a broader set of perspectives on Gen AI in cybersecurity. Fourth, the number of participants was relatively small. While the sample size is appropriate for an exploratory qualitative interview study aimed at gaining in-depth insights, the findings should not be interpreted as statistically generalizable. The study does not aim to provide quantitative evidence or prevalence estimates, but rather to identify themes, perspectives, and requirements that can inform future research and design efforts.

Finally, the study relies on self-reported data from interviews, reflecting participants' individual experiences and subjective perceptions. Given that Gen AI is a highly topical and sometimes controversial subject, participants may hold strong opinions that influence their responses. Despite these limitations, the study provides valuable exploratory insights into cybersecurity specialists' perspectives, particularly regarding offensive security practices. To the best of our knowledge, this study is among the first to explicitly examine practitioners' perspectives on the use of Gen AI in offensive cybersecurity contexts.

9 Conclusion

Our study found that all 11 participants use Gen AI to some extent, but not necessarily in their work. From cybersecurity specialists, this limited adoption was somewhat expected due to many security and privacy concerns. The most frequently mentioned concern is leaking company secrets to third parties, which could pose significant security risks. As some participants mentioned, to fully leverage Gen AI in their work, they need a local solution that gives them control over their data. However, some participants acknowledged both the challenges of setting up a local solution and the drawbacks of many open-source models, which often perform worse than web-based models. Other concerns include hallucinations and a general scepticism about the use of Gen AI for advanced tasks.

The findings from our interviews suggest that greater adoption and trust of Gen AI in cybersecurity tasks requires meeting certain conditions: (1) there needs to be an easier setup of locally deployed models, and the local solution should provide performance comparable to web-based models, (2) there needs to be a comprehensive explainability of the Gen AI model steps and choices for it to be included and trusted for automated tasks, such as pentesting, so a pentester should be able to easily follow what the tool does and understand its reasoning behind its choices, (3) hallucinations need to eliminated or at least mitigated.

The participants mentioned that Gen AI tools could be used as path advisors and guide for new students and beginners, and for endpoint analysis, code generation and analysis, information gathering, and report generation. Among these solutions, report generation was the most frequently mentioned use case, which was not surprising, as reporting was the most disliked task among participants.

Future Directions. Our study provided valuable insights into the types of functionalities that cybersecurity professionals would find useful and interesting in future cybersecurity Gen AI applications. Several participants suggested the potential usefulness of generating POC applications. From a technical perspective, this functionality is already feasible to some extent and could be integrated into pentesting tools. Others suggested that Gen AI serve as a path advisor in pentesting, assisting with mapping web applications, performing endpoint analysis, or supporting student training. Source code analysis was also mentioned. As with many of these suggestions, implementation is possible with current technology, but participants emphasized that it should be reliable and operate using local tools to ensure security. Another often mentioned feature is reporting. A potential Gen AI tool could collect data throughout a pentest and automatically generate a structured final report, following predefined templates.

Participants also highlighted several requirements and constraints that would influence their willingness to adopt AI tools. The most frequently cited was that no sensitive data should be leaked or shared with third parties. Other important factors included comprehensive explainability, an intuitive and user-friendly interface, and reduced hallucination or inaccurate outputs. Future research and tool development should consider these requirements carefully to ensure adoption and trustworthiness in professional cybersecurity workflows.

Acknowledgment. This research was funded in part by the Swedish Industrial Graduate School on Cybersecurity (SIGS-CyberSec), a project of the Knowledge Foundation of Sweden (KKS). The authors would like to express their gratitude to all participants in the studies reported in this paper and also thank the reviewers for their time and invaluable feedback.

References

1. Al Balawi, M., Alnabhan, M., Atoum, M.S.: Generative AI for advanced malware detection. In: Proceedings of 4th Intelligent Cybersecurity Conference (ICSC). IEEE (2024). https://doi.org/10.1109/ICSC63108.2024.10895965
2. Al Siam, A., Alazab, M., Awajan, A., Faruqui, N.: A comprehensive review of AI's current impact and future prospects in cybersecurity. IEEE Access **13** (2025). https://doi.org/10.1109/ACCESS.2025.3528114

3. Alshaikh, O., Parkinson, S., Khan, S.: Exploring perceptions of decision-makers and specialists in defensive machine learning cybersecurity applications: the need for a standardised approach. Comput. Secur. **139** (2024). https://doi.org/10.1016/j.cose.2023.103694
4. Capuano, N., Fenza, G., Loia, V., Stanzione, C.: Explainable artificial intelligence in cybersecurity: a survey. IEEE Access **10**, 93575–93600 (2022). https://doi.org/10.1109/ACCESS.2022.3204171
5. Goyal, D., Subramanian, S., Peela, A.: Hacking, The Lazy Way: LLM Augmented Pentesting. arXiv preprint arXiv:2409.09493 (2024). https://doi.org/10.48550/arXiv.2409.09493
6. Hubinger, E., et al.: Sleeper Agents: Training Deceptive LLMs that Persist Through Safety Training. arXiv preprint arXiv:2401.05566 (2024). https://doi.org/10.48550/arXiv.2401.05566
7. ISO/IEC: 25010 - systems and software engineering - systems and software quality requirements and evaluation (square) - system and software quality models (2023)
8. Konstantinou, A., et al.: Leveraging LLMs for non-security experts in threat hunting. Mach. Learn. Knowl. Extr. **7**(2), 31 (2025). https://doi.org/10.3390/make7020031
9. Liu, F., et al.: Exploring and Evaluating Hallucinations in LLM-Powered Code Generation. arXiv preprint: arXiv:2404.00971 (2024). https://doi.org/10.48550/arXiv.2404.00971
10. Lyon, G.F.: Nmap Network Scanning: The Official Nmap Project Guide to Network Discovery and Security Scanning. Insecure. Com, LLC (2008)
11. Miles, M.B., Huberman, A.M., Saldaña, J.: Qualitative Data Analysis: A Methods Sourcebook, 3 edn. Sage (2014)
12. Shao, M., et al.: An Empirical Evaluation of LLMs for Solving Offensive Security Challenges. arXiv preprint: arXiv:2402.11814 (2024). https://doi.org/10.48550/arXiv.2402.11814
13. Stratton, S.J.: Population research: convenience sampling strategies. Prehosp. Disaster Med. (2021). https://doi.org/10.1017/S1049023X21000649
14. Utbildningsdepartementet (Sverige): Lag (2003:460) om etikprövning av forskning som avser människor (2023)
15. Zhao, W.X., et al.: A Survey of Large Language Models, v.16. arXiv preprint: arXiv:2303.18223 (2023). https://doi.org/10.48550/ARXIV.2303.18223

Hardware and IoT Security

From RAM to Topology: Forensic Analysis of ZeroTier One-Systems

Sven-Maurice Althoff[1(✉)] and Daniel Spiekermann[2]

[1] FernUniversität Hagen, Hagen, Germany
althoff@me.com
[2] FH Dortmund, Dortmund, Germany
daniel.spiekermann@fh-dortmund.de

Abstract. ZeroTier One (ZTO) is a decentralized, peer-to-peer software-defined networking platform that enables secure virtual networks across heterogeneous systems and NAT boundaries. While its strong encryption and minimal persistent logging enhance privacy and resilience, they also limit the effectiveness of traditional forensic approaches based on disk artifacts or network traffic analysis. As a consequence, volatile memory remains the primary source of forensic evidence. This paper presents the first systematic cross-platform memory forensic analysis of ZTO on Linux, Windows, and macOS. Using hypervisor-assisted RAM snapshots, Volatility, and structured byte-level inspection, we reliably identify ZTO client presence, recover runtime configuration and authentication material, and extract peer-related metadata. Our results show that volatile memory exposes key forensic artifacts, including network identifiers, tokens, and peer information, enabling partial reconstruction of overlay membership and topology despite encrypted communication and the absence of logs. To support practical investigations, we introduce *MemZTector*, a lightweight proof-of-concept tool that automates the extraction of ZTO-specific artifacts from raw memory dumps. Our findings demonstrate that memory forensics is not merely complementary but often the only viable approach for investigating decentralized ZTO deployments.

Keywords: memory forensics · ZeroTier One · software-defined networks · SD-WAN

1 Introduction

Decentralized overlay networks are increasingly used to interconnect endpoints across heterogeneous platforms and administrative domains. ZeroTier One (ZTO) is a prominent example: it provides software-defined, peer-to-peer virtual networking across NAT boundaries without relying on a centralized controller. While this design improves flexibility and resilience, it complicates digital investigations. End-to-end encryption renders payload inspection ineffective, and ZTO minimizes persistent logging and disk-resident configuration traces. As a result,

© The Author(s), under exclusive license to Springer Nature Switzerland AG 2027
K. Renaud et al. (Eds.): EICC 2026, CCIS 3012, pp. 169–184, 2027.
https://doi.org/10.1007/978-3-032-28957-5_10

conventional evidence sources such as log files, controller state, or packet captures often provide limited forensic value.

In contrast, a running ZTO instance must maintain transient but forensically relevant state in volatile memory, including network identifiers, runtime configuration, authentication material, and peer relationship metadata. To our knowledge, the forensic footprint of ZTO in RAM has not yet been systematically characterized across operating systems. This gap is operationally relevant for incident response and forensic readiness in modern, decentralized network environments [6].

This paper presents the first cross-platform memory forensic study of ZTO, covering Linux, Windows, and macOS. Using controlled experiments, we investigate four central research questions: (RQ1) Can we detect the presence of a ZTO client in memory? (RQ2) What sensitive cryptographic materials can we recover? (RQ3) What can we observe from network communication sessions? (RQ4) How far can we reconstruct peer relationships or network structures?

The main contributions of this paper are as follows.

- We present the first systematic analysis of ZTO memory artifacts across Linux, Windows, and macOS platforms. Our experiments reveal the potential for data extraction in conjunction with the recoverability of client presence, authentication tokens as well as communication and peer information.
- We derive stable in-memory anchors and offsets that enable reliable extraction of network configuration and node/peer information from raw dumps.
- We introduce *MemZTector*, a lightweight proof-of-concept tool that automates the extraction of ZTO-specific evidence from memory images.

We structure the remainder of the paper as follows. Section 2 discusses related work, while Sect. 3 summarizes background information. Section 4 describes our testbed and the scenario used for our initial investigations. Section 5 presents the analysis of ZTO memory forensics. Based on these results, we show in Sect. 6 *MemZTector*, a proof-of-concept (PoC) for the extraction and analysis of network information from a memory dump, which is evaluated in Sect. 7. Section 8 concludes and gives an outlook on future work.

2 Related Work

Research on software-defined networking (SDN) has extensively examined security threats and architectural weaknesses. Kreutz et al. [10] provide a foundational survey of SDN security, identifying attack vectors across control and data planes using the STRIDE model. Their work highlights the risks of centralized control, which has motivated subsequent studies on SDN resilience and forensic preparedness. Extending this perspective, Iqbal et al. [8] analyze attacks against SDN application layers and management interfaces, demonstrating how compromised control components can be leveraged for information disclosure and privilege escalation.

Several approaches address forensic investigations in SDN environments. Chin et al. [6] propose a forensic methodology for SDN switches that leverages live memory and southbound communication analysis. Zhang et al. [17] introduce a multi-plane evidence collection framework spanning data, control, and management layers. These works assume centrally managed SDN architectures and rely on persistent control-plane artifacts, limiting their applicability to decentralized, peer-to-peer overlay networks such as ZeroTier One. Research focusing on ZTO is limited to case studies such as Sood et al. [16], where ZTO is used as an ad-hoc VPN-like network. In IoT deployments, Hriţcan et al. implement ZTO as a lightweight alternative to traditional site-to-site VPNs [7].

Memory forensics has emerged as a key technique for analyzing systems where critical state information is not persistently stored. It enables the recovery of transient artifacts such as process state, encryption material, and network metadata from volatile memory [5]. Tools such as *Volatility* are widely used for malware analysis, incident response, and digital investigations [14]. A detailed discussion of Volatility, its plugins, and analysis techniques can be found in [11]. Beyond classical memory forensics, related work has also examined forensic traces in privacy-preserving overlay systems. Angeli et al. [4] analyze Tor-related artifacts in Windows environments across disk, network, and memory sources, showing that overlay-based communication systems can leave recoverable traces despite their privacy-oriented design. Prior work by Kauser et al. [9] on Tor and I2P similarly demonstrates that P2P networks leave recoverable artifacts in volatile memory, including routing metadata and session identifiers, even when designed to limit persistent state.

The intersection of SDN and memory forensics has received limited attention. Da Luz Lemoz et al. [13] analyze memory artifacts of OpenFlow-based SDN switches, while Alshaya et al. [3] apply YARA-based memory scanning to an SDN controller. Both approaches target controller-centric architectures and do not address peer-to-peer SDN platforms that deliberately minimize persistent logging and centralized state.

In contrast to existing work, this paper focuses on the volatile memory footprint of ZeroTier One, a decentralized peer-to-peer SDN platform. To the best of our knowledge, no prior study has systematically analyzed ZTO across operating systems to identify, categorize, and extract forensic artifacts from volatile memory. Our work bridges this gap by combining memory forensics with structured analysis of ZTO's runtime data structures, enabling forensic investigations in environments where traditional SDN assumptions do not hold.

3 Background

3.1 ZeroTier One

ZeroTier One (ZTO) is an open-source, peer-to-peer (P2P) overlay networking platform that creates software-defined virtual networks across heterogeneous hosts and network boundaries (e.g., NAT) as shown in Fig. 1. Each node communicates via its underlay interface (IPv4/IPv6) while simultaneously participating in one or more virtual networks that assign overlay addresses and enforce membership-based access control.

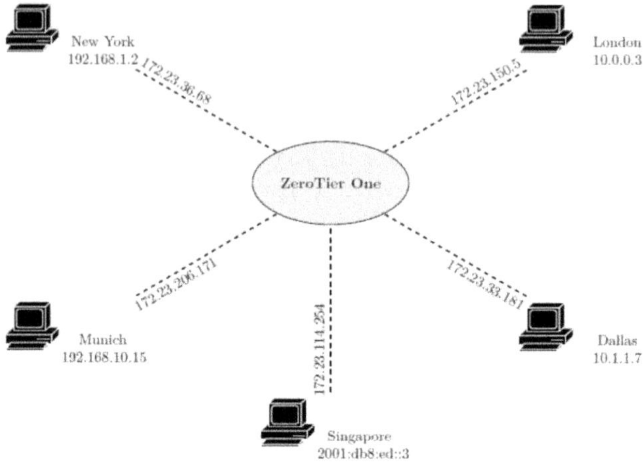

Fig. 1. Exemplary distributed network infrastructure of ZTO.

Unlike conventional SDN architectures that rely on centralized controllers, ZTO distributes control and state among participating nodes, allowing devices to securely communicate across NAT boundaries and disparate networks without manual configuration or dedicated infrastructure.

Conceptually, ZTO can be described using two logical layers as shown in Fig. 2.

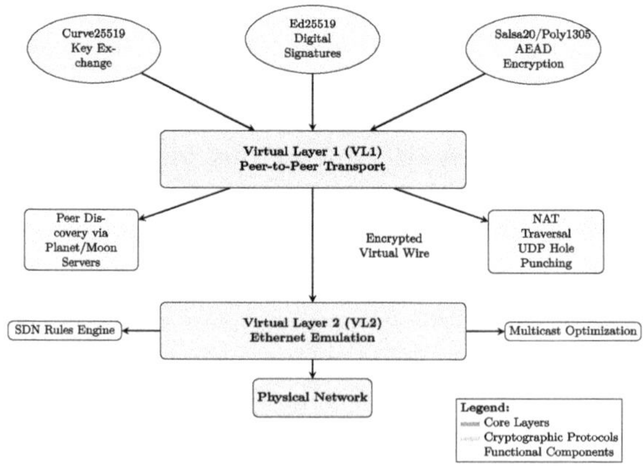

Fig. 2. ZTO Protocol Architecture.

Virtual Layer 1 (VL1) provides encrypted P2P transport, peer discovery (via public root infrastructure and optional privately hosted nodes, *moons*), and NAT traversal (e.g., UDP hole punching). **Virtual Layer 2 (VL2)** emulates Ethernet-like connectivity on top of VL1 and implements network membership and configuration enforcement. Each virtual network is identified by a network identifier (NWID). During runtime, nodes maintain network configuration and membership state (e.g., assigned overlay addresses and policy flags) as well as peer/path metadata required for connectivity management.

These runtime structures are central to our analysis, as they constitute stable anchors for identifying ZTO presence and extracting configuration and peer-related artifacts from volatile memory.

ZTO's encrypted transport and decentralized state reduce the availability of conventional disk- and traffic-based evidence sources. Consequently, runtime state in volatile memory becomes a primary target for forensic analysis.

3.2 Memory Forensics

Memory forensics is the acquisition and analysis of volatile memory (RAM) to recover transient digital evidence that exists only during system operation [5]. In contrast to disk-based forensics, RAM may contain process state, in-memory configuration data, active network metadata, and short-lived authentication material that is not reliably persisted to storage [2]. This is particularly relevant for modern systems that minimize local logs or protect network traffic using end-to-end encryption. Volatile memory is also the only location where encryption material like session keys or pre-shared tokens may be accessible in unencrypted form [12].

From a forensic analysis perspective, the interpretation of RAM artifacts is subject to several limitations. Modern operating systems rely on paging and demand paging, so parts of a process' virtual address space may not be resident in physical memory at acquisition time. In addition, swapping may move pages to disk (e.g., swap files or pagefiles), shifting relevant evidence outside the RAM snapshot [1]. Finally, memory acquisition can suffer from smearing, i.e., inconsistencies introduced by concurrent system activity, resulting in torn or mixed data structures [15]. While pausing virtual machines during hypervisor-assisted acquisition mitigates smearing, paging-related incompleteness and runtime churn can still affect artifact completeness.

In practice, memory analysis commonly relies on frameworks such as *Volatility*, which provide operating-system-specific plugins for process enumeration, module inspection, and network/socket analysis, thereby enabling a structured workflow for extracting and correlating volatile artifacts [11,14]. However, the recoverability and representation of artifacts can vary across operating systems due to differences in memory management and data encoding. This motivates cross-platform validation when characterizing forensic traces of the same application family.

In this work, memory forensics provides the methodological basis to identify ZTO runtime artifacts and to recover network configuration, authentication tokens, and peer/path metadata from raw memory snapshots.

4 Experimental Setup and Data Acquisition

From a forensic perspective, ZTO poses distinct challenges. Its end-to-end encryption and limited persistent logging reduce the availability of conventional evidence sources such as log files and packet captures. In addition, its dynamic peer relationships and largely memory-resident runtime state make volatile memory a primary source of forensic artifacts, including configuration data, peer information, and communication metadata. Reliable investigation therefore depends on timely memory acquisition.

To examine these artifacts under controlled conditions, we created a simplified test environment with multiple nodes in a shared subnet. Within this setup, we used ICMP traffic as a minimal and reproducible communication baseline for identifying ZTO-related activity in memory.

4.1 Testbed and Platforms

ZTO enables network segmentation through unique identifiers, allowing the creation of isolated virtual networks. For this experiment, we configured a dedicated network with ZeroTier 1.160.0, named *test_network*, with the network ID *ebe7fbd4450c4966*. This identifier is essential for client authentication and network joining procedures.

The configuration ensures seamless communication between all registered clients while maintaining strict traffic management. Furthermore, the configuration automatically assigns to each client a unique IPv4 address, supporting a flexible and consistent network topology across diverse platforms. The evaluation was performed on three major operating systems, namely Kali Linux 2024.4 with Kernel 6.5.0-kali2-amd64, Microsoft Windows 11 Pro (64bit) 24H2 (Build 22631.2861), and macOS Monterey 12.6 (Build 21G115).

4.2 Data Acquisition

Due to ZTO's decentralized design and the absence of persistent logs or unencrypted traffic traces, the acquisition of volatile memory is essential for forensic analysis. We therefore focus on capturing the runtime state of ZTO clients under controlled and reproducible conditions.

Memory snapshots were acquired using QEMU's `dump-guest-memory` functionality, which produces a complete raw image of a virtual machine's RAM. To ensure forensic soundness, all acquisitions were performed with the virtual machines paused, preventing memory modifications during capture. This approach preserves active processes, runtime configuration, communication buffers, and transient authentication material at the time of acquisition.

The resulting memory images were stored in raw format and analyzed offline. By relying on hypervisor-assisted acquisition, we avoided the deployment of in-guest collection agents, thereby minimizing additional memory footprint and ensuring a consistent and platform-independent acquisition process across all experiments.

5 Analysis

To validate the presence of ZTO artifacts in volatile memory, we first conducted a structured examination of the acquired dumps using the Volatility framework. We used Volatility 3 Framework 2.27.0 on Kali Linux 6.12.33-1 (2025-06-25), which enables process- and network-centric inspection of memory images and provides a well-established workflow for digital forensic investigations.

The analysis starts with process enumeration with *windows.pslist*[1], which revealed active instances of the `zerotier-one` process as shown in Listing 1.1. This finding confirmed client activity with the two processes *zerotier-one_x*[2] (service) and *zerotier_deskt*[3] (UI) in memory and established a baseline for further analysis.

Listing 1.1. Volatility process listing for ZeroTier One

```
$ python3 vol.py -f zto_win.dmp windows.pslist | grep -i zero

6552    8481    zerotier-one_x    0xa006dd874080    31    -    0    False
    2025-01-04 17:42:07.000000 UTC    N/A    Disabled
6580    5760    zerotier_deskt    0xa006de8a7080     3    -    1    False
    2025-01-04 18:14:04.000000 UTC    N/A    Disabled
```

Network analysis with the Volatility plugin *windows.netstat* identified open UDP sockets maintained by *zerotier-one_x64.exe*. Additionally, some connections corresponded to the internal network 172.23.0.0/16, initiated by the *svchost.exe* and *SYSTEM* processes (which were displayed, but do not indicate a connection between svchost.exe and ZTO as shown in Listing 1.2.)

Listing 1.2. Volatility network connections (shortened)

```
python3 vol.py -f zto_win.dmp windows.netstat

0xa006ded66688  UDPv4   172.23.36.68    137     *       0
    4       System  2025-01-04 18:19:19.000000 UTC
0xa006dcf54e38  UDPv4   20.20.20.21     138     *       0
    4       System  2025-01-04 16:43:29.000000 UTC
0xa006ded65988  UDPv4   172.23.36.68    138     *       0
    4       System  2025-01-04 18:19:19.000000 UTC
0xa006df77d788  UDPv4   172.23.36.68    1900    *       0
    3868    svchost.exe     2025-01-04 18:19:19.000000 UTC
```

[1] or the according *linux.pslist* in case of Linux memory dumps.
[2] Which is shortened from zerotier-one_x64.exe.
[3] Which is shortened from zerotier-desktop.exe.

```
0xa006e52549e8  UDPv4    20.20.20.21      9993    *       0
    6552    zerotier-one_x  2025-01-04 17:42:11.000000 UTC
0xa006e52570e8  UDPv4    20.20.20.21      35929   *       0
    6552    zerotier-one_x  2025-01-04 17:42:11.000000 UTC
```

To determine further details of the process, we dumped the entire process memory using Volatility's *memmap*-plugin.

5.1 Structured Memory Analysis and Struct Mapping

Conventional string or regex searches on raw memory, while useful for initial detection, are insufficient for a rigorous scientific forensic analysis. Because of this, we performed an internal structure reconstruction to ensure the reliability and specificity of the string based artifact extraction.

Since ZTO is open source, we analyzed its source code[4] to identify the C++ classes and data structures that store critical runtime state information within the *zerotier-one* process. The relevant network information is processed in *services/OneService.cpp* within objects corresponding to the following core structures:

- **NetworkConfig:** Holds network setup details, including the Network ID (NWID), assigned overlay IP addresses ('assignedAddresses'), routing entries ('routes'), and administrative flags ('allowManaged').
- **Peer/Path:** Stores metadata for known or active nodes, including the Node ID, real-time latency statistics, and external reachability endpoints ('surfaceAddresses').

Based on these results, we identified specific byte sequences and offsets that reliably mark the beginning of a serialized ZTO configuration structure in the application's process memory. The NWID can be reliably identified using the hexadecimal string `0x763D3030303030303030303030303030370A`, which serves as a strong forensic identifier and enables further analysis once located. By mapping these data to their relative positions within the detected memory object, we established reliable byte offsets for additional data fields like the name of the network following the NWID after 170 bytes as shown in Fig. 3a. This allows extraction of the critical data even if the surrounding memory is fragmented.

Based on this, detailed information, such as network configuration data or communication secrets, can be extracted.

5.2 Network Information

The NWID can be found at multiple locations within the memory dump. By analyzing the surrounding data preceding and following each occurrence, additional details about the configuration can be derived. If the data located 211 bytes before the identified offset is not set to `0x00`, this position provides further insight into the network configuration as shown in Fig. 3b.

[4] The source code is available at: https://github.com/zerotier/ZeroTierOne.

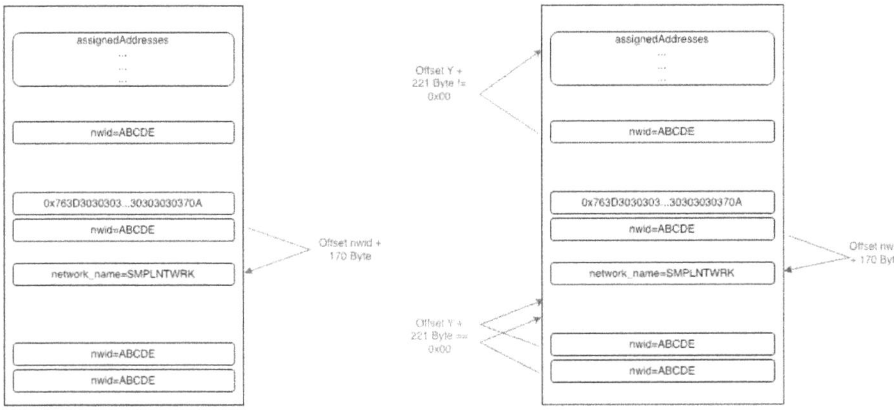

(a) Finding NWID and network name based on the identification hexadecimal string (b) Offset calculation to detect network information (red: miss, green: hit)

Fig. 3. Process dump memory.

The data at this position can be be extracted in JSON syntax as follows:


```
[
  {
    "nwid": "ebe7fbd4450c4966",
    "name": "test_network",
    "assignedAddresses": ["172.23.36.68/16"],
    "routes": [
      { "target": "172.23.0.0/16", "via": null }
    ],
    "allowManaged": true,
    "allowGlobal": false,
    "allowDefault": false,
    "allowDNS": false,
    "mtu": 2800,
    "portDeviceName": "ethernet_32770",
    "status": "OK"
    /* additional fields omitted for brevity */
  }
]
```


The extracted information covers the internal network configurations, including MAC addresses and IP addresses of our test-network as well as details about the network that matches the chosen options as shown in Fig. 4. As the MAC addresses appear to be ephemeral, the IP addresses and subnet information align with our network configuration we used for the clients, which is hidden from the existing volatility plugins *windows.netscan* and *windows.netstat*.

Fig. 4. Configuration options.

5.3 Node Information

By analyzing the memory dump, we detected hard-coded strings such as node_id or zt_peer_packet, which provide valuable data access, including metrics or details about peers. An analysis of the string node_id and the used offsets provides details about devices inside the virtual network as shown in Listing 1.3. Each entry is separated by `0x0A` and between 54 and 56 Bytes long, containing information about the node, its status or packet counts and error rates between nodes including directional traffic (tx/rx).

Listing 1.3. Extracted peer metrics from ZTO memory

```
zt_peer_latency_count{node_id="778cde7190"} 9
zt_peer_latency_sum{node_id="778cde7190"} 15032449914
zt_peer_path_count{node_id="62f865ae71",status="dead"}
zt_peer_path_count{node_id="cafe9efeb9",status="alive"}
```

The extracted artifacts show that peers like `778cde7190` and `cafe9efeb9` were active and maintained communication with the client, as indicated by their latency distributions and non-zero counters. Other nodes such as `62f865ae71` were recorded as `dead`, reflecting unavailable or terminated connections at the time of acquisition.

5.4 Node Authentication

ZTO enforces access control through a cryptographically protected membership model. During runtime, clients store network-specific authorization material in volatile memory to establish and maintain encrypted peer-to-peer connections. These credentials are transient and typically not persisted to disk, making them a relevant target for memory forensics.

In all analyzed memory dumps, we identified the file `authtoken.secret` mapped into the address space of the ZTO process. On Windows systems, the file is normally located in the user-specific ZeroTier directory and could be recovered from memory using Volatility's *filescan* and *dumpfiles* plugins:

```
\Users\Docker\AppData\Local\ZeroTier\authtoken.secret
```

The extracted file contained the authorization token *qzkjof7pxfiayoa 5q1p968oj* in clear text, as shown in Listing 1.4. The presence of this token in volatile memory provides direct evidence of active or recent network membership at the time of acquisition.

Listing 1.4. Extracted authtoken from ZTO memory

```
python3 vol.py -f zto_win.dmp  windows.dumpfiles --virtaddr
    0xa006dfdc1c10
<snip>
DataSectionObject 0xa006dfdc1c10 authtoken.secret

cat file.0xa006dfdc1c10.DataSectionObject.authtoken.secret.dat
qzkjof7pxfiayoa5q1p968oj
```

From a forensic perspective, recovered authentication tokens enable (i) attribution of a node to a specific NWID, (ii) correlation with other in-memory artifacts such as peer metadata or network configuration, and (iii) temporal reconstruction of access privileges even if credentials were later revoked. Although the token alone does not allow node impersonation without the corresponding private key, its recovery constitutes a high-value forensic artifact in decentralized overlay networks.

6 Implementation

To assess the practical feasibility of extracting ZTO-related forensic artifacts from volatile memory, we implemented *MemZTector*[5], a lightweight Python proof of concept that operates directly on raw process memory dumps. It first locates stable ZTO-specific configuration markers to recover the NWID, network name, and subnet, then identifies IP address strings in memory and correlates them with the recovered subnet to distinguish overlay-internal addresses from unrelated artifacts. In addition, *MemZTector* extracts candidate public endpoints from peer-path structures by retaining only syntactically valid IP/-port pairs. The prototype comprises fewer than 400 lines of code and depends exclusively on Python's standard library. Although not intended as a production-ready forensic tool, *MemZTector* demonstrates that reliable and repeatable extraction of ZTO-specific artifacts from volatile memory is feasible and can be largely automated.

```
python3 memztector.py pid.6552.dmp
=== Results ===
Network name: test_network
Network ID: ebe7fbd4450c4966
Subnet: 172.23.0.0/16
```

[5] The PoC is available at https://github.com/dfeu/memZTector.

```
Found IPs in subnet (unique): 3 (scanned: 60)
        172.23.36.3
        172.23.36.68
        172.23.68.181

-- Public endpoints --
  20.20.20.21:9993
  20.20.20.21:35929
  46.5.229.48:44807
  46.5.229.48:44815
  46.5.229.48:44817
  46.5.229.48:44821
  46.5.229.48:44826
  46.5.229.48:44834
  46.5.229.48:44838
Unique public endpoints: 9 (read: 241)
```

The discrepancy between the public endpoints and the addresses found inside the ZTO overlay is expected and follows directly from ZTO's transport model and NAT traversal:

- **NAT Traversal and Ephemeral Ports**
  ZTO performs UDP hole punching and maintains multiple candidate paths per peer [7]. NAT assigns *ephemeral* external ports, which explains the many port variations for the same public IP. The well-known port `9993/UDP` coexists with these ephemeral ports used during path negotiation and keep-alives.
- **Temporal Effects in RAM**
  Volatile memory may hold *stale* path entries from previous sessions (e.g., old NAT bindings) alongside current ones. Thus, a single dump can show more public endpoints than are currently in use.

Based on the extracted artifacts, an in-depth forensic investigation is feasible. The recovered overlay identifiers and in-subnet addresses enable the reconstruction of the virtual topology and peer relationships. At the same time, the discovered public endpoints provide further details for external sources, such as firewall and flow logs.

7 Evaluation

To validate our results and confirm the reliability of the identified artifacts, we conducted a structured evaluation across two different network protocols. The evaluation design followed two principles: (i) to reproduce realistic usage contexts of ZTO in controlled laboratory settings, and (ii) to compare the consistency of extracted artifacts across different operating systems and communication workloads. We derived the environments from the scenario used in our initial research as described in Sect. 4, but each was completely newly created (e.g. new network names and NWIDs, new IP addresses, operating system). This cross-platform consistency confirms that the identified ZTO traces are not artifacts of single executions but persist across independent runs. The reproducibility of these results under controlled conditions reinforces the reliability of volatile memory as a dependable evidence source in decentralized network environments.

- **Scenario 1**
  Building on our baseline, this scenario introduced Hypertext Transfer Protocol (HTTP) transactions using a lightweight Python-based web server. HTTP, as an unencrypted application-layer protocol, enabled the observation of higher-level communication artifacts, such as request/response structures, session identifiers, and data payloads.
- **Scenario 2**
  The second and most advanced scenario examined file transfers using the SSH File Transfer Protocol (SFTP). Here, encrypted communication channels introduced additional complexity, as payload contents were no longer directly observable.

This methodological approach, derived from the layered protocol analysis framework in [17], enables a detailed evaluation of our results described in Sect. 5. We defined four stages, each stage focusing on one specific research question listed in Sect. 1.

7.1 Stage 1: Baseline Client Presence

The first evaluation case verified the ability to detect the active ZTO process in volatile memory. Using `pslist` in Volatility, the presence of the `zerotier-one` binary was consistently identified across Windows, Linux, and macOS. The recovered process metadata (process ID, execution path, timestamps) matched the expected runtime state of the client, confirming that memory forensics provides a reliable baseline indicator of ZTO activity.

7.2 Stage 2: Network Communication Metadata

In the second stage, we evaluated whether network communication could be observed in volatile memory. Using *windows.netscan*, *linux.sockStat* and *mac.netstat*, open sockets associated with ZTO were recovered, including the well-known default port `9993/UDP`. This shows that forensic investigators can reliably link active communication channels to the ZTO process, despite end-to-end encryption. Using *MemZTector*, the tool provides in-depth insights that reveal internal node information not previously visible with standard Volatility plugins.

7.3 Stage 3: Token and Configuration Recovery

The third step focused on recovering ephemeral authentication tokens and runtime configuration data from RAM. By scanning the dumps with the *filescan*-plugin and manual string extraction, we identified tokens and configuration options, assigned overlay IP addresses, and network membership descriptors in JSON format. The artifacts were cross-validated against the experimental setup and confirmed to be accurate.

7.4 Stage 4: Peer Node Detection

The final investigation examined whether peer relationships could be reconstructed. A manual string search could extract the metrics that Volatility could not access. We were able to extract metrics embedded in memory that reveal multiple peer node identifiers, their latency data, and path status values (`alive`/`dead`). In addition to this, *MemZTector* can extract both internal and public IP addresses, providing an additional source for investigations. These results allowed us to partially reconstruct the overlay topology and correlate it with the controlled testbed configuration. The ability to detect remote peers from a single memory dump significantly enhances the evidentiary value of volatile artifacts.

7.5 Summary

Across our two scenarios, the methodology consistently reproduced expected artifacts and validated our results. Process presence was confirmed as a baseline indicator, network sockets revealed active communication, tokens exposed authentication state, and peer information enabled partial topology reconstruction. Together, these findings demonstrate that memory forensics offers a reliable and repeatable approach to investigating ZTO deployments, even in the absence of persistent logs or accessible network captures.

8 Conclusion

In this work, we presented the first systematic investigation of volatile memory artifacts in ZTO, a decentralized peer-to-peer SDN platform. To achieve this, we designed a cross-platform experimental framework that covers both Linux and Windows systems, enabling us to analyze the presence and structure of forensic evidence in volatile memory. Our methodology focused on three key areas: detection of client presence, reconstruction of peer information, and recovery of authentication material.

Our analysis demonstrates that ZTO leaves identifiable traces in memory, including process signatures, join tokens, and partial peer connection metadata. While forensic tools such as Volatility and bulk_extractor are partially effective in locating these artifacts, they also reveal limitations when analyzing overlay networks, where many conventional evidence sources are unavailable. However, our analysis partially depends on version-sensitive memory artifacts, including fixed offsets and characteristic strings. Its robustness across software versions, build configurations, and operating system changes remains to be evaluated.

We argue that memory forensics is not simply a complementary technique, but rather frequently the only feasible approach for investigating ZTO deployments, since persistent logs and communication records are often scarce or intentionally absent. In this way, incident responders can still uncover client activity, while law enforcement investigators may extract peer metadata to reconstruct covert communication networks. In summary, our investigation successfully addresses the research questions posed in the introduction:

1. **RQ1 - Process and Artifact Presence:** We confirmed the consistent presence of the active `zerotier-one` process and application-specific artifacts (NWIDs, Node IDs) across Windows, Linux, and macOS.
2. **RQ2 - Authentication State Recovery:** We established a repeatable method for recovering ephemeral authentication tokens (`authtoken.secret`) from RAM, providing conclusive evidence of network membership.
3. **RQ3 - Session Extraction:** The extraction of network communication session data is possible by determining relevant offsets in the memory dump. The implementation of *MemZTector* validated the feasibility of automating the extraction of network data like network identifiers or internal addresses.
4. **RQ4 - Topology Reconstruction:** We demonstrated that structured memory analysis enables the reliable extraction of internal overlay IP subnets and a partial reconstruction of the virtual topology via peer connection metrics.

While the methodology we present enables consistent recovery of volatile artifacts, certain constraints remain. The analysis depends on controlled environments and static snapshots, which may not fully capture the dynamics of live or large-scale ZTO deployments. Furthermore, variations in operating system memory management can obscure or fragment critical data structures, thereby limiting the completeness of the artifacts. Future work should extend this approach to continuous memory acquisition, automated pattern detection, and the correlation of volatile artifacts with external telemetry such as controller logs or endpoint traces. These extensions may improve robustness against fragmentation and noise in real-world memory dumps, thereby further enhancing forensic reliability and operational applicability in decentralized real-world networks.

References

1. Abrossimov, E., Rozier, M., Shapiro, M.: Generic virtual memory management for operating system kernels. In: Proceedings of the Twelfth ACM Symposium on Operating Systems Principles, pp. 123–136 (1989). https://doi.org/10.1145/74851.74863
2. Al-Sharif, Z.A., Al-Senjalawi, R., Alzoubi, O.A.: The effects of document's format, size, and storage media on memory forensics. Forensic Sci. Int. Digit. Investig. **48**, 301692 (2024). https://doi.org/10.1016/j.fsidi.2024.301692
3. Alshaya, A., Kardorff, A., Facundus, C., Baggili, I., Richard III, G.: Memory forensics of the opendaylight software-defined networking (SDN) controller. In: Proceedings of the 18th International Conference on Availability, Reliability and Security, pp. 1–8 (2023). https://doi.org/10.1145/3600160.3600196

4. Angeli, V.M., Atamli, A., Karafili, E.: Forensic analysis of tor in windows environment: a case study. In: Proceedings of the 17th International Conference on Availability, Reliability and Security. ARES 2022. Association for Computing Machinery, New York, NY, USA (2022). https://doi.org/10.1145/3538969.3543808
5. Case, A., Richard III, G.G.: Memory forensics: the path forward. Digit. Investig. **20**, 23–33 (2017). https://doi.org/10.1016/j.diin.2016.12.004
6. Chin, T., Xiong, K.: A forensic methodology for software-defined network switches. In: DigitalForensics 2017. IAICT, vol. 511, pp. 97–110. Springer, Cham (2017). https://doi.org/10.1007/978-3-319-67208-3_6
7. Hriţcan, D.F., Graur, A., Balan, D.: Securing IoT environments using zerotier and opnsense. In: 2024 23rd RoEduNet Conference: Networking in Education and Research (RoEduNet), pp. 1–4. IEEE (2024). https://doi.org/10.1109/roedunet64292.2024.10722755
8. Iqbal, M., Iqbal, F., Mohsin, F., Rizwan, M., Ahmad, F.: Security issues in software defined networking (SDN): risks, challenges and potential solutions. Int. J. Adv. Comput. Sci. Appl. **10**(10), 298–303 (2019). https://doi.org/10.14569/IJACSA.2019.0101042
9. Kauser, S., Safdar, T., Hasan, M.H., Akashah, E., Kazmi, S.: Windows 10's browser forensic analysis for tracing p2p networks' anonymous attacks. Comput. Mater. Continua **72**, 1251–1273 (2022). https://doi.org/10.32604/cmc.2022.022475
10. Kreutz, D., Ramos, F., Veríssimo, P., Esteve Rothenberg, C., Azodolmolky, S., Uhlig, S.: Software-defined networking: a comprehensive survey. arXiv e-prints **103** (2014). https://doi.org/10.1109/JPROC.2014.2371999
11. Ligh, M.H., Case, A., Levy, J., Walters, A.: The Art of Memory Forensics: Detecting Malware and Threats in Windows, Linux, and Mac Memory, 1st edn. Wiley Publishing (2014)
12. Lindenmeier, C., Hammer, A., Gruber, J., Röckl, J., Freiling, F.: Key extraction-based lawful access to encrypted data: taxonomy and survey. Forensic Sci. Int. Digit. Invest. **50**, 301796 (2024). https://doi.org/10.1016/j.fsidi.2024.301796
13. da Luz Lemos, F.A., dos Santos Cavali, T., Fonseca, K.V.O., Fonseca, M.S.P., de Faria, R.A.: Enhancing the security of software-defined networking through forensic memory analysis. J. Netw. Syst. Manag. **32**(4), 82 (2024). https://doi.org/10.1007/s10922-024-09862-4
14. Nyholm, H., et al.: The evolution of volatile memory forensics. J. Cybersecur. Priv. **2**(3), 556–572 (2022). https://doi.org/10.3390/jcp2030028
15. Parida, T., Nath, K., Das, S.: Sam: a mechanism to facilitate smear-aware forensic analysis of volatile system memory. J. Appl. Secur. Res. **19**(2), 300–329 (2024)
16. Sood, K., Nosouhi, M.R., Kumar, N., Gaddam, A., Feng, B., Yu, S.: Accurate detection of IoT sensor behaviors in legitimate, faulty and compromised scenarios. IEEE Trans. Dependable Secure Comput. **20**(1), 288–300 (2021). https://doi.org/10.1109/tdsc.2021.3131991
17. Zhang, S.H., Meng, X.X., Wang, L.H.: SDNForensics: a comprehensive forensics framework for software defined network. In: Proceedings of the International Conference on Computer Networks and Communication Technology (CNCT 2016), pp. 92–99. Atlantis Press (2016). https://doi.org/10.2991/cnct-16.2017.13

From Scores to Alarms: NP-Calibrated Event-Level Detection with a Multi-horizon Digital Twin for ICS

Hoda Mehrpouyan(✉)

Boise State University, Boise, ID 83702, USA
hodamehrpouyan@boisestate.edu

Abstract. Digital-twin-based detection in industrial control systems (ICS) is frequently evaluated with heuristic thresholds, obscuring the false-alarm budget and making operating points brittle under background shift. We propose a deployable hybrid detector for the Secure Water Treatment (SWaT) testbed: a Gated Recurrent Unit (GRU) digital twin is trained on normal operation, multi-horizon forecast residuals are mapped to structured residual features, and a Random Forest (RF) produces a residual-space attack score. Operating points are selected *without* tuning on held-out attack days by calibrating a threshold τ_α via a Neyman–Pearson (NP) quantile rule in a design exceedance budget $\alpha = 10^{-2}$ on a designated negative calibration stream. Pointwise exceedances are then converted into operator-facing alarm *events* using a fixed persistence, minimum-duration, merge, and cooldown policy, allowing workload reporting in alarms/day, false alarms/day, and alarm occupancy.

Under a strict day-to-day temporal protocol with held-out attack days, the hybrid twin+RF achieves strong ranking performance (AUC = 0.921, AP = 0.896). At deployable operating points, NP-normal calibration yields event recall 0.429 with 1.524 false events/day (occupancy 0.157), while background-matched calibration using out-of-bag scores on early-deployment negatives reduces nuisance workload to 0.381 false events/day (occupancy 0.152) at the same event recall. A residual-energy baseline is substantially weaker (AUC ≈ 0.68, AP ≈ 0.27) and incurred false events 24–26/day under background calibration. Overall, the results show that auditable calibration and event-level workload are essential to deployable ICS detection beyond pointwise separability.

Keywords: Digital twin · Industrial control systems · Cyber-attack detection · Event-level alarming

1 Introduction

Industrial control systems (ICS) are the foundation for critical infrastructure such as water treatment, power generation, and manufacturing. Growing connectivity and remote access expand the attack surface, allowing cyber–physical

© The Author(s), under exclusive license to Springer Nature Switzerland AG 2027
K. Renaud et al. (Eds.): EICC 2026, CCIS 3012, pp. 185–200, 2027.
https://doi.org/10.1007/978-3-032-28957-5_11

attacks and faults that manifest as subtle, time-correlated deviations in multivariate sensor and actuator streams [14,15,18]. In deployment, the core challenge is not only to *rank* anomalous behavior, but to (i) choose a *reproducible operating point* with an explicit false-alarm budget and (ii) present alarms as *operator-actionable events* rather than isolated threshold crossings.

Digital twins, data-driven or physics-informed predictors of closed-loop plant behavior, provide a natural basis for detection. A twin trained on nominal operation forecasts expected trajectories, and prediction residuals encode structured evidence of abnormal behavior. However, many twin-based detectors retain ad hoc thresholding (e.g. fixed multiples of standard deviation, tuned cutoffs, or best-F_1 on a test split), making operating points unstable under day-to-day background variation and obscuring false-alarm budgets and workload. In contrast, end-to-end sequence classifiers can achieve strong discrimination, but their thresholds are often tuned on evaluation data, and their outputs are not naturally aligned with event-level incident triage.

This paper targets this *operationalization gap*: given a pointwise score stream, how do we produce (i) an auditable threshold selected from negative data under a specified exceedance budget and (ii) event-level alarms that reflect how operators respond? We instantiate this question on the Secure Water Treatment (SWaT) dataset using a hybrid residual-space attack detector. A GRU digital twin is trained on normal operation to produce multi-horizon residuals, and a supervised Random Forest (RF) maps structured residual summaries to a score. Importantly, we do *not* select thresholds by maximizing the set of tests F_1. Instead, we select thresholds using an explicit Neyman–Pearson (NP)-style quantile rule on a designated negative calibration stream, and we convert exceedances into alarm *events* via a documented persistence/minimum-duration/merge/cooldown policy. This yields operating points that are reproducible and auditable and that can be reported directly in workload units (alarms/day, false alarms/day, alarm occupancy) under a strict day-wise temporal protocol.

Our contributions include: 1- **Auditable selection of operating-points under temporal shift.** We formalize NP-style calibration by setting a target exceedance budget α on a specified negative calibration stream, avoiding threshold tuning on the evaluation-period attack labels. We study two deployable choices: *NP* normal (held-out normal-only calibration) and *NP-bg* background-matched calibration using early-deployment negatives (with out-of-bag scoring to reduce optimistic bias). 2- **Operator-facing alarm semantics and workload reporting;** We define an explicit eventization rule, persistence, minimum duration, merge gap, and cooldown that maps pointwise exceedances to alarm events. We report event recall and workload metrics (false events/day, alarms/day, occupancy) alongside pointwise AUC/AP and precision/recall to expose the practical precision–workload trade-off. 3- **A hybrid residual space detector to instantiate the protocol on SWaT.** We use a normal-trained GRU predictor to generate multi-horizon residual evidence and a supervised RF head to learn a decision rule in residual space, providing an interpretable and reproducible testbed for studying calibration and eventization under a chronological split. 4-

Baselines in matched protocol and label access. We include supervised baselines trained under the same day-wise split (RF-RAW, Ridge/Logistic-RAW, GRU-CLS), a normal-only baseline (GRU-AE) and a twin-only residual-energy baseline, all evaluated with the same NP calibration and eventization to ensure comparisons reflect operating-point selection and alarm semantics rather than temporal leakage or inconsistent label access.

The remainder of the paper is organized as follows. Section 2 summarizes related work. Section 3 presents our GRU-twin residual pipeline, NP-style quantile calibration, and the eventization policy used to produce operator-facing alarms. Section 4 evaluates the approach on SWaT under a strict day-wise protocol using both pointwise (AUC/AP) and event-level workload metrics. Section 5 concludes.

2 Related Work

ICS anomaly and attack detection spans classical model/observer residual methods, learning-based sequence detectors, and digital-twin forecasting. Digital twins range from physics-based and hybrid models to purely data-driven predictors used for monitoring and decision support [6,17,22]. In ICS settings, RNN, GRU and LSTM forecasters trained in nominal operation are commonly paired with residual or reconstruction-error scoring and then thresholded for alarms; this aligns with classical fault diagnosis and attack-detection formulations based on residual generation [16]. SWaT studies confirm that prediction residuals are informative for cyber–physical attacks [7,8,10,13]. However, many reported pipelines directly threshold residual magnitudes or rely on tuned cutoffs (e.g., best-F_1 on a test split) and temporally entangled evaluation protocols, which can confound generalization with post hoc operating-point selection. In contrast, we adopt a strict chronological (day-wise) protocol and treat thresholding and post-processing as explicit design choices, reporting operator-facing event metrics in addition to pointwise ranking measures.

Selecting thresholds as score quantiles on negative data is closely related to quantile calibration and conformal prediction, which provide finite-sample coverage guarantees under the assumptions of exchangeability or that samples are independent and identically distributed [1,20,23]. Extensions to time series (e.g., blocking, weighted, or covariate-shift adaptations) highlight that dependence and nonstationarity make the calibration background consequential [21,24]. We therefore take an NP-style engineering perspective: the operating point is defined as an empirical quantile of scores on a *designated negative stream*, and we empirically audit realized exceedance and workload when calibrating on normal-only negatives versus background-matched early-deployment negatives (using out-of-bag scoring to reduce optimistic bias). Separately, alarm management practice evaluates incidents as *events* rather than isolated threshold crossings, using persistence, deadbands/hysteresis, suppression and aggregation to prevent alarm floods, with workload tracked by alarm rates and occupancy [5,11].

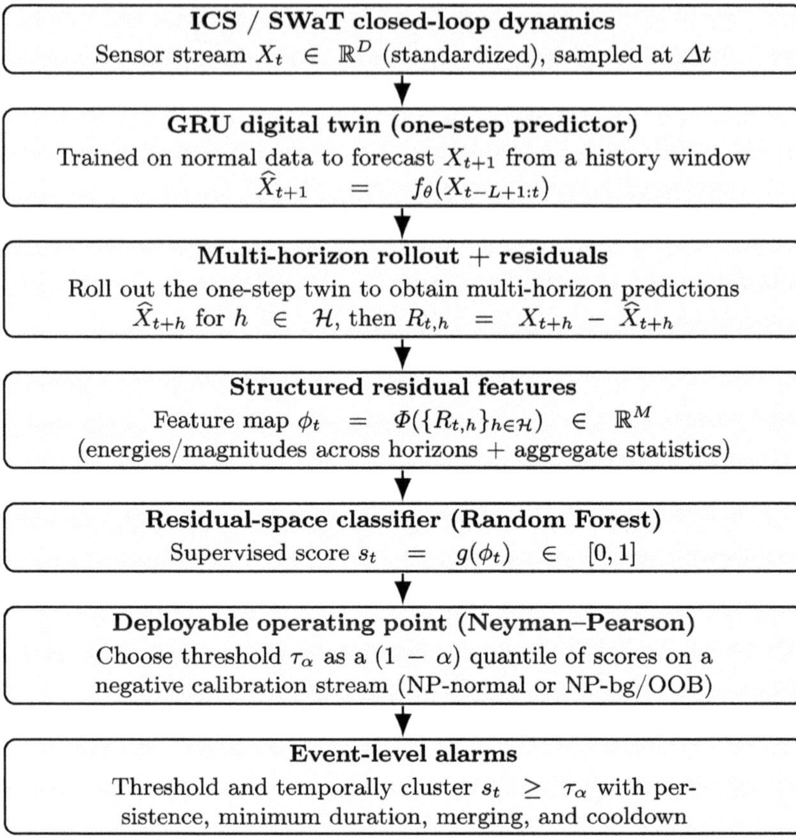

Fig. 1. Formalization of the hybrid detector as a residual-based pipeline with deployable calibration and eventization.

Consistent with these principles, we apply an explicit persistence/minimum-duration/merge/cooldown eventization policy and report false events/day and occupancy alongside AUC/AP and precision/recall, linking deployable calibration to operator-facing workload.

3 Methodology

We formalize the proposed detector as an observer–residual–decision pipeline: a learned predictor (digital twin) produces multi-step forecast residuals, residuals are compressed into structured features, and a supervised classifier assigns an anomaly score that is converted to alarms via an auditable operating-point rule. Figure 1 summarizes the data flow implemented in our code.

The primary contribution of this work is *operationalization* rather than a new classifier architecture. *NP-style calibration* to select operating thresholds by an explicit exceedance budget on a designated negative calibration stream (avoiding test-label tuning), (ii) *eventization* to transform sample-wise threshold exceedances into operator-facing alarm events via a fixed policy (persistence, minimum duration, merge gap, cooldown), and (iii) a *workload audit* that reports interpretable rates (false events/day, alarms/day, occupancy) alongside conventional ranking metrics (AUC/AP) and event recall. We use a hybrid GRU-twin + RF detector as a representative instantiation of this protocol, but the calibration/eventization/workload machinery is model-agnostic and can be applied to alternative scoringfunctions (e.g., residual-energy scores or normal-only autoencoders) under the same temporal split.

3.1 Setup

Let $X_t \in \mathbb{R}^D$ denote the standardized sensor vector at time t sampled at interval Δt (in SWaT, $\Delta t \approx 1\,\text{s}$). We test

$$H_0 : \text{normal operation,} \tag{1}$$

$$H_1 : \text{attack/abnormal operation.} \tag{2}$$

We assume that the conditional distribution of future trajectories differs between hypotheses. For a history length L and a finite horizon set $\mathcal{H}$ be

$$X_{t+h} \mid X_{t-L+1:t} \sim \begin{cases} P_0(\cdot \mid X_{t-L+1:t}), & H_0, \\ P_1(\cdot \mid X_{t-L+1:t}), & H_1 . \end{cases}$$

for all $h \in \mathcal{H}$. In our SWaT runs, we use $L = 10$ and $\mathcal{H} = \{1, 5, 15\}$; after sensor alignment, D is the number of shared numeric channels (e.g. $D = 51$ in the reported run).

3.2 Digital Twin and Multi-horizon Residuals

The GRU twin is trained on normal data to minimize one-step mean-squared error:

$$\widehat{X}_{t+1} = f_{\theta^\star}(X_{t-L+1:t}), \qquad \theta^\star \in \arg\min_{\theta} \mathbb{E}_{H_0}\left[\|X_{t+1} - f_\theta(X_{t-L+1:t})\|_2^2\right]. \tag{3}$$

To obtain predictions at multiple horizons, we roll out the one-step model (iteratively feeding predicted states) to produce $\widehat{X}_{t+h}$ for each $h \in \mathcal{H}$. The residuals are then

$$R_{t,h} := X_{t+h} - \widehat{X}_{t+h} \in \mathbb{R}^D, \qquad h \in \mathcal{H}. \tag{4}$$

It is convenient to view the collection of multi-horizon residuals as a tensor $R_t := \{R_{t,h}\}_{h\in\mathcal{H}} \in \mathbb{R}^{D\times|\mathcal{H}|}$, and its vectorization as

$$r_t := \text{vec}(R_t) \in \mathbb{R}^{DK}, \qquad K := |\mathcal{H}|. \tag{5}$$

Working Assumptions. (i) (*Nominal boundedness*) Under H_0, the residuals are approximately centered and have a bounded scale (relative to standardization). (ii) (*Detectable deviation*) Under H_1, at least one horizon exhibits a persistent change in residual magnitude/energy over a nontrivial interval.

3.3 The Residual Space is Natural

For fixed $(X_{t-L+1:t}, f_{\theta^\star})$, each future value can be written as $X_{t+h} = \widehat{X}_{t+h} + R_{t,h}$ for $h \in \mathcal{H}$. Thus, compared to a fixed trained twin, anomaly detection can be analyzed in the residual domain R_t rather than directly in the observation domain. We use this observation only to motivate residual-based monitoring; the detector does *not* retain the full residual tensor and, therefore, does not claim losslessness or sufficiency.

3.4 Structured Residual Features: a Deliberately Lossy Summary

In practice, we do not feed the full residual tensor R_t to the classifier. Instead, we compute a finite-dimensional summary

$$\phi_t = \Phi(R_t) = \Phi(\{R_{t,h}\}_{h\in\mathcal{H}}) \in \mathbb{R}^M, \tag{6}$$

where Φ aggregates per-sensor and cross-horizon statistics designed to capture common ICS attack signatures: distributed drifts (energy growth), localized manipulations (peaks) and horizon-dependent deviations (rollout error that grows with h). This representation is intentionally *lossy* (not injective), chosen to reduce variance and stabilize learning; any optimality discussion is therefore *relative to the induced feature space.*

A canonical set of components (all computable from $\{R_{t,h}\}_{h\in\mathcal{H}}$) includes:

(Per-horizon energy)
$$E_t^{(h)} := \|R_{t,h}\|_2^2, \quad h \in \mathcal{H}, \tag{7}$$

(Global multi-horizon energy)
$$E_t := \sum_{h\in\mathcal{H}} \|R_{t,h}\|_2^2, \tag{8}$$

(Per-channel across-horizon energy)
$$E_{t,i} := \sum_{h\in\mathcal{H}} \left(R_{t,h}[i]\right)^2, \quad i = 1,\ldots,D, \tag{9}$$

(Cross-horizon peak statistics)
$$M_{t,i} := \max_{h\in\mathcal{H}} |R_{t,h}[i]|, \quad M_t := \max_{1\le i\le D} M_{t,i}, \tag{10}$$

(Aggregate summaries)
$$\bar{E}_t := \frac{1}{D}\sum_{i=1}^{D} E_{t,i}, \quad E_t^{\max} := \max_{1\le i\le D} E_{t,i},$$
$$E_t^{\text{top-}k} := \sum_{i\in \text{Top}k(E_{t,1:D})} E_{t,i}. \tag{11}$$

In our implementation, Φ instantiates the energy/magnitude features across $\mathcal{H}$ together with a small set of aggregate statistics, generating M on the order of a few hundred dimensions.

Design Rationale for Φ: Residual energies and magnitudes are standard when attack directions are unknown: they detect both distributed drifts (energy growth) and localized manipulations (peaks on a subset of channels). We therefore summarize multi-horizon residuals using per-horizon energies, per-channel energies across horizons, and simple order statistics (max/top-k). This is intentionally lossy to reduce variance and stabilize the residual-space classifier; its utility is empirically validated by ablation Φ in Table 3.

Twin-Only Residual-Energy Baseline: Because Eq. (8) already defines a natural scalar anomaly score, we also evaluate a reviewer-expected *residual* energy baseline that sets $s_t^{\text{eng}} := \sum_{h\in\mathcal{H}} \|R_{t,h}\|_2^2$ (or $s_t^{\max} := \max_{h\in\mathcal{H}} \|R_{t,h}\|_2^2$) and then applies the *same* NP calibration and the *same* eventization policy as the hybrid Twin+RF score. The results are reported in Sect. 4.2.

3.5 Supervised Scoring in the Induced Feature Space

Let Q_0 and Q_1 denote the distributions of ϕ_t under H_0 and H_1. The Bayes-optimal rule in the ϕ-space is

$$T^\star(\phi) = \mathbf{1}\left[\Pr(Y = 1 \mid \phi) \geq \tfrac{1}{2}\right]. \tag{12}$$

We learn a score $s_t = g(\phi_t) \in [0, 1]$ using a Random Forest classifier. We do *not* claim global optimality for the original time-series problem; instead, forests provide a practical nonparametric mechanism for learning nonlinear decision boundaries in the chosen residual feature space. This choice is further motivated by theoretical analyzes that establish consistency for certain randomized forest constructions under explicit assumptions and model restrictions (e.g. [2,3,19]). We validate the generalization empirically using chronological (day-wise) splits.

3.6 NP-Style Quantile Calibration for Deployable Operating Points

A central design choice is that the operating point is not selected by maximizing the set of tests F_1, but by imposing an explicit exceedance *budget* on a designated negative calibration stream. This yields a reproducible and auditable rule for choosing thresholds without using attack labels from the evaluation period.

Let $\mathcal{C}$ be a set (or stream) of negative calibration indices, and let $\{s_i\}_{i\in\mathcal{C}}$ be their anomaly scores. For a target *design* exceedance budget $\alpha \in (0, 1)$, we define the calibrated threshold

$$\tau_\alpha \;=\; Q_{1-\alpha}(\{s_i : i \in \mathcal{C}\}), \tag{13}$$

i.e., the empirical $(1 - \alpha)$ quantile of negative calibration scores. Under i.i.d. or exchangeable negatives, (13) can be interpreted as an approximate level-α type-I operating point. In our time-series setting, the negatives are temporally dependent and can shift over days; therefore, we do *not* treat (13) as ensuring type-I test-time control. Instead, α serves as an operational budgeting parameter anchored to the chosen negative background, and we report the exceeded behavior realized in disjoint negative evaluation periods.

NP-Normal vs. NP-bg/OOB. We consider two choices of negative calibration background: (i) **NP-normal**, where $\mathcal{C}$ is a held-out normal-only segment (*normal-cal*); and (ii) **NP-bg**, where $\mathcal{C}$ is the negative subset of an early deployment background (e.g., early attack-period negatives). To reduce optimism when $\mathcal{C}$ overlaps the RF training period, NP-bg is calculated using out-of-bag (OOB) scores on those negatives.

Audit: Realized Exceedance on Disjoint Negatives. To quantify how calibration transfers under dependence and background shift, we audit the *realized* exceedance rate on a disjoint negative evaluation set $\mathcal{E}$ that is not used to set τ_α:

$$\widehat{\alpha}_{\mathcal{E}}(\tau_\alpha) = \frac{1}{|\mathcal{E}|}\sum_{t\in\mathcal{E}} \mathbf{1}[s_t \geq \tau_\alpha]. \tag{14}$$

We report both the design budget α and the realized rate $\widehat{\alpha}_{\mathcal{E}}$, which acts as a background-shift diagnostic rather than a "violated guaranty" in the nonstationary setting.

Block-Level Audit for Temporally Dependent Streams. Because scores are temporally dependent, we optionally calibrate and audit operating points on block-aggregated scores to align false-alarm budgeting with operator cadence. For non-overlapping blocks $\{\mathcal{B}_j\}$ of length W seconds, define $b_j = \max_{t\in\mathcal{B}_j} s_t$ (default). We set a block threshold by $\tau^{(W)}_{\alpha_{\text{blk}}} = Q_{1-\alpha_{\text{blk}}}(\{b_j : j \in \mathcal{C}_{\text{blk}}\})$ and report the observed exceedance rate on disjoint negative blocks as a drift diagnostic. The event workload (false events/day, occupancy) is calculated after applying the fixed eventization policy (Sect. 3.7).

3.7 From Pointwise Alarms to Events

Eventization. Given timestamps $\{\theta_t\}_{t=1}^T$ and anomaly scores $\{s_t\}_{t=1}^T$, with a calibrated threshold τ, we form a binary exceedance stream $z_t = \mathbf{1}[s_t \geq \tau]$. To reduce chatter, we apply a persistence rule: an alarm is eligible at time t only if at least k of the most recent m samples exceedances, i.e., $\sum_{i=\max(1,t-m+1)}^{t} z_i \geq k$. Maximum contiguous runs of eligible samples define candidate alarm intervals, which are filtered to enforce a minimum duration of samples $\ell_{\min}$. Candidate intervals separated by gaps of at most g samples are merged. Finally, we apply a cooldown of the c samples by merging any event whose start occurs within the c samples of the previous event end, reflecting the operator's handling time. The resulting disjoint intervals constitute the event-level alarm stream used for event recall and workload metrics (alarms/day, false alarms/day, occupancy).

Parameters. Eventization is parameterized by $(k, m, \ell_{\min}, g, c)$, corresponding to persistence, persistence window, minimum event length, merge gap, and cooldown, respectively; in experiments, we convert fixed durations in seconds to samples using the sampling interval Δt (Sect. 3.1).

Pointwise vs. Block Scores. Unless stated otherwise, eventization is applied to pointwise scores s_t. When using block-aggregated scores b_j (e.g., $b_j = \max_{t\in B_j} s_t$ for non-overlapping W-second blocks), we apply the same procedure to (T_j, b_j) using block-end timestamps T_j and interpret $(m, \ell_{\min}, g, c)$ in seconds.

4 Case Study

The Secure Water Treatment (SWaT) testbed is a scaled-down but fully operational industrial water treatment plant constructed for cyber–physical security research. The plant comprises six interconnected stages: raw water storage, ultra-filtration, chemical dosing, reverse osmosis, and two clean-water storage/distribution stages, each instrumented with PLC-controlled actuators (pumps and valves) and sensors measuring level, flow, pressure, and conductivity. Sensors and actuators are networked through industrial fieldbuses and monitored through an SCADA/HMI.

We use the public SWaT dataset consisting of a multi-day *normal* log followed by a multi-day *attack* log containing time-stamped, labeled cyberphysical attacks (e.g. sensor spoofing, actuator manipulation, and coordinated multi-stage attacks) [7,13]. This structure supports a deployable protocol: learning nominal dynamics from the normal-only period and evaluating detection on temporally held-out attack days.

4.1 Experimental Setup

While the GRU digital twin is trained exclusively in normal operation, the downstream Random Forest (RF) head is trained using labeled samples from the early attack days in the designated training period. Accordingly, our method is best viewed not as purely unsupervised anomaly detection, but as a *hybrid residual space attack detector*: the twin provides a normal-trained, process-consistent residual representation, and the RF learns an attack-vs-normal decision rule in that residual space. Our *deployability* claim therefore concerns the *operating-point selection and alarm semantics*, not label-free learning: thresholds are set by an NP-style negative-quantile rule on a designated calibration stream without tuning on held-out evaluation attack labels, and alarms are reported as operator-facing events using a fixed eventization policy (persistence/min-length/merge/cooldown). We automatically detect the timestamp column and the attack label column in the attack log (`Normal/Attack`), strip header/column whitespace, and retain the intersection of numeric channels shared between the normal and attack logs. In this run, this yields **51 shared sensor channels**. All channels are standardized using mean and variance computed from the normal log only. The sampling interval is approximately $\Delta t \approx 1$ s.

We train a GRU-based predictor [4] f_θ on the normal log for next-step forecasting. Using overlapping history windows of length $L = 10$, the model predicts the next state $\hat{x}_{t+1} = f_\theta(x_{t-L+1:t})$ and is trained with mean-squared error using Adam [12] for 10 epochs (batch size 256). Residuals are computed as

$$R_{t,h} = x_{t+h} - \hat{x}_{t+h}, \qquad h \in \mathcal{H}, \tag{15}$$

where $\hat{x}_{t+h} = f_\theta(x_{t-L+1:t})_h$ denotes the h-step-ahead prediction produced by the normal-trained multi-horizon twin f_θ from the length-L history window. We denote the residual tensor at time t by

$$R_t = \{R_{t,h}\}_{h\in\mathcal{H}} \in \mathbb{R}^{|\mathcal{H}|\times D}. \tag{16}$$

After timestamp alignment and channel intersection, we use 51 channels with 495,000 normal samples and 449,919 attack-period samples at $\Delta t \approx 1$s, and a strict day-wise split (Dec. 28–30 train, Dec. 31–Jan. 2 test) with seed 123.

Residual Feature Map and Hybrid Classifier. We construct $\phi_t \in \mathbb{R}^{273}$ as in Sect. 3 and train an RF classifier with `n_estimators`=300, `max_samples`=0.632, `min_samples_leaf`=5 and `class_weight`=`balanced_subsample`, producing $s_t = \Pr(y_t = 1 \mid \phi_t)$.

Chronological Split. We split the normal data into N-fit (first 80%) for negative RF training and N-cal (last 20%) for NP-normal calibration. We split the attack log day-wise: Dec. 28–30 for RF training and Dec. 31–Jan. 2 for held-out evaluation. N-cal is never used for RF training.

NP-Normal Operating Point and Negative Stream for False Alarms. We set a Neyman–Pearson operating point at the design level $\alpha = 10^{-2}$ by calibrating the threshold τ_α on the *normal calibration stream* N-cal: $\tau_\alpha = Q_{1-\alpha}(s_t \mid t \in \text{N-cal})$. Unless otherwise stated, realized false-alarm rates are reported with respect to N-cal, i.e. $\text{FPR}_{\text{N-cal}}(\tau) = \Pr(s_t \geq \tau \mid t \in \text{N-cal})$. As a diagnostic for background shift, we also report the fraction of labeled-normal samples in the A-test that exceed τ_α ("test-neg"), but this does not define the NP calibration.

Robustness Check. We also ran a strictly forward-in-time cross-fit that separates twin training, RF training, NP calibration, and negative auditing into disjoint blocks; the same trend for NP-normal vs. NP-bg workload persists.

For NP-normal, E is the held-out normal-eval block; for NP-bg, E is the held-out bg-E block from the background day.

4.2 Results on Held-Out Attack Days

Table 1 reports pointwise scoring. In addition to pointwise scoring, we audit a block-level mode that aggregates per-second scores into non-overlapping B=10 s blocks via $s_j^{\text{blk}} = \max_{t\in\text{block } j} s_t$ and applies an NP-calibrated block threshold τ_{blk}. Because block aggregation changes the score distribution and threshold-crossing dynamics, the induced alarm-event stream (and thus workload) is not directly comparable to pointwise results; In this block mode, NP-normal yields $\tau_{\text{blk}} = 0.141545$ with 3.048 total events/day and 0.762 false events/day, while NP-bg/OOB yields $\tau_{\text{blk}} = 0.216375$ with 1.143 total events/day and 0.000 false

Table 1. Pointwise performance and event-level workload on held-out attack-test days.

Operating point	τ	AUC	AP	F1	evtRec	false/day	total/day
RF NP-norm	0.1059	0.921	0.90	0.867	0.429	1.524	3.810
RF NP-bg/OOB	0.1279	0.921	0.90	0.874	0.429	0.381	2.667
Eng-SUM NP-bg	calib.	0.677	0.27	0.044	0.429	24.764	26.669
Eng-MAX NP-bg	calib.	0.679	0.27	0.044	0.429	25.526	27.431

Note. For energy baselines the NP-calibrated threshold τ is on the raw residual-energy scale and is therefore not numerically comparable to the RF score scale; we report it as "calib." and focus on the induced event-level workload.

events/day on held-out days. *False/day* is computed on the held-out attack-test days as the number of predicted alarm events with zero temporal overlap with any labeled attack interval, normalized by the evaluation duration in days:

$$\text{false/day} \;=\; \frac{1}{T_{\text{eval}}} \sum_{\hat{e}\in\hat{\mathcal{E}}} \mathbf{1}(\forall a \in \mathcal{A} :\; |\hat{e} \cap a| = 0)\,, \tag{17}$$

where $\hat{\mathcal{E}}$ is the set of predicted alarm events after eventization, $\mathcal{A}$ is the set of ground-truth attack intervals, and T_{eval} is the evaluation span in days.

Simple residual-energy thresholding is substantially weaker than learning a decision rule in the residual space. Under identical calibration and eventization, residual-energy scores exhibit poor ranking quality (AUC $\approx$ 0.68, AP $\approx$ 0.27) compared to Twin+RF (AUC $\approx$ 0.92, AP $\approx$ 0.90). More importantly, in the recall of the matched events in this split (evtRec = 0.429), the energy baselines incur dramatically higher operator workload (24–26 false events/day) than Twin+RF with NP-bg/OOB (0.381 false events/day), indicating that RF exploits residual structure beyond aggregate energy and yields a materially better workload–coverage trade-off under temporal generalization.

Comparison of NP-normal across the two granularities illustrates the effect of score aggregation: block-max scoring calibrates a higher threshold in the block domain (τ_{blk}) and yields a different alarm-event stream than pointwise thresholding (Table 1, τ_α), which can materially change false reported /day and total alarms/day even when ranking performance (AUC/AP) is similar. Within a *fixed* granularity, NP-bg/OOB provides an auditable operating-point selection that does not use held-out evaluation attack labels: it calibrates τ using a designated negative calibration background (here, OOB scores on attack-train negatives) and can reduce false/day (e.g. Table 1: 1.524→0.381) while maintaining comparable event recall (evtRec) in this split. Accordingly, we do not interpret cross-table workload differences as contradictions; they arise from different operational granularities, whereas within-table comparisons isolate the effect of the calibration background under a fixed granularity.

Table 2. Baseline performance.

Method	τ_α	Prec	Rec	F1	AUC	AP	Event Rec.	False/day
RF-RAW	0.203333	0.543	0.886	0.673	0.917	0.889	0.929	20.57
GRU-CLS	0.976113	0.743	0.033	0.064	0.936	0.861	0.214	0.38
GRU-AE(err)	0.055924	0.669	0.873	0.758	0.933	0.884	0.929	32.76
Ridge-RAW	0.222929	0.979	0.307	0.467	0.596	0.548	0.429	0.00

4.3 Supervised and Normal-Only Baselines

To make the learning regime explicit, we separate *normal-only unsupervised* baselines from *supervised* detectors. The GRU autoencoder (GRU-AE) is trained on normal data only and serves as a normal-only unsupervised reference. In contrast, the hybrid twin+RF and the supervised baselines in Table 2 are trained using labeled attack samples from the early days of attack. All methods are evaluated under the same temporal split and are calibrated to deployable operating points using the same NP-style thresholding and the same eventization procedure. For each baseline, we calibrate τ_α on the same negative stream (NP-normal unless otherwise stated) and apply identical eventization parameters $(k, m, \ell_{\min}, g, c)$.

We therefore evaluate four baselines using the same chronological split and the same NP-normal thresholding rule ($\alpha = 10^{-2}$, calibrated on normal-cal negatives):

- **RF-RAW:** a Random Forest trained on raw lag features (history taps) augmented with simple deltas and window statistics, capturing short-term temporal structure without an explicit twin.
- **Ridge-RAW** [9]: a regularized linear classifier trained on the same raw lag feature map, representing a strong linear baseline with low variance.
- **GRU-CLS:** a supervised GRU sequence classifier trained directly on raw sequences, producing a probability score.
- **GRU-AE (normal-only):** a GRU autoencoder trained on normal data only; anomaly scores are reconstruction errors. This baseline represents a deployable unsupervised regime without attack labels.

RF-RAW and GRU-AE achieve high event recall, but do so with substantially higher false-alarm rates, indicating that their high sensitivity translates into heavy operator burden under the NP-normal operating point. GRU-CLS attains high ranking performance but is poorly calibrated for NP-thresholding in this run (very low recall at the $\alpha = 10^{-2}$ threshold), yielding few alarms but missing most attacks. Ridge-RAW provides a conservative alternative with near-zero false-alarm events but significantly reduced recall and weaker ranking (AUC/AP), consistent with limited linear expressiveness. Overall, the proposed hybrid residual-space classifier provides a stronger balance of pointwise accuracy and deployable event workload: it preserves high discrimination (AUC/AP)

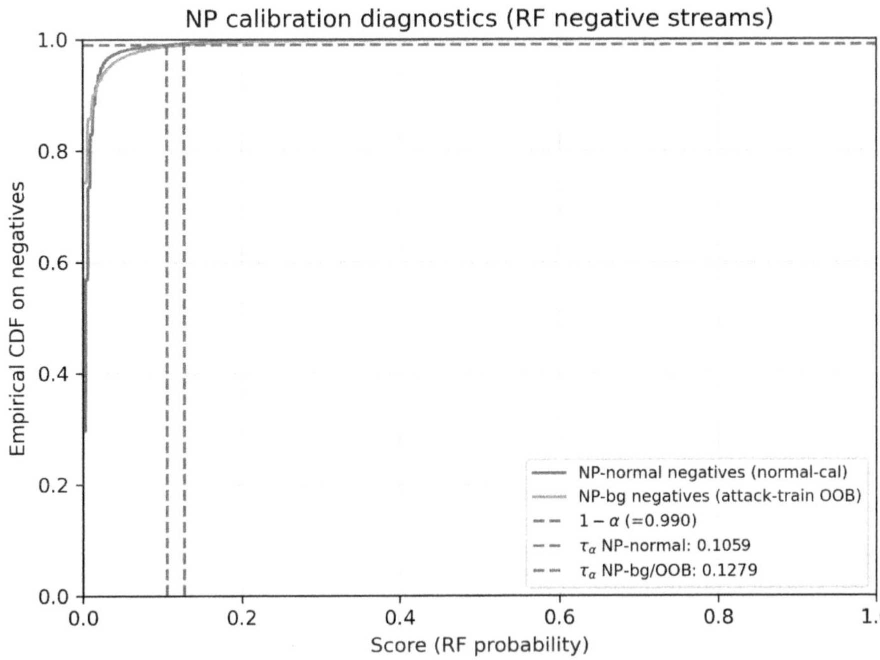

Fig. 2. NP calibration diagnostics for the Twin+RF score on the two negative calibration streams. We plot the empirical CDF of scores on (i) held-out normal calibration windows (NP-normal) and (ii) early-deployment attack-train negatives scored out-of-bag (NP-bg/OOB). The horizontal dashed line marks the target quantile $1 - \alpha$ ($\alpha = 10^{-2}$); the vertical dashed lines show the resulting thresholds τ_α on each stream. The shift in τ_α explains why NP-normal and NP-bg/OOB produce different alarm workloads under the same eventization policy.

while maintaining a substantially lower false-alarm rate than the most sensitive baselines.

Figure 2 visualizes the selection of the NP threshold on the two negative streams and highlights how the choice of the calibration background shifts τ_α, which in turn changes the workload at the event-level.

Feature-Map Φ Ablation. Table 3 ablates the residual feature map $\phi_t = \Phi(\{R_{t,h}\}_{h\in\mathcal{H}})$ while maintaining the GRU twin, the RF hyperparameters, the day-wise train/test split and the eventization policy fixed. Residuals are computed for horizons $\mathcal{H} = \{1, 5, 15\}$ and then summarized by Φ using families of statistics designed to capture (i) *localized spikes* (per-sensor magnitudes) and (ii) *distributed drifts* (per-sensor energy and horizon-level aggregates). Concretely, our `full` feature map concatenates per-sensor per-horizon $|r|$ and r^2, plus a small set of horizon-wise aggregates (e.g., max/mean and ℓ_2 energy). The `noG` variant removes global/horizon-wide summary terms, retaining only per-sensor statistics; `Eonly` keeps only energy-like terms (e.g., r^2-based features); `h1`

Table 3. Ablation of the residual feature map Φ (NP-bg/OOB calibration, $\alpha = 10^{-2}$). Evaluation on held-out attack-test days (Dec 31–Jan 2).

Φ variant	dim	τ	AUC	AP	Prec.	Rec.	F1	evtRec	false/day
ext	359	0.136919	0.920	0.858	0.939	0.815	0.873	0.571	1.52
full	315	0.138335	0.910	0.826	0.940	0.813	0.872	0.500	1.14
noG	306	0.140626	0.917	0.842	0.952	0.790	0.864	0.429	0.76
h1	105	0.163861	0.912	0.872	0.936	0.801	0.863	0.429	1.52
Eonly	153	0.143288	0.926	0.846	0.943	0.795	0.863	0.571	0.76

restricts Φ to a single horizon ($h = 1$); and `ext` expands `fully` with additional summary statistics intended to improve separability under heterogeneous attack manifestations.

Deployable Calibration and Reported Metrics. For each Φ variant we train the RF on normal-fit negatives and attack-train samples, then set a deployable threshold τ using the Neyman–Pearson (NP) calibration with *background* matched negatives: specifically, τ is the empirical $(1 - \alpha)$ quantile of *out-of-bag* (OOB) scores on the attack-train negative subset (NP-bg/OOB), with $\alpha = 10^{-2}$. Pointwise precision/recall/F1 are computed on held-out attack-test days. Event-level metrics are computed by thresholding scores at τ and applying a fixed persistence-and-merging alarm policy; we report event recall (evtRec, the fraction of ground-truth attack episodes overlapped by at least one alarm event) and false events/day (alarms not overlapping any ground-truth attack, normalized by test duration). Overall, the ablation indicates that performance is relatively stable across reasonable summaries, while using only a single horizon (`h1`) or removing global summaries (`noG`) can reduce event-level coverage.

We use a default event policy with persistence $(k, m) = (20, 60)$ s, minimum event length 60 s, merge gap 300 s, and cooldown 300 s. For a stream with negative-only that is disjoint from the training, we sweep the decision threshold τ and select $\tau^\star$ as the most sensitive threshold for achieving ≤ 5 false alarm events/day. We then report the recall and occupancy of events on held-out attack days at $\tau^\star$.

5 Conclusion and Future Work

We presented a hybrid digital-twin detector for the SWaT testbed that couples a normal-trained gated recurrent unit (GRU) predictor with structured multi-horizon residual features and a residual-space Random Forest (RF) score. The central contribution is operationalization for deployment: operating points are set by a Neyman–Pearson (NP) quantile rule on a specified negative calibration stream (without tuning on held-out attack days), and pointwise exceedances are converted into operator-facing alarm *events* via a fixed persistence/minimum-duration/merge/cooldown policy. Under a strict day-wise temporal protocol with held-out attack days, the detector achieves strong ranking performance (AUC = 0.921, AP = 0.896), and substantially outperforms a residual-energy baseline. At deployable operating points, NP-normal calibration yields event recall 0.429 with 1.524 false events/day (occupancy 0.157), while background-matched calibration using out-of-bag scores on early-deployment negatives reduces nuisance workload to 0.381 false events/day (occupancy 0.152) at the same event recall.

Three lessons emerge. First, pointwise separability (AUC/AP) does not determine deployable behavior because threshold selection and eventization dominate operator workload. Second, the calibration background must be stated explicitly: negative score distributions shift across days and regimes, and this shift directly changes the induced false-event rate under a fixed event policy. Third, event post-processing can suppress short or intermittent attacks even when pointwise recall is high, underscoring the need to report event-level metrics alongside sample-wise metrics.

Future work will focus on drift-robust calibration for nonstationary ICS telemetry and on event-level error control that targets workload directly (e.g. false events/day and occupancy), as well as adaptive eventization policies that better preserve detection of brief attack episodes without increasing nuisance alarms.

References

1. Angelopoulos, A.N., Bates, S., et al.: Conformal prediction: a gentle introduction. Found. Trends® Mach. Learn. **16**(4), 494–591 (2023)
2. Biau, G.: Analysis of a random forests model. J. Mach. Learn. Res. **13**(1), 1063–1095 (2012)
3. Breiman, L.: Random forests. Mach. Learn. **45**(1), 5–32 (2001)
4. Chung, J., Gulcehre, C., Cho, K., Bengio, Y.: Empirical evaluation of gated recurrent neural networks on sequence modeling (2014). https://arxiv.org/abs/1412.3555
5. Engineering Equipment and Materials Users' Association (EEMUA): Eemua publication 191: Alarm systems – a guide to design, management and procurement. EEMUA Guidance Publication (2024). ISBN 978-0-85931-243-1
6. Fuller, A., Fan, Z., Day, C., Barlow, C.: Digital twin: enabling technologies, challenges and open research. IEEE Access **8**, 108952–108971 (2020)
7. Goh, J., Adepu, S., Junejo, K.N., Mathur, A.P.: A dataset to support research in the design of secure water treatment systems. In: Critical Information Infrastructures Security (2016). https://api.semanticscholar.org/CorpusID:3075307
8. Goh, J., Adepu, S., Tan, M., Lee, Z.S.: Anomaly detection in cyber physical systems using recurrent neural networks. In: IEEE 18th International Symposium on High Assurance Systems Engineering (HASE), pp. 140–145. IEEE (2017)
9. Hoerl, A.E., Kennard, R.W.: Ridge regression: biased estimation for nonorthogonal problems. Technometrics **12**(1), 55–67 (1970). http://www.jstor.org/stable/1267351
10. Inoue, J., Yamagata, Y., Chen, Y., Poskitt, C.M., Sun, J.: Anomaly detection for a water treatment system using unsupervised machine learning. In: 2017 IEEE International Conference on Data Mining Workshops (ICDMW), pp. 1058–1065. IEEE (2017)

11. International Society of Automation (ISA): Ansi/isa-18.2-2016: Management of alarm systems for the process industries. ANSI/ISA Standard, American National Standard, approved March 17, 2016 (2016)
12. Kingma, D.P., Ba, J.: Adam: a method for stochastic optimization (2017). https://arxiv.org/abs/1412.6980
13. Mathur, A.P., Tippenhauer, N.O.: Swat: a water treatment testbed for research and training on ICS security. In: 2016 International Workshop on Cyber-Physical Systems for Smart Water Networks (CySWater), pp. 31–36 (2016). https://doi.org/10.1109/CySWater.2016.7469060
14. Mehrpouyan, H.: Fault tolerance vs. attack detection in industrial control systems: a deep learning approach. In: 2025 IEEE International Conference on Cyber Security and Resilience (CSR), pp. 556–561 (2025). https://doi.org/10.1109/CSR64739.2025.11130080
15. O'Toole, S., Mehrpouyan, H.: Towards cyber-physical representation and cyber-resilience against attack and failure within a hydraulic network simulation toolkit. In: 2024 IEEE Security and Privacy Workshops (SPW), pp. 246–252 (2024). https://doi.org/10.1109/SPW63631.2024.00029
16. Pasqualetti, F., Dörfler, F., Bullo, F.: Attack detection and identification in cyber-physical systems. IEEE Trans. Autom. Control **58**(11), 2715–2729 (2013)
17. Qi, Q., Tao, F., Zuo, Y., Zhao, D.: Digital twin service towards smart manufacturing. Procedia CIRP **72**, 237–242 (2018)
18. Rasapour, F., Serra, E., Mehrpouyan, H.: Framework for detecting control command injection attacks on industrial control systems (ICS). In: Seventh International Symposium on Computing and Networking (CANDAR), pp. 211–217 (2019). https://doi.org/10.1109/CANDAR.2019.00035
19. Scornet, E., Biau, G., Vert, J.P.: Consistency of random forests (2015)
20. Shafer, G., Vovk, V.: A tutorial on conformal prediction. J. Mach. Learn. Res. **9**(3) (2008)
21. Stankeviciute, K., Alaa, A.M., van der Schaar, M.: Conformal time-series forecasting. In: Ranzato, M., Beygelzimer, A., Dauphin, Y., Liang, P., Vaughan, J.W. (eds.) Advances in Neural Information Processing Systems, vol. 34, pp. 6216–6228. Curran Associates, Inc. (2021)
22. Tao, F., Zhang, M.: Digital twin shop-floor: a new shop-floor paradigm towards smart manufacturing. IEEE Access **5**, 20418–20427 (2017)
23. Vovk, V., Gammerman, A., Shafer, G.: Algorithmic Learning in a Random World. Springer (2005)
24. Zaffran, M., Féron, O., Goude, Y., Josse, J., Dieuleveut, A.: Adaptive conformal predictions for time series. In: International Conference on Machine Learning, pp. 25834–25866. PMLR (2022)

Spectre Attacks Detection in RISC-V Processors Using Hardware-Level Machine Learning Models

Mustafa Şirin(✉) and Ebu Yusuf Güven

Computer Engineering, Istanbul University-Cerrahpasa, Istanbul, Turkey
{mustafa.sirin,eyguven}@iuc.edu.tr

Abstract. Modern processors used features such as speculative execution and branch prediction to enhance performance; however, these mechanisms leave microarchitectural traces when mispredictions occur, enabling attacks such as Spectre. The Spectre attack has had a significant impact on the hardware security domain and has emerged as a vulnerability affecting almost all contemporary processors. To mitigate its effects, various methods have been developed to detect and defend against Spectre attacks. Owing to their capability to observe hardware-level events, hardware performance counters (HPC) stand out as a valuable resource for detecting Spectre attacks. In this study, we investigate the detection of Spectre-v2 attacks using HPC-based machine learning models. With our models trained with data collected on RISC-V architecture, we achieved Spectre attack detection with over 99% accuracy. Our findings demonstrate that traces of Spectre-v2 are distinguishable within HPC data and that HPC-based models are effective in attack detection.

Keywords: Hardware Attacks · Spectre Detection · Hardware Performance Counters · Attack Detection

1 Introduction

To enable modern processors to operate faster and more efficiently, hardware developers have implemented microarchitectural optimizations in processors to achieve high performance. While techniques such as speculative execution and branch prediction enhance processor performance, the speculative execution mechanism irreversibly alters the microarchitectural state following prediction errors, leading to serious security vulnerabilities [6]. The Spectre attack, which exploits this microarchitectural effect in the processor, executes speculative instructions by mistraining the branch prediction unit to access secret data. This execution can cause data leakage by leaving traces in the cache [19]. The Spectre attack exploits mispredictions in speculative execution within the processor to obtain secret information. For this reason, it has become an important research topic in the field of hardware security, and many different variants have

© The Author(s), under exclusive license to Springer Nature Switzerland AG 2027
K. Renaud et al. (Eds.): EICC 2026, CCIS 3012, pp. 201–211, 2027.
https://doi.org/10.1007/978-3-032-28957-5_12

emerged. Studies have shown that hardware based on Intel, ARM, and RISC-V architectures are vulnerable to this attack [13].

Since Spectre is a security vulnerability that affects all processors, defense techniques have been developed. In studies conducted on both the detection and defense sides of the attack, attack detection studies based on software patches in particular cause high system load [11]. Software-based techniques developed for defense against the attack attempt to strengthen the system by preventing speculative execution effects. However, software-based techniques cause significant performance loss because they interfere with the structure added to the processor for performance [6,13,20]. Attack detection and prevention efforts have also brought about new attack variants. For this reason, existing detection and defense systems against Spectre attacks may be insufficient to address all possible attack types [19]. There are various variants of the Spectre attack, including V1, V2, and V4 [20]. Due to its impact on processors with many different architectures, it is important to develop attack detection techniques specific to both the attack variant and the target architecture.

Microarchitectural events such as branch mispredictions affect Hardware Performance Counters (HPC), which are used as special-purpose counters in all modern processors [2]. HPCs enable the monitoring of hardware events with relatively little impact on system performance [12]. In this context, HPC analysis, which records microarchitectural events during speculative execution of processors, has emerged as an important research area for detecting malicious behavior and identifying Spectre-like attacks [4]. Studies show that machine learning techniques yield effective results in detecting Spectre attacks from HPC data. Andreu et al. [3] presented a detection and classification study for hardware attacks and benign applications by training a machine learning model using HPC information obtained from Intel x86 architectures. Using Naive Bayes, Support Vector Machine (SVM), Random Forest, and Decision Tree methods for detection and classification, the study achieved over 99% accuracy. In Aktas-Aydin et al. [1], who proposes a hybrid method using LSTM and CNN, achieved a 95% accuracy rate with their model trained on HPC data. However, it is important to note that other programs running concurrently on the system may influence the same HPC events triggered by a Spectre attack, which could increase the risk of false positives. Addressing the false positive issue, AL-Zu'bi et al. [2] examined nearly 40 HPCs under various benign and attack scenarios to determine the most suitable HPCs for attack detection and the best performing ML algorithms. The studies also show that attackers mimic benign applications during spectre attacks, thereby reducing the detection rates of machine learning models using HPC [14].

This paper presents a controlled analysis examining the HPC-level distinguishability of the Spectre-v2 attack. We focus on Spectre-v2 detection using machine learning algorithms on RISC-V based systems. We used the HARPY-V dataset [17], which contains HPC data collected during attacks and benign applications on a RISC-V-based processor. We extracted Spectre-v2 and benign application data from the dataset and evaluated it using machine learning models. We trained our model for attack detection using Logistic Regression, Support Vector Machine (SVM), Random Forest, and XGBoost methods. We achieved an accuracy rate above 99% in all four models.

In the following section of the article, we explained the Spectre attack, machine learning, the dataset we used, and related studies in the Background section. In the Method section, we explained the method we presented in the study and described our preprocessing and experimental setup steps. In the Results section, we discussed our results in detail. In the Conclusion section, we discussed the results and mentioned future works.

2 Background

Spectre attacks, which exploit techniques implemented to improve processor performance, pose a security threat to processors [10]. This section details the Spectre attack, describes the machine learning models and dataset used in the study, and reviews related works in the literature.

2.1 Spectre

Modern processors use techniques such as branch prediction and speculative execution to achieve high performance. When the target of a branch depends on a memory value being read, the processor predicts the target and attempts to execute it in advance. This situation brings with it security problems that could cause data leakage in the processor [9]. The Spectre attack removes memory isolation by exploiting speculative execution in the processor. The processor speculatively executes certain instructions by predicting the outcome of branches. Even if changes are rolled back in case of a wrong prediction, this leaves traces in units like the cache and branch predictor, leading to data leakage [10].

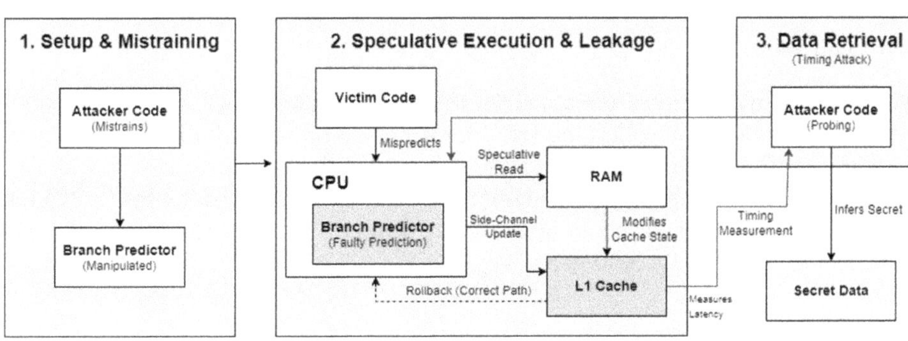

Fig. 1. Spectre attack: schematic overview

The attack typically consists of three stages: first, the necessary prerequisites are prepared, such as mistraining the branch predictor to force it into an incorrect path and loading the target memory location into a register. Next, features such as out-of-order execution or speculative execution are exploited to steer secret data toward a covert side channel. In the final stage, the attacker retrieves the secret data through the side channel [15]. An overview of Spectre attacks is presented in Fig. 1. There are numerous variants of the Spectre attack based on the different branch prediction units employed in modern processors. Spectre-v1 targets the pattern history table branch predictor through conditional branches, whereas Spectre-v2 exploits forward indirect branches by attacking the branch target buffer structure. The Spectre-RSB attack impacts the system by targeting the return stack buffer predictor [9].

2.2 Machine Learning

Machine learning creates mathematical models that can make predictions and decisions without being specifically programmed, by learning from past data or training data. The accuracy of the model depends largely on the amount of data used; more data generally leads to more accurate predictions [18]. Machine learning models emerge as a powerful method for attack detection by defining system data using statistical models.

Logistic Regression, which has high speed and interpretability, is a method that works by mapping data to a probability space and works effectively in binary classification problems. SVM is a maximum interval-based method that classifies data by mapping it to a high-dimensional space. SVM can give effective results in nonlinear classification problems with its high accuracy and generalization capabilities. Random Forest is a decision tree-based method that works with multiple decision trees and builds them by randomly selecting features and examples. It is a model with high accuracy and robustness. XGBoost is a method that incrementally improves model accuracy by optimizing the loss function based on gradient boosting decision trees [7].

2.3 Dataset

In this study, we used the open source HARPY-V dataset [17]. The HARPY-V dataset was created by collecting HPC information on the Xuantie C910 processor with RISC-V architecture by performing 16 different hardware attacks and running 16 different benign applications. For each process, data was collected by reading 20 counter data available on the Xuantie C910 processor. Each class contains approximately 2000 data [17].

2.4 Related Works

In the literature, methods such as software-based monitoring, microarchitectural trace analysis and hardware performance counters can be used to detect Spectre attacks. Experiments using machine learning show high detection accuracy.

Spectre Detection Approaches. Aiming to detect speculative memory accesses with control flow extraction, address analysis and taint analysis methods, oo7 has detected fifteen purpose-built code patterns affected by Spectre [21]. Cheang et al. [5] introduced the concept of trace property-dependent observational determinism (TPOD) for the detection of new leaks introduced by speculative execution. Guarnieri et al. [8] analyzed the semantic notion of security against speculative execution attacks. Researchers developed Spectector, an algorithm based on symbolic execution to automatically detect the presence of speculative leaks [8]. Ting et al. [19] presented a method to automatically detect Spectre attacks at the RTL level using two phases: speculative load data manipulation and cache state alteration during the speculative time window. In addition, it has been shown that increasing the number of samples in the dataset with synthetic data generation increases the accuracy rate in spectre attack detection [1].

HPC-Based Machine Learning Approaches for Spectre Detection. SpecDefender, which provides both Spectre attack detection and defense using HPC information, has shown that it can detect the attack in 2 s with over 90% model accuracy [6]. Le et al. [13] trained a neural network model using data from 4 hardware performance counters and achieved over 99% accuracy in attack detection. Mushtaq et al. [16] analyzed behavioral data of concurrent processes collected through HPCs and introduced the WHISPER tool which uses multiple machine learning models to detect side channel attacks at runtime. Li et al. [15] improved the evasion robustness of the detection model by randomly switching between detectors with different settings to prevent attackers from mimicking benign signatures and evading detection systems, and showed that it can maintain detection accuracy above 80%. AL-Zu'bi et al. [2] shown a systematic evaluation of the best performing combinations of HPC knowledge and ML algorithms for the detection of Spectre attacks using HPC. Tong et al. [20] trained a machine learning model using 4 HPC metrics collected on an Intel processor detected Spectre and Meltdown attacks with variants with over 99% accuracy. HPC measurements can be affected by different workloads. Taking this issue into account Kotha et al. [11] analyzed the impact of system noise on intrusion detection by analyzing branch and cache miss distributions in different CPU modes. Andreu et al. [3] presented an intrusion detection study on Intel and AMD platforms in 3 different scenarios using HPC information. The model proposed in [3] detected many different hardware attacks with over 99% accuracy.

In this study, we present a model that detects spectre attacks by training HPC data with machine learning. In the literature, studies on spectre detection with HPC data have presented evaluations with different processor architectures and different HPC information. In our study, we evaluate HPC data obtained from a RISC-V architecture processor. We consider the spectre-v2 variant and present its comparison with HPC data against a benign application.

3 Method

In this work, we specifically focus on the Spectre-v2 attack and a single benign application to create a controlled experimental environment. Spectre-v2 is chosen because its attack structure can leave microarchitectural traces on HPCs during branching and cache-related events. Stress_m, which generates high system activity by creating load such as CPU, memory and I/O with configuration, is chosen as a reference application because it can leave an impact on HPCs during this heavy work. This type of application can create a heavy load, causing HPC profiles that superficially resemble attack behavior. This controlled analysis allows us to assess the distinctive capability of HPCs before extending the approach to a wider range of benign applications. The workflow of the proposed approach is shown in Fig. 1.

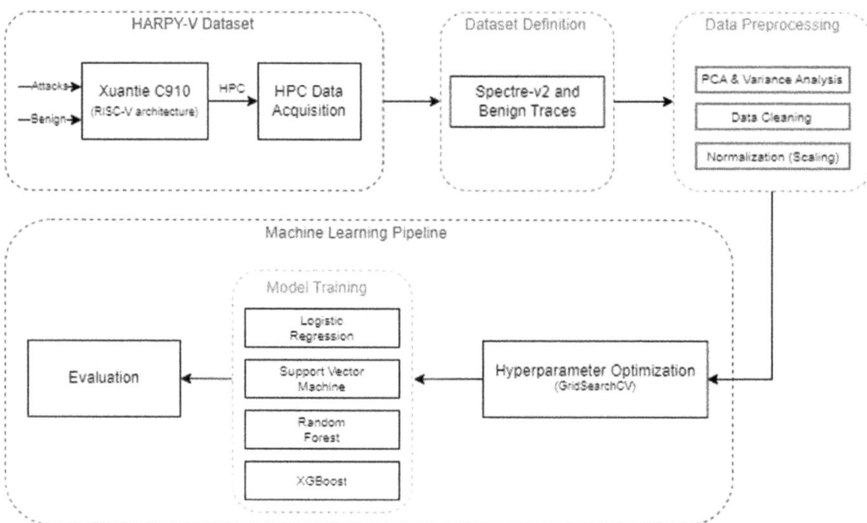

Fig. 2. The machine learning pipeline implemented to classify Spectre-v2 attacks using HPC data collected from a RISC-V architecture.

In this study, the HARPY-V dataset, created by collecting HPC information from the XuanTie C910 processor with RISC-V architecture under specific attack and benign scenarios, was used. Data related to the Spectre-v2 attack and the benign application named stress_m were extracted from this dataset, and a new dataset consisting only of these two classes was used for model training (see Fig. 2).

Preprocessing steps were applied to the created dataset to make it suitable for model training, and then the machine learning pipeline consisting of hyperparameter optimization, model training, and evaluation of results was performed.

3.1 Preprocessing

The raw dataset contains a time column. This column is not required for classification, so it was removed from the dataset before training. The dataset contains 20 different HPC measurements for each example. Seven of these HPCs were selected based on PCA and variance assessments and provided to the model as features. The HPCs selected as features are r002, r005, r007, r00E, r00F, r010 and r013. The selected HPCs represent the Instruction retired counter, I-UTLB and JTLB miss counters, L1 DCache Read Access/Miss counters, L1 DCache Write Access counter and L2 Cache Read Miss counter [17]. Including different hardware events as features in the model can also provide a comprehensive perspective.

To reduce the effect of different numerical scales and scale each feature to a zero mean and unit variance, 'StandardScaler' was applied to all features. In this way, the aim was to reduce scale-related bias in model training. The preprocessed dataset was used for model training and model evaluation.

3.2 Experimental Settings

All experiments in this study were performed on a system equipped with an Intel® Core^{TM} i7-11800H processor (2.30 GHz), 16 GB RAM, and an NVIDIA GeForce RTX 3050 Ti Laptop GPU. Since the classes in the dataset contained different numbers of values, stratified sampling was used to preserve class distribution, and the dataset was randomly split into 70% training and 30% testing. Four different machine learning models were used for model training: Logistic Regression, Support Vector Machine, Random Forest, XGBoost.

For each model, parameters such as kernel type, tree depth, learning rate, etc., were added to the search space, and hyperparameter optimization was performed using the GridSearchCV method.

The cross-validation value *cv=StratifiedKFold(n_splits=5,shuffle=True)* was defined to maintain class balance within each fold. Cross-validation accuracy was used as the criterion for selecting the best model parameters. The best model parameters obtained for the Logistic Regression model, the parameters were set as *C = 100, class_weight = balanced, max_iter = 2000, penalty = l1*, and *solver = liblinear*. For the Support Vector Machine model, *C = 10, gamma = scale, and kernel = rbf* were used. In the Random Forest model, the parameters were *criterion = gini, max_depth = None, min_samples_leaf = 2, min_samples_split = 2, and n_estimators = 200*. For the XGBoost model, the parameters were set as *colsample_bytree = 0.8, learning_rate = 0.2, max_depth = 3, n_estimators = 100*, and *subsample = 1.0*.

4 Result

The experiments first analyzed the values associated with the spectre-v2 attack and benign application in the dataset. Four machine learning models evaluations were then performed.

Exploratory Analysis with Principal Component Analysis (PCA): To examine whether the dataset could be classified into classes, principal component analysis was performed by reflecting the seven HPC features used in training and testing into two components, PC1 and PC2. The PCA graph is shown in Fig. 3.

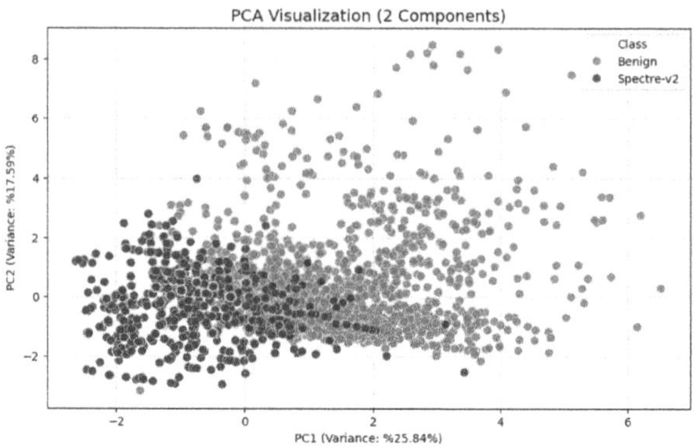

Fig. 3. PCA Projection.

PC1 accounts for 25.84% of the total variance, while PC1 and PC2 together explain 43.43% of the total variance. Analyzing the PCA graph shows that the different distribution of attack and benign points indicates that the selected features increase the discriminability in classification. When examining the regions close to the (0,0) point in particular, it can be seen that the points overlap. This indicates that it is not possible to draw a perfect linear separation line by looking at only two components. Looking at the graph, it can be predicted that the classification model results will yield good accuracy results with the separation of points spread across different regions.

Cross-Validation Performance of Classification Models: In Table 1 , the cross-validated performance metrics of the four classification models are provided. Table 1 summarizes the best model results obtained with GridSearchCV. The results show that all four models can detect the spectre attack with high accuracy against benign applications. In particular, ensemble models demonstrated nearly perfect performance with 99.9% accuracy and F1 score.

The lowest performance was shown by a linear model, logistic regression, with an accuracy value of 99.13%. The fact that the model's precision value is slightly higher than its recall value indicates that the model is cautious when deciding on the attack class. The highest performance was shown by the Random Forest model with an accuracy value of 99.95%. Similarly, the F1 score result was also 99.95%. The XGBoost model followed with a similar performance, achieving an accuracy value of 99.92%. The SVM model showed an accuracy success of 99.35% and is noteworthy for its cross-validation recall result of 1.0. This indicates that the model did not miss any Spectre-v2 examples.

Table 1. Cross-Validation Performance Metrics with Standard Deviations

Model	Accuracy	F1 Score	Precision	Recall
Logistic Regression	0.9913 *Std: 0.0025*	0.9925 *Std: 0.0022*	0.9958 *Std: 0.0040*	0.9892 *Std: 0.0041*
Support Vector Machine	0.9935 *Std: 0.0043*	0.9944 *Std: 0.0038*	0.9944 *Std: 0.0024*	1.0000 *Std: 0.0000*
Random Forest	0.9995 *Std: 0.0007*	0.9995 *Std: 0.0006*	0.9991 *Std: 0.0011*	1.0000 *Std: 0.0000*
XGBoost	0.9992 *Std: 0.0016*	0.9993 *Std: 0.0014*	0.9991 *Std: 0.0019*	0.9995 *Std: 0.0009*

Evaluation of the Best Model Parameters: For each model, the models were retrained on the training set using hyperparameters that maximize cross-validation accuracy, and the results were evaluated. The resulting confusion matrices are shown in Fig. 4.

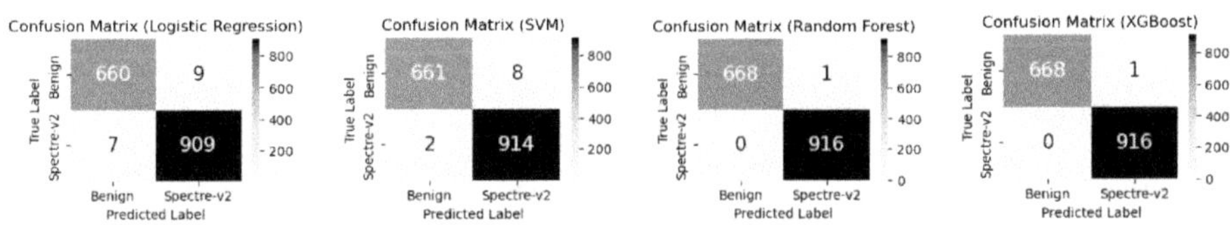

(a) (b) (c) (d)

Fig. 4. Confusion matrices of the highest-accuracy models: (a) Logistic Regression. (b) SVM. (c) Random Forest. (d) XGBoost

When analyzed in Fig. 4a, it is observed that despite the high success rate of Logistic regression, the rate of missing Spectre-v2 samples is higher than other models. Although the SVM model (Fig. 4b) produces better results than logistic regression, it is seen to produce a small number of false positives and false negatives. The best results were obtained with the Random Forest (Fig. 4c) and XGBoost (Fig. 4d) models, and these two models produced only one false positive each. Furthermore, it was observed that these two models did not miss any Spectre-v2 samples.

5 Discussion and Conclusion

In the studies conducted on Intel-based platforms, which constitute most of the studies in the literature; Kotha et al. [11] achieved over 99% accuracy with 100000 samples by extracting 12 features from 20 HPCs, while Andreu et al. [3] achieved over 99% accuracy with 28000 samples. Li et al. [15] achieved 99% accuracy in attack detection, but the detection accuracy of the model decreased when evasive attacks were included. In RISC-V based work, Le et al. [13] achieved 99% accuracy with 4 HPC information and 16000 samples. In this study, over 99% accuracy was achieved with 7 HPC information and 5282 samples. It can be evaluated that the results achieved in this study are due to the fact that both the attack traces on RISC-V are evident and the selected features reflect the attack.

The results we obtained in this study show that Spectre-v2 detection on RISC-V architecture can be realized with high accuracy with a small number of data. All 4 different machine learning models we used in this study achieved over 99% accuracy. The result shows that an advanced attack like Spectre-v2 can be effectively detected with a real-time behavior-based analysis. Our results provide a practical baseline for the development of real-time defense approaches on RISC-V based systems.

Although the proposed approach achieves high accuracy under controlled conditions, model robustness must be ensured in real-world environments due to noise that may arise from system activities. In future work, the evaluation will be expanded to include a wide range of benign applications and attack types, and detection performance will be assessed under noisy conditions to better reflect practical RISC-V environments.

References

1. Aktas-Aydin, H., Yalcin, G.: Spec17tre: a new dataset in hardware security and using deep learning for detecting spectre attacks. Arab. J. Sci. Eng., 1–12 (2025)
2. AL-Zu'bi, M., Weissenbacher, G.: Statistical profiling of micro-architectural traces and machine learning for spectre detection: a systematic evaluation. In: 2024 Design, Automation & Test in Europe Conference & Exhibition (DATE), pp. 1–6 (2024). https://doi.org/10.23919/DATE58400.2024.10546539
3. Andreu, D., Otero, B., Canal, R.: Multi hardware-attack dataset and ML-based detection using processor stress patterns on x86. WiPiEC J.-Works Prog. Embed. Comput. J. **11**(1), 7–7 (2025).https://doi.org/10.64552/wipiec.v11i1.94
4. Bhade, P., Paturel, J., Sentieys, O., Sinha, S.: Lightweight hardware-based cache side-channel attack detection for edge devices (edge-cascade). ACM Trans. Embed. Comput. Syst. **23**(4) (2024). https://doi.org/10.1145/3663673
5. Cheang, K., Rasmussen, C., Seshia, S., Subramanyan, P.: A formal approach to secure speculation. In: IEEE 32nd Computer Security Foundations Symposium (CSF), pp. 288–28815 (2019). https://doi.org/10.1109/CSF.2019.00027
6. Choudhari, A., Guilley, S., Karray, K.: SpecDefender: transient execution attack defender using performance counters. In: Proceedings of the 2022 Workshop on Attacks and Solutions in Hardware Security, pp. 15–24 (2022)
7. Fan, Z., You, Z.: Research on network intrusion detection based on XGBoost algorithm and multiple machine learning algorithms. Theor. Nat. Sci. **31**(1), 161–166 (2024)
8. Guarnieri, M., Köpf, B., Morales, J.F., Reineke, J., Sánchez, A.: Spectector: principled detection of speculative information flows. In: 2020 IEEE Symposium on Security and Privacy (SP), pp. 1–19 (2020).https://doi.org/10.1109/SP40000.2020.00011
9. Jang, H., Shin, Y.: MicroCFI: microarchitecture-level control-flow restrictions for spectre mitigation. IEEE Access **11**, 138699–138711 (2023). https://doi.org/10.1109/ACCESS.2023.3340680
10. Kocher, P., et al.: Spectre attacks: exploiting speculative execution. In: 2019 IEEE Symposium on Security and Privacy (SP), pp. 1–19 (2019). https://doi.org/10.1109/SP.2019.00002

11. Kotha, J.K.C., Gaudiot, J.L.: A data-driven approach using hardware performance counters to detect micro-architectural attacks. In: 2025 IEEE 49th Annual Computers, Software, and Applications Conference (COMPSAC), pp. 2301–2306 (2025). https://doi.org/10.1109/COMPSAC65507.2025.00324
12. Langehaug, T.J., Graham, S.R., Kabban, C.M.S., Borghetti, B.J.: MADFAM: microarchitectural data framework and methodology. IEEE Access **10**, 23511–23531 (2022). https://doi.org/10.1109/ACCESS.2022.3153313
13. Le, A.T., Hoang, T.T., Dao, B.A., Tsukamoto, A., Suzaki, K., Pham, C.K.: Spectre attack detection with neutral network on RISC-V processor. In: 2022 IEEE International Symposium on Circuits and Systems (ISCAS), pp. 2467–2471 (2022). https://doi.org/10.1109/ISCAS48785.2022.9937212
14. Li, C., Gaudiot, J.L.: Challenges in detecting an "evasive spectre." IEEE Comput. Archit. Lett. **19**(1), 18–21 (2020). https://doi.org/10.1109/LCA.2020.2976069
15. Li, C., Gaudiot, J.L.: Detecting spectre attacks using hardware performance counters. IEEE Trans. Comput. **71**(6), 1320–1331 (2022). https://doi.org/10.1109/TC.2021.3082471
16. Mushtaq, M., et al.: WHISPER: a tool for run-time detection of side-channel attacks. IEEE Access **8**, 83871–83900 (2020). https://doi.org/10.1109/ACCESS.2020.2988370
17. Pou Freixas, A., Otero, B., Canal, R.: RISC-V hardware attack traces on on-chip hardware performance counters (HARPY-V Dataset) (2025).https://doi.org/10.34810/data2538
18. Shaveta: A review on machine learning. Int. J. Sci. Res. Archive **9**(1), 281–285 (2023). https://doi.org/10.30574/ijsra.2023.9.1.0410
19. Ting, C.C., Huang, Y.T., Chen, Y.T., Chen, Y.R., Lin, E.H.: Spectre attack detection with formal method on RISC-V processor at RTL design level. In: 2025 International VLSI Symposium on Technology, Systems and Applications (VLSI TSA), pp. 01–04 (2025). https://doi.org/10.1109/VLSITSA64674.2025.11046778
20. Tong, Z., Zhu, Z., Zhang, Y., Liu, Y., Meng, D.: KSM: killer of spectre and meltdown attacks. In: 2024 27th International Conference on Computer Supported Cooperative Work in Design (CSCWD), pp. 157–162 (2024). https://doi.org/10.1109/CSCWD61410.2024.10580256
21. Wang, G., Chattopadhyay, S., Gotovchits, I., Mitra, T., Roychoudhury, A.: oo7: low-overhead defense against spectre attacks via program analysis. IEEE Trans. Software Eng. **47**(11), 2504–2519 (2021). https://doi.org/10.1109/TSE.2019.2953709

Code and Binary Analysis

Static Detection of Compromised Java Components by Topological Anomalies

Oscar Z. de Paiva(✉), Wilson V. Ruggiero, and Marcos A. Simplicio Jr.

Polytechnic School, Universidade de São Paulo, São Paulo, Brazil
{ozpaiva,wilson,msimplicio}@larc.usp.br

Abstract. Open-source components are assets of utmost importance in software development nowadays. However, they also create new attack venues in modern software supply-chains: attacks in the last five years have shown that one can inject malicious code into these components with relative ease. To address this issue, this paper presents a novel protection mechanism aimed at detecting, in open-source components and applications, the varieties of malicious code typically disseminated through supply-chain attacks. As the technology focus of our case study, we selected Java due to its relevance in contemporary corporate environments. The mechanism employs a topological static analysis approach that first clusters the call graph of the program or component under analysis, following the definitions of namespaces contained within it and the relations between these namespaces. Connections between security-critical methods within and across clusters are then used to flag the presence of malicious insertions, as the normal occurrence of these connections is correlated, as we demonstrate, with minimum numbers of method calls from within the clusters to methods of the same APIs as the security-critical ones. In our tests, this approach demonstrated high accuracy, ranging from 86.3% to 99.98% in a publicly available dataset. This illustrates that the proposed technique is robust and useful in practice, especially for the low number of false-positive classifications observed.

Keywords: SSC Security · Java Security · Malware Detection

1 Introduction

Most of the code of modern software applications consists of direct or indirect third-party dependencies, especially Open-Source Components (OSCs) [11]. This demonstrates how essential OSCs have become in reducing software development costs and promoting innovation around the world. However, the ubiquity of these components has unfortunately opened new attack venues for malicious actors to compromise systems via the Software Supply Chains (SSCs) on which these depend. Indeed, in the last decade, attackers have been able to infiltrate communities involved in OSC development, or compromise systems used by them, to inject malicious code into OSCs and "trojanize" them with relative ease [7,9,11,12].

© The Author(s), under exclusive license to Springer Nature Switzerland AG 2027
K. Renaud et al. (Eds.): EICC 2026, CCIS 3012, pp. 215–233, 2027.
https://doi.org/10.1007/978-3-032-28957-5_13

Cybersecurity specialists have proposed various solutions for such attacks, and also mandated/encouraged their adoption [13]. Examples include stronger authentication, authorization and intrusion detection for source and package repositories, SCA (Software Composition Analysis) and SBOMs (Software Bill of Materials), in-process isolation, dependency sandboxing, and static or dynamic analysis techniques. Despite those advances, the risk associated with SSC attacks in OSCs remains largely unmitigated.

In this paper, we contribute to this effort by proposing a novel static analysis technique for detecting malicious code excerpts injected into Java components through SSC attacks. We choose Java as the focus of our study due to a combination of: (1) limited attention dedicated to the technology in the SSC security literature; (2) widespread adoption in enterprise environments, which increases the risk of SSC attacks on Java OSCs (as shown by security incidents involving Log4j [4] and Struts2 [8]). Similarly to the few works in the literature that address the security of Java SSCs [10,16,18], our technique relies on analyzing relations between security-critical methods (e.g., involving I/O or command interpretation) within the components. However, unlike previous works, ours focuses on high-level structural properties of Java programs as defined by their components, as well as on the dependency relations between these components, as defined by Java packages (the technology's namespace system).

The benefits of this approach are: (1) the components and their relationships provide natural structures whose shared features characterize the expected relations between security-critical methods within and across structures, revealing anomalies due to malicious insertions; (2) relying on such structural properties increases resistance to simple obfuscation techniques, an important step toward countering evasion tactics. The features considered in this work are the incidence of components onto Java APIs containing security-critical methods, measured by the number of calls made within the components to methods from those APIs. Based on our empirical findings, in a component, the presence of close connections between security-critical methods (a common characteristic of malicious excerpts injected through SSC attacks) is associated with minimal values of incidences of that component onto the APIs that contain these methods. In other words, such close connections are concentrated in program parts that make higher use of the APIs that contain the security-critical methods. The minimal values of incidence can then be obtained from a dataset with uncontaminated programs through an Unsupervised Machine Learning approach.

We restrict our analysis to excerpts that adhere to the typical structures and behaviors represented in the "Backstabber's Knife Collection" dataset [12], which is a collection of malicious code excerpts disseminated through SSC attacks. We implemented our proposal using the WALA library [15] to build call graphs from Java bytecode and extract program components and dependency relations between them, and used the XCorpus dataset [3] of Java programs to adjust the values of minimal incidence. The results obtained indicate detection accuracies ranging between 86.3% and 99.98%, depending on the malicious excerpt type.

The rest of this paper is organized as follows. Section 2 presents the behaviors and structures typically exhibited by malicious code excerpts injected into OSCs through SSC attacks, highlighting our general threat model. Section 3 presents our proposal, and Sect. 4 describes our experimental analysis. Section 5 discusses limitations of our technique and possible adaptations to other technologies. Section 6 reviews the main related works, and, finally, Sect. 7 concludes the document and presents future research directions.

2 Threat Model and Types of Malicious Code at Issue

There are few restrictions on the form or behavior of malicious code that can be injected into an application via SSC attacks on OSCs. These restrictions are mainly related to how the infected component is integrated into the application's control-flow, which determines how the malicious code is triggered and if/how it can trigger application code. In turn, these aspects are determined by the underlying attack vector and dynamic linking mechanisms provided by different programming environments. Besides inserting malicious code into components, attackers can also *remove* code from them (e.g., access control checks). Since that would be equivalent to *inserting* a vulnerability into the target component, detecting such malicious modifications by analyzing only the applications' binaries or source code is an undecidable problem.

Due to the undecidability issue, the design of static analysis techniques for detecting malicious code must restrict its targeted construct types. In this work, we focused on the capability of detecting at least the most common malicious constructs inserted into software applications through SSC attacks on OSCs. Those constructs were extracted from the "Backstabber's Knife Collection" (BKC) [12], a dataset of contaminated OSCs from open-source ecosystems of various technologies (mainly Python, Javascript, Ruby and Java). We analyzed the dataset version with the commit identifier prefixed by , from Sept. 24, 2025, restricting ourselves to instances that had been previously analyzed by the dataset maintainers and are accompanied by code samples in which the location of the malicious construct was identified (totaling 2467 instances). We also restricted the analysis to instances of which the code sample implements at least one of the following malicious behaviors, as defined in [12]: *data exfiltrators*, *droppers*, *backdoors* and *reverse shells* (1648 instances in total). Lastly, instances of which the malicious constructs were designed to compromise only desktop environments (for example, by relying on browser features, or on the capture of data from the clipboard or from keyboard events) were excluded from the analysis.

The total number of instances from the BKC dataset satisfying our final selection criteria was 1583. We analyzed the malicious constructs in the code samples of each one of the selected instances, searching for common implementation patterns. Unfortunately, many malicious constructs appear in multiple dataset instances (an indication that they were part of dissemination campaigns), causing some statistical distortions. Despite this issue, there are clear implementation patterns found in the selected instances: in 1475 of these (93,2%), the malicious behavior is implemented in small code excerpts of a few dozen commands that rely on the usage of methods from at least two of the following categories of security-critical standard APIs: ( ) network access, ( ) file read, ( ) file write, ( ) command interpretation (or dynamic manipulation), and ( ) inspection of system properties or environment variables. Data exfiltrators are thus implemented with methods from and , or and categories; droppers, with methods from , and , or and categories; and backdoors and reverse shells, with methods from and categories.

Listing 1.1 provides an example of a malicious code excerpt in Java (a dropper), selected from the BKC dataset following our criteria. We highlight that, in most of the instances not following the criteria, the malicious construct either is implemented with methods solely from the category, or employs methods from user libraries (i.e. not from standard APIs) for accessing at least one of the functionality categories. Notably, despite being a common characteristic of malwares, only about 100 of the selected instances from the BKC dataset employ obfuscation techniques – albeit all very simple, such as base 64 encoding.

```
URLClassLoader loader = (URLClassLoader) ClassLoader.getSystemClassLoader();
Class urlclassloaderClass = URLClassLoader.class;
URL url = null;
try{
    url = new URL("...");
    Method m = urlclassloaderClass.getDeclaredMethod("addURL", new Class[]{URL.class});
    m.setAccessible(true);
    m.invoke(loader, new Object[]{url});
    Class.forName("...", true, loader);
} catch (Exception e) {
    e.printStackTrace();
}
```

Listing 1.1. Example of a malicious Java code excerpt (a dropper) extracted from the "Backstabber's Knife Collection" dataset [12].

Threat Model: Based on the behavioral and implementation patterns of malicious code observed in the BKC dataset, it is possible to define the threat model contemplated in this work. We consider attackers capable of using any attack vector to inject malicious code into OSCs and the applications that depend on these components. The malicious code behave as *data exfiltrators*, *droppers*, *backdoors* or *reverse shells*, and consist of small code excerpts that use methods from the combinations of standard APIs as previously defined. These assumptions are compatible with attackers that aim to maximize the number of infected applications, in order to penetrate the networks in which they run or extracting sensitive data from them. Attacks targeting specific applications or institutions (for example, through the injection of application-specific malicious code) are considered outside the scope of analysis, as well attacks that employ sophisticated code obfuscation techniques.

Java specifics and Threat Model: There is a significant underrepresentation of malicious code excerpts injected into Java OSCs, not only in the BKC dataset (which has 3 instances) but broadly in the specialized literature. For this reason, in this work we assume that behavioral and structural patterns of malicious code excerpts targeting other technologies are also found in excerpts targeting Java. Naturally, this assumption creates a risk of misrepresenting patterns found in real-world malicious Java code, as most malicious code examples in the BKC dataset are injected into installation scripts – something not available in Java. While this may simply indicate that attackers prefer to rely on this feature for simplicity (as it provides an easy-to-use hook to the control-flow and trigger malicious code), it could also explain the predominance of implementations relying solely on standard APIs (as user libraries may not be available for executing installation scripts). However, we claim that these limitations do not invalidate the analysis, as implementations relying solely on standard APIs are compatible with the assumed threat model: an attacker using only standard APIs avoids issues with dependencies, thus maximizing contamination rate. We also observe that, since Java's execution environment relies on semi-interpretation, it does not provide the same convenient features for natively executing code directly from memory variables as scripting languages do. Therefore, malicious code relying on the category of security-critical methods may not be easily translatable to Java. Lastly, the set of methods in each category of security-critical functionalities may vary according to the standard Java API version, but an attacker seeking to maximize infection would likely choose older versions for better compatibility – a choice we also made by targeting Java 1.6.

3 Description of the Technique

For detecting the types of malicious code excerpts defined in Sect. 2, our technique relies on the analysis of statically observable topological features typically exhibited by uninfected Java programs and components. The analysis is performed in three steps: (1) hierarchical division of the program into parts; (2) investigation on how security-critical methods, from the categories defined in Sect. 2, are connected in the control-flow within (and through) the program parts; (3) examination of how program parts are connected to external parts (i.e. parts from the Java standard libraries), and how those connections are associated with the connections between security-critical methods. Further details and the rationale for each step are given as follows (illustrative examples of the concepts presented in this section can be found on our repository[1]

1. The division of programs and their components into parts is based on Java packages, the language's mechanism for organizing code into namespaces in which there can be no name conflicts, and over which are enforced (non-security-related) access-control rules. One useful characteristic of this mechanism is the relation between package names. These names are defined by simple strings, i.e., strings that do not need to obey any grammatical rule representing relationships between any two packages. Despite this lack of restrictions, developers by convention name packages following an "inverted" URI grammar rule that typically relates the components' name, project, developer

[1] at https://github.com/ozpaiva/static-topol-analysis):.

organization, country, and other details, hierarchically, providing a measure of the degree of relatedness between packages. We use this measure to cluster the program's Call Graph (CG), producing to other graphs: the PDG (Package Dependency Graph), a digraph whose vertices represent packages from the analyzed program or from Java standard APIs, and whose edges represent a dependency relation between two packages based on caller-callee relations denoted on the CG; and the PGDG (Package Group Dependency Graph), a digraph whose vertices represent groups of packages from the analyzed program or from the Java standard API's (the grouping being determined by the similarity between packages names), and whose edges represent the dependency relations between groups as defined in the caller-callee relations denoted on the CG.

2. The connections between methods from the security-sensitive API categories are represented by *junctions*, which are special vertices from the CG representing the roots of trees, i.e., subgraphs of the CG connecting two or more Security-Critical Methods (SCMs) that correspond to the trees' leafs. Each tree has a corresponding projection onto the PDG and the PGDG. Based on the height of the tree and their projections, different classes of junctions can be defined, representing connections that occur within a single method, package, or package group, or across different package groups.
3. The relation between a program part (user-defined package or package group) with parts from the Java standard APIs is represented by the concept of *incidence*, which can be defined by the number (absolute or relative) of calls between methods from the program part and the external part (in absolute and relative numbers). The concept of incidence denotes how much a program part depends on the external part to implement their intended functionalities, providing a form of "code smell". The normal occurrence of different classes of junctions within a program part can thus be associated with minimal values of incidence from the part onto the external parts that contain each method from a combination of SCMs (also denoted as SCMC). That is the main hypothesis of our detection method, and the minimal values of incidence can be learned through Unsupervised Machine Learning from a dataset of uncontaminated programs.

In summary, the robustness of our method is based on the fact that, for implementing the types of malicious code represented in Sect. 2, the attacker cannot avoid inserting, into the CG of the infected program or component, junctions between a SCMC – and, for the insertion to be imperceptible, it must occur into a part that exhibits minimal values of incidence. If the parts exhibiting these minimal values are sufficiently rare – and they are, as we demonstrate in Sect. 4 –, a detector based on parameters of minimal incidence can exhibit good detection performance. The details of each of the three method steps are presented in more formal terms in what follows.

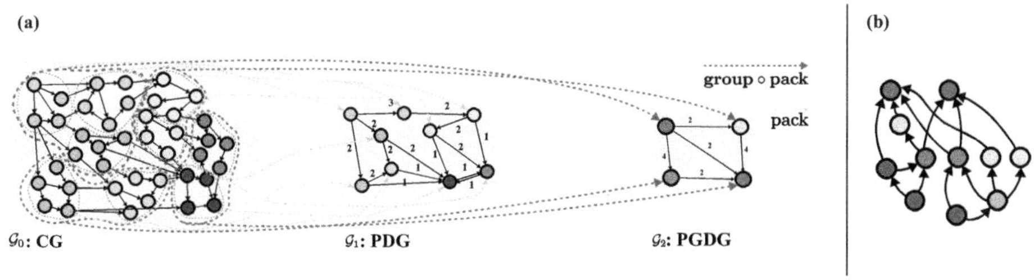

Fig. 1. **(a)** Relations between a hypothetical CG and the PDG and PGDG built from it, according to the functions **pack** and **group ∘ pack** that cluster the methods corresponding to every vertex of the CG. The PDG and PGDG are both weighted digraphs; an edge weight correspond to the number of calls between the methods mapped to the vertex from which the edge leaves, and the methods mapped to the vertex onto which the edge arrives. **(b)** A simple graph illustrating the definitions of junction: relatively to the security-critical methods in blue, the vertices in red are first order junctions, and the orange vertex is a second-order junction. Vertices in light gray are incomplete, and vertices in dark gray are complete, but do not fit the definition of junction.

3.1 Three Layers of Topological Representation: CG ($\mathcal{G}^{(0)}$), PDG ($\mathcal{G}^{(1)}$) and PGDG ($\mathcal{G}^{(2)}$)

In the form of a directed graph, a call graph is a representation, of the caller-callee relations between methods[2] within a program $\mathcal{P}$, denoting the high-level aspects of $\mathcal{P}$'s interprocedural control-flow. In the simplest formulation possible, the call graph, denoted as $\mathcal{G}^{(0)} = (V^{(0)}, E^{(0)})$, has its set of vertices $V^{(0)}$ mapped by a bijection $\mathbf{b}^{(0)}$ to the methods in $\mathcal{P}$. The set of edges $E^{(0)}$ is formed by pairs $e^{(0)} = (v^{(0)}, v'^{(0)})$, $v^{(0)}, v'^{(0)} \in V^{(0)}$, such that the method m, which satisfies $v^{(0)} = \mathbf{b}^{(0)}(m)$, calls the method m' that satisfies $v'^{(0)} = \mathbf{b}^{(0)}(m')$ from anywhere in the sequence of instructions that forms it.

We take the $\mathcal{G}^{(0)}$ as the lowest level of possible topological representations of $\mathcal{P}$, since it defines its topology as defined by pairwise relationships between its methods – hence the superscript 0 in $\mathcal{G}^{(0)}$. The two other layers of topological representation, represented by the PDG and the PGDG, are thus denoted as $\mathcal{G}^{(1)}$ and $\mathcal{G}^{(2)}$, and can be defined as follows:

- The PDG is a weighted digraph, denoted as $\mathcal{G}^{(1)} = (V^{(1)}, E^{(1)})$, whose set of vertices $V^{(1)}$ is composed of elements denoted as $v^{(1)}$, such that there is a bijection $\mathbf{b}^{(1)}$ between these elements and the set of packages in $\mathcal{P}$. Let **pack** be a function mapping each method m in $\mathcal{P}$ to the package containing the class where m is defined; the set of vertices $E^{(1)}$ consists of pairs $e^{(1)} = (v^{(1)}, v'^{(1)})$, where $v^{(1)}, v'^{(1)} \in V^{(1)}$, such that, for some method m satisfying $v^{(1)} = \mathbf{b}^{(1)} \circ \mathbf{pack}(m)$, there is a method m' that is called by m and satisfies

[2] or procedures/subroutines, or functions, depending on the programming paradigm under consideration; as the focus of this work is Java, which is an object-oriented language, we adopted the term *method*.

$v'^{(1)} = \mathbf{b}^{(1)} \circ \mathbf{pack}(m')$. The weight associated to $e^{(1)}$ is given by a function $w^{(1)} : E^{(1)} \longmapsto \mathbb{Z}$ that maps it to the number of distinct pairs (m, m') for which the aforementioned condition is satisfied.

- The PGDG, in turn, is a also weighted digraph, denoted as $\mathcal{G}^{(2)} = (V^{(2)}, E^{(2)})$, whose set of vertices $V^{(2)}$ is composed of elements, denoted as $v^{(2)}$, such that there is a bijection $\mathbf{b}^{(2)}$ between these elements and the set of package groups to which the packages in $\mathcal{P}$ belong. Let **group** be the function that maps each package in $\mathcal{P}$ to the package group to which it belongs, and let **pack** be the same function as previously defined; the set of edges $E^{(2)}$ is composed of pairs $e^{(2)} = (v^{(2)}, v'^{(2)})$, where $v^{(2)}, v'^{(2)} \in V^{(2)}$, such that, for some method m satisfying $v^{(2)} = \mathbf{b}^{(2)} \circ \mathbf{group} \circ \mathbf{pack}(m)$, there is a method m' that is called by m and satisfies $v'^{(2)} = \mathbf{b}^{(2)} \circ \mathbf{group} \circ \mathbf{pack}(m')$. The weight associated to $e^{(2)}$ is given by a function $w^{(2)} : E^{(2)} \longmapsto \mathbb{Z}$ that maps it to the number of distinct pairs (m, m') for which the forementioned condition is satisfied.

The PDG- and PGDG-based clustering scheme can be implemented with an efficient algorithm of $O(|V^{(0)}| + |E^{(0)}|)$ complexity. Figure 1 (a) shows a visual summary of the scheme, and the relations between $\mathcal{G}^{(0)}$, $\mathcal{G}^{(1)}$ and $\mathcal{G}^{(2)}$.

3.2 Junction Vertices

Junction vertices are special vertices of $\mathcal{G}_0$ that can serve as markers for how the methods of a particular SCMC can be connected throughout the interprocedural control-flow of a program $\mathcal{P}$. In more formal terms, a junction vertex j in the call graph $\mathcal{G}^{(0)}$ of program $\mathcal{P}$ is defined in relation to the set M of a SCMC under consideration, whose elements are denoted as m_i, with $0 < i < n$ and $n = |M|$. Let a path between any two nodes $v^{(i)}, v'^{(i)}$ of graph $\mathcal{G}^{(i)}$ be denoted as a sequence $p^{(i)} = \left(v_0^{(i)}, v_1^{(i)}, ..., v_{l-1}^{(i)}\right)$, where $|p^{(i)}| = l$ denotes the length of the sequence, $v_0^{(i)} = v^{(i)}$, $v_{l-1}^{(i)} = v'^{(i)}$ and, for $j \in [0, l]$, $\left(v_j^{(i)}, v_{j+1}^{(i)}\right) \in E^{(i)}$. Let also **dijkstra** be a function that maps every pair $v^{(i)}, v'^{(i)}$ of vertices either to a path $p^{(i)}$ of minimal length connecting them, or to **nil** if there is no such path in $\mathcal{G}^{(i)}$. Lastly, let a vertex $v^{(0)}$ be denoted as **complete**, relatively to a set M, if, for every $m_i \in M$, $\mathbf{dijkstra}(v^{(0)}, \mathbf{b}^{(0)}(m_i)) \neq \mathbf{nil}$. Two classes of junctions can be then be described in the Definitions 1 and 2.

Definition 1. *Relatively to a set M of a SCMC, a junction of first order j is a vertex in $\mathcal{G}^{(0)}$ that: (1) is complete; (2) has no complete neighbors – i.e. for any vertex $v^{(0)}$ such that $(j, v^{(0)}) \in E^{(0)}$, $v^{(0)}$ is not complete;*

Definition 2. *Relatively to a set M of a SCMC, a junction of second order j is a vertex in $\mathcal{G}^{(0)}$ that: (1) is complete; (2) has complete neighbors of first or second order; (3) would be a junction of first order relatively to M if all edges $(j, v^{(0)})$ were removed from $E^{(0)}$, where $v^{(0)}$ is a complete neighbor of j.*

Figure 1 (b) provides, for a simple hypothetical graph, an illustration of junctions of first and second order according to the definitions presented. Regardless of their order, junction vertices are referred to simply as junctions henceforth, for simplicity. More informally, a junction can be defined as a vertex on which first met all the reverse paths – i.e. paths that follow edges in reverse –, starting from the vertices corresponding to methods in the SCMC under consideration. In other words, every junction j is the root of a tree $\mathcal{T}_j^{(0)}$, which is a subgraph of $\mathcal{G}^{(0)}$ formed by (and only by) the vertices and edges of the smallest paths from j to every node $\mathbf{b}^{(0)}(m_i)$, where $m_i \in M$. Each path is thus a branch of the tree, starting on j and ending on a node associated with a method from the SCMC.

Associated to every tree $\mathcal{T}_j^{(0)}$ of a junction j are $\mathcal{T}_j^{(1)}$ and $\mathcal{T}_j^{(2)}$, both trees that are subgraphs of $\mathcal{G}_j^{(1)}$ and $\mathcal{G}_j^{(2)}$, respectively, and whose roots are determined by the expressions $\mathbf{b}^{(1)} \circ \mathbf{pack}(j)$ and $\mathbf{b}^{(2)} \circ \mathbf{group} \circ \mathbf{pack}(j)$. Let the projection of a path $p^{(0)}$ onto $\mathcal{G}^{(i)}$, for $i \in [0,1]$, be defined as the path $p^{(i)}$, the sequence of vertices obtained once all chains of equal elements are removed from (leaving only the first element of each chain on) the sequence $s^{(i)} = (v_0^{(i)}, v_1^{(i)}, ..., v_{l-1}^{(i)})$, where $l = |p^{(0)}|$, $v_j^{(i)} = \mathbf{pf}(v_j^{(0)})$, $j \in [0,l]$, and $\mathbf{pf}$ is a function mapping elements from $V^{(0)}$ onto $V^{(i)}$. The branches of $\mathcal{T}_j^{(1)}$ and $\mathcal{T}_j^{(2)}$ are defined by the projections of the branches from $\mathcal{T}_j^{(0)}$ respectively onto $\mathcal{G}^{(1)}$, with $\mathbf{pf} = \mathbf{b}^{(1)} \circ \mathbf{pack} \circ \mathbf{b}^{(0)^{-1}}$, and onto $G^{(2)}$, with $\mathbf{pf} = \mathbf{b}^{(2)} \circ \mathbf{group} \circ \mathbf{pack} \circ \mathbf{b}^{(0)^{-1}}$.

Defining $h(\mathcal{T})$ as the height of a tree, i.e., the length of its biggest branch, a junction j can be classified into one of the four categories according to the values of $h(\mathcal{T}_j^{(0)})$, $h(\mathcal{T}_j^{(1)})$ and $h(\mathcal{T}_j^{(2)})$ as follows:

- **J0** ($h(\mathcal{T}_j^{(0)}) = 1$): the method associated to j is directly connected to, i.e., directly calls, all the methods in M.
- **J1** ($h(\mathcal{T}_j^{(0)}) > 1$, $h(\mathcal{T}_j^{(1)}) = 1$): the method associated to j is indirectly connected to the methods in M through other methods, but all are contained in the same package as the junction.
- **J2** ($h(\mathcal{T}_j^{(0)}) > 1$, $h(\mathcal{T}_j^{(1)}) > 1$, $h(\mathcal{T}_j^{(2)}) = 1$): the method associated to j is indirectly connected to the methods in M through other methods, all of which belonging to the same package group as the former. However, at least one of the methods belong to a different package as the others.
- **J+** ($h(\mathcal{T}_j^{(2)}) > 1$): the method associated to j is indirectly connected to the methods in M through at least one method that belongs to a different package group as the former method.

The junction search procedure described in this section can be implemented, for a given SCMC, by an algorithm of complexity $O(|E^{(0)}| + |V^{(0)}| * log(|V^{(0)}|))$.

3.3 Incidence Analysis

For each method m in the set M of a given SCMC, an application package p and the package group g that contains it as defined by $g = \mathbf{group}(p)$, the following incidence indicators can be calculated (by an algorithm of complexity $O(|E^{(0)}| + |V^{(0)}|)$):

$$I^{(1)} = w^{(1)}(p, \mathbf{pack}(m)) \qquad R^{(1)} = \frac{I^{(1)}}{\sum_{(p,p')\in E^{(1)}} w^{(1)}(p,p')}$$

$$I^{(2)} = w^{(2)}(g, \mathbf{group} \circ \mathbf{pack}(m)) \qquad R^{(2)} = \frac{I^{(2)}}{\sum_{(g,g')\in E^{(2)}} w^{(2)}(g,g')}$$

For detecting malicious excerpts that introduce junctions of class **J0** and **J1** into compromised programs, our method relies on two assumptions regarding the natural occurrence of such junctions within a package p:

- it is correlated with the incidences of p onto every package that contains a method in M (the Security-Critical Package Combination, or SCPC), as denoted by $I^{(1)}$ (in absolute terms) and $R^{(1)}$ (relatively to the number of calls of methods in p to methods contained in other packages);
- it is also correlated with the incidences of $g = \mathbf{group}(p)$ onto every package group that contains packages comprising the methods in M (the Security-Critical Package Group Combination, or SCPGC), as denoted by $I^{(2)}$ (in absolute terms) and $R^{(2)}$ (relatively to the number of calls of methods in g to methods in other package groups).

With these assumptions, and also with an adequate – and, mostly importantly, non-compromised – dataset, it is possible to empirically determine minimal values of $I^{(1)}$, $R^{(1)}$, $I^{(2)}$, and $R^{(2)}$ for a package to be expected to contain junctions of class **J0** and **J1** connecting the methods in M. That is, in effect, an Unsupervised Machine Learning approach employing a model composed of $4 * n$ parameters, organized in tuples $\left(\theta_I^{(1)}, \theta_R^{(1)}, \theta_I^{(2)}, \theta_R^{(2)}\right)$, one for each method in M. These parameters define an infinite subspace $\mathcal{R} \in \mathbb{R}^{4\times n}$, and they can be determined by a training process that aims at maximizing the number of packages in the dataset containing the junctions under consideration that fall within $\mathcal{R}$, while minimizing the number of packages not containing these junctions that fall within $\mathcal{R}$. Later, during an inference phase, the model can be used to classify as anomalies (i.e. as contaminated) the packages that contain junctions connecting the methods in M, but fall outside $\mathcal{R}$.

For detecting malicious excerpts that introduce junctions of class **J2** into compromised programs, we employ a simplified version of the same technique. It is based on the analogous assumption that the natural occurrence of these junctions within a package group g is correlated with incidences $I^{(2)}$ and $R^{(2)}$ of g onto every package group containing the methods in M (the SCPGC). The Machine Learning process used has fewer parameters, only two for each method M: $\left(\theta_I^{(2)}, \theta_R^{(2)}\right)$, organized in pairs. These parameters define an infinite subspace $\mathcal{R} \in \mathbb{R}^{2\times n}$, which, as in the previous case, delimits the region in which a package group is expected (not) to contain the aforementioned junctions.

4 Experimental Evaluation

To validate the effectiveness of the method presented in Sect. 3, we implemented and tested it in a five-step semi-automatic process, using a public dataset of Java

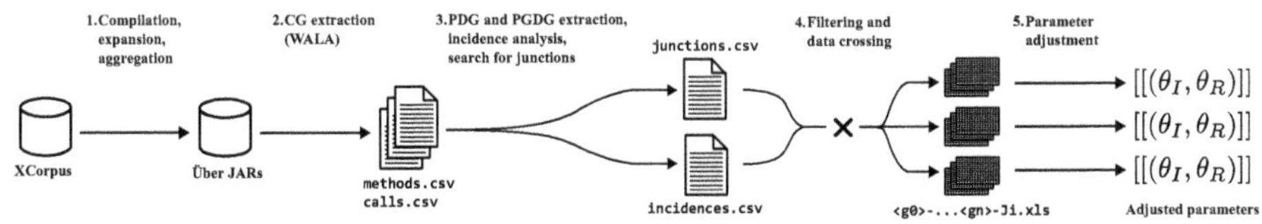

Fig. 2. Our 5-step experimental procedure. Step 1 involved the compilation of the XCorpus dataset, expansion of instances related to more than one executable, and aggregation of all instances, including their dependencies, into Über JARs. In Step 2, we extracted the CG from every JAR, and saved each CG in two files (and). In Step 3, we extracted the PDG and CDG from every CG, and the incidences and junctions from them (and). In Step 4 we filter repeated occurrences, crossing the data in the files to produce spreadsheets named (where correspond to the package groups represented in the file, and correspond to the junction class to which the refer). Step 5 is the parameter adjustment process detailed in Sect. 3.3.

programs as main input. Figure 2 illustrates the process. The first three steps were performed using specially developed software tools, while the last two, which involved data analysis, were executed manually in common spreadsheet software. For reproducibility, we made available, in our repository, the code and executables of all specially developed software tools, and the data produced in each step of the process. The non-manual steps were executed in approximately 120 min on a computer with an Intel 11800H CPU and 32GB of RAM, running Ubuntu 22.04 and Java HotSpot version 1.8.0-371.

The dataset chosen for this experiment was XCorpus [3], a curated, compilable and executable dataset of Java programs. For being a curated and versioned dataset, XCorpus contributes to the reproducibility and analysis of our results. Moreover, as its instances are compilable and executable, all their dependencies are resolved, which allows us to construct complete call graphs of the instances. Despite these advantages, one drawback of using XCorpus is that its instances may correspond to obsolete or, at least, old versions of still used Java programs, libraries and frameworks. However, to the best of our knowledge, there was no dataset exhibiting all the advantages of XCorpus at the time of writing.

In step 2, the CG of XCorpus' instances were extracted with the WALA library for program analysis [15]. According to the options provided by the library, the analysis was configured as follows: 0-1-CFA (instance-based) control-flow analysis algorithm, interprocedural-only exception analysis, instance-based points-to-precision resolution algorithm, object graph field propagation, every method considered as possible entry point, and Java 1.6 stubs (this configuration was selected mainly for reducing computational cost of the analysis, and may not correspond to the most precise analysis available in WALA).

The junction search performed in step 3 was based on a list of SCMs based on the lists provided by [10] and [14]. We merged those lists, removed from the result methods that do not immediately trigger a security-critical functionality (for example, the constructor of the class), and also removed methods absent in the Java 1.6 standard API (the most recent version compatible with XCorpus' instances). Lastly, we also added to the list methods that were missing (for example, and), and adopted a search procedure that uses the name of each method in the list as the prefix of a search string, which allows accounting for additional methods.

4.1 Findings

In the whole XCorpus dataset, there were found 6694 unique Java packages (roughly – as there were some name collisions and packages repackaged with different names). Those packages were divided into 371 package groups, according to a map manually crafted (and which can be found in our repository). The junction search procedure found 10608 non-unique junctions across all XCorpus instances, and the ones belonging to **J0**, **J1** and **J2** were selected (as **J+** was out-of-scope). Values of incidence were extracted considering SCPs and SCPGs defined over the categories of Java APIs: , (equivalent to), and (reuniting families). The manual parameter adjustment procedure performed in step 5 produced the values represented in Tables 1 and 2.

The values represented in Tables 1 and 2 show that, in most cases, junctions of class **J0** are linked to higher relative incidences (and also, to a lesser extent, absolute incidences) than junctions of class **J1** are. These adjusted parameters also show that relative incidences in the PDG are generally higher than in the PGDG, indicating that packages tend to be more specialized in functionalities that more frequently use the APIs related to the SCPCs. The same is not true for absolute incidences, because package groups naturally contain more classes. Lastly, these parameters also show that junctions of classes **J2** are associated, with some exceptions, to lower values of relative incidences in comparison to junctions of classes **J1** and **J0**.

To analyze the possible efficacy of detectors based on the parameters adjusted in step 5, we can proceed as follows. Let *in* and *out* be, respectively, the number of elements (packages or package groups) that contain junctions and respectively fall inside and outside $\mathcal{R}$, the region selected by the adjusted parameters. Let *sel* be the total number of elements contained in $\mathcal{R}$. Considering a hypothetical scenario in which an attacker randomly contaminates a percentage r of packages (or package groups) from the dataset with malicious excerpts based on a given SCPC (SCPGC), every contaminated package (or package group) outside $\mathcal{R}$ corresponds to a true positive. Likewise, every contaminated package (or package group) inside $\mathcal{R}$ corresponds to a false negative. Every non-contaminated package or package group inside $\mathcal{R}$ corresponds to a true negative; and outside $\mathcal{R}$, also to a true negative, unless it contains natural, benign junctions – i.e. it corresponds to a false positive. The TP and TN (true positive and true negative) numbers can then be calculated as $TP = (n - sel) * r$ and $TN = (1 - r) * (n - out)$, where n is either equal to 6694 or 371 (the total number of packages and package groups in the dataset, respectively). The detector's accuracy, acc, calculated as $acc = (TP + TN)/n$, achieves its minimal value when $r = 1$, and we consider this minimal value in our analysis. We also considered the detector's $\overline{F_1}$ score,

Table 1. Parameters of minimal incidences associated with the presence of junctions connecting different combinations of methods (**J0** or **J0+J1** within packages, or **J0+J1+J2** in package groups), each from and APIs.

	io+net combinations						lang+net combinations					
	J0		**J1+0**		**J2+1+0**		**J0**		**J1+0**		**J2+1+0**	
	io	net	io	net	io	net	lang	net	lang	net	lang	net
$\theta_I^{(1)}$	14	7	26	6	–	–	48	13	48	10	–	–
$\theta_R^{(1)}$	4.6%	1.6%	0.6%	1.65%	–	–	8.5%	0.5%	4.2%	0.5%	–	–
$\theta_I^{(2)}$	41	17	41	6	62	39	1554	70	583	22	1329	68
$\theta_R^{(2)}$	2.2%	0.6%	0.1%	1.94%	1.7%	0.35%	9.8%	0.4%	7.8%	0.2%	10%	0.17%

Table 2. Parameters of minimal incidences associated with the presence of junctions connecting different combinations of methods (**J0** or **J0+J1** within packages, or **J0+J1+J2** in package groups), each from APIs.

	J0			**J1+0**			**J2+1+0**		
	lang	net	io	lang	net	io	lang	net	io
$\theta_I^{(1)}$	144	24	28	144	24	28	–	–	–
$\theta_R^{(1)}$	33%	2.5%	6.8%	11%	0.3%	1.73%	–	–	–
$\theta_I^{(2)}$	11224	725	1618	1329	88	312	1329	40	189
$\theta_R^{(2)}$	18.6%	1.1%	2.4%	13.6%	0.3%	2.4%	9.1%	0.3%	1%

i.e. its average F_1 score, defined as the integral between 0 and 1 of the function $F_1(r) = 2 * (n - sel) * r / (r * (2 * n - sel - out) + out)$.

The number of natural occurrences of junctions for each SCPC and junction class, along with the detection accuracy for each class and combination, are displayed in Table 3. We can see that the natural occurrence of junctions of classes **J0**, **J1** and **J2** is itself relatively rare, with a few dozen occurrences in thousands of packages or hundreds of package groups. The detection accuracies achieved are very high, ranging from 86.3% to 99.98%. The lowest value of $\overline{F_1}$ obtained was 0.806. The false positive rate, FP, calculated as $FP = out * (1 - r)$, is at most equal to *out* and, thus, very low. That is an important performance quality for a detector when processing high volumes of instances, as too many false positives may overwhelm security analysts. In this test, the statistical significance achieved may not actually be enough for a rigorous evaluation of our technique and the parameters adjusted can be overfit to XCorpus instances, but we believe these results already provide good evidence of the potential of our technique.

Ablation Study. We conducted a simple ablation exercise on the adjusted parameters, by zeroing them one by one in the models and observing the accuracy degradation. It was possible to see that there is no great difference between the individual contribution of each parameter to the accuracy of the models, as no single parameter contributes more than 8% to the accuracy of any model.

Table 3. Resulting detection accuracies and $\overline{F_1}$ for all combinations of junction classes and package groups

	io+net			lang+net			lang+net+io		
Class	**J0**	**J1+0**	**J2+1+0**	**J0**	**J1+0**	**J2+1+0**	**J0**	**J1+0**	**J2+1+0**
sel	51	143	44	112	284	51	1	26	50
in	10	16	15	12	28	32	1	4	16
out	6	6	7	15	41	43	0	6	11
Accuracy	99.2%	97.8%	88.1%	98.3%	95.8%	86.3%	99.98%	99.6%	86.5%
$\overline{F_1}$	0.993	0.986	0.902	0.985	0.960	0.806	1.00	0.995	0.880

The most important parameters are the ones related to incidences onto the group, followed by and, lastly, . This was expected, since the group reunites methods that, compared to the methods from other groups, are less effective indicators of functionality.

5 Discussion: Resistance to Evasion

If our technique is eventually used to detect SSC attacks in real-world settings, it is expected – as always is in adversarial settings – that attackers will employ methods to try to evade detection. We next discuss some of the methods that may be employed for this purpose, and the counteracting mechanisms that may be put in place as possible improvements or complements to our technique.

Insertion of Spurious Calls: perhaps the most basic technique an attacker may employ to avoid detection is inserting spurious calls to methods in the SCPs to achieve, in the contaminated package, the thresholds of relative and absolute incidences associated to the normal occurrence of junctions as represented in our model. By spurious calls, we mean calls that, in the caller method, have their produced values discarded in dead local variables, for example. To counteract the insertion of spurious calls, data and control flow analysis techniques could be employed to search for dead-end code.

Insertion of Spurious Methods: another manipulation technique related to the previous one is the insertion, into a package or package group, of methods (possibly originally legitimate) connected by caller-callee relations, constituting whole call graphs. With this, the attacker could thus subvert the most basic techniques of detection of spurious calls. To detect the insertion of whole spurious call graphs into compromised components, other topological analysis techniques could be employed, for example, in the search of subgraphs weakly connected to the complete call graph of the package or package group. Anomalies in the package or package group class hierarchy, possibly associated to the malicious insertion of spurious sets of methods, can also be used to detect these methods.

Insertion of Spurious Packages or Package Groups: in addition to possibly trying to insert spurious calls or methods within a single package or package group, attackers may also try to insert whole spurious packages and package groups into a Java component to indirectly reach security-critical functionalities. These packages and package groups may originally be associated with legitimate, benign libraries, for example, command interpreters or HTTP clients. In practical terms, this may be equivalent of inserting junction of class **J+** in the compromised program – and junctions of this class are currently not detectable by our technique. To overcome this limitation, the list of security-sensitive methods used in our technique could be expanded with methods from non-standard libraries. Another alternative would be the extraction of "signatures" of intended functionality from package and package groups, which would allow the creation of models of expected connectivity between components depending on their signatures. Graph learning and other deep learning techniques could seem promising to build these models, with which abnormal connections could be detected and used as indicators of compromised components. Techniques of detection of spurious methods and calls could be used to complement those models, as low connectivity between packages and package groups or spurious calls between methods may serve as additional indications connections inserted irregularly. These mechanisms may also be employed to counteract namespace obfuscation.

Subversions of the Definitions of Junction: attackers can try to subvert the classification of methods as second order junctions. To do so, from a compromised method, they could call another complete (and legitimate) method for reaching the security-critical method required to implement the inserted code's malicious behavior. A more robust definition of junction method would be needed to detect this malicious manipulation, possibly involving data flow analysis to determine if any legitimate connection is used to indirectly reach a security-critical method. That analysis could complement the ones used to detect malicious calls and components, as previously discussed. An easier alternative for the attacker in this realm would be to implement malicious excerpts using methods solely from the category – in this case, techniques such as data and control-flow analysis or graph learning techniques could be applied to detect them.

Subversion of the Static Analysis Accuracy: as static analysis is inherently inaccurate, attackers could take advantage of its imprecisions to hide malicious insertions. Moreover, Java frameworks frequently use features such as dependency injection, which makes the static construction of complete call graphs very difficult. In those cases, tools such as Jasmine [2], if available for the framework in question, could be used to guide the static analysis process (however, note that these problems affect mostly junctions of class **J+**, which are out-of-scope in the current version of our technique). Dynamic analysis approaches could also be used to complement the static analysis, in general.

Ultimately, with all those countermeasures in place, we argue that attackers would be forced to use malicious excerpts that do not adhere to the assumptions detailed in Sect. 2. These excerpts would, thus, necessarily be built to exploit higher-level functionality – e.g., target methods and data structures implementing access control functionalities in downstream applications. Their resulting behavior would be less general, but instead coupled to specific applications and libraries, increasing the cost of SSC attacks and limiting their scale and impact.

6 Related Works

A seminal work focused on detecting malicious Java components is Jarhead [16]. In that study, the authors rely on static indicators, such as the usage of security-critical APIs and the presence of code obfuscation, to identify malicious components (JARs, specifically) in Java Applets – an obsolete technology nowadays. In the context of SSC security, [14] employs static call graph analysis to evaluate the reliance of Java libraries on security-critical methods of the standard API. The authors relate that usage to the presence of vulnerabilities in the CVE database, to extract security robustness indicators that could assist developers in selecting a library for a specific functionality. Applying concepts of ZTA (Zero-Trust Architecture), specifically the ideas of continuous monitoring and automatic restriction of authorization policies, the authors of ZTD [1] propose an in-process isolation mechanism for Java components. The mechanism is based on the observation that, under attack, those components utilize security-critical APIs that they do not access during normal execution. All the these three works – even the last two, which were designed for the protection of legitimately vulnerable components (or, in principle, malicious code that is indistinguishable from vulnerabilities) –, rely on notions of normal usage of security-critical APIs, and these notions are related to the concept of incidence that we propose in this work.

Aiming to detect malicious Java code in OSCs, [10] investigates simple possible indicators of malicious behavior in JARs. One of the studied indicators, which the authors conclude to perform well, is the usage of security-critical methods from Java standard APIs in a single method. The authors also propose intra-procedural data-flow analysis to detect suspicious data flows between the security-critical methods, but leave the analysis of interprocedural data flows for future work. Shear [18] is another work focused on the detection of malicious code excerpts in Java OSCs, employing static analysis and deep learning techniques. Potentially malicious code excerpts are identified in the components under analysis through their use of security-sensitive APIs, belonging to a set of categories similar to those defined in Sect. 2 of this work. A program slicing technique is then employed to extract the code excerpts from the components. Next, embeddings of the code excerpt are then used to train a deep learning classifier, which ultimately demonstrated high detection accuracy. Unlike our work, neither Shear nor the previous utilizes higher-level topological features from analyzed programs to train an anomaly detector – which we believe to be a more robust approach, especially in a scenario of scarce training data. However, the program features used by the supervised detection techniques in both studies may complement ours, enabling higher detection accuracy.

The analysis of topological properties of Java programs has been used in some works (such as [5]) with the aim of defining metrics for software engineering studies. Those works conclude, in summary, that the topologies of Java programs exhibit properties, such as locality, that distance them from randomly built structures. In the context of Java security, works that employ concepts related to topological properties of programs are found in the field of Android security, applying these properties specifically for the detection of "piggybacked" applications. These applications are repackaged versions of legitimate ones, into which malicious code has been embedded – an attack setting that is similar to SSC attacks on OSCs. To identify piggybacked Android applications, the authors of Dapasa [6] rely on differences, between malicious and piggybacked applications, in the usage of security-sensitive APIs. They define the concept of sensitive subgraphs, that are built using an algorithm that propagates a security-sensitive metric (derived from the notion of TF-IDF) through the methods in the call graph. The piggybacked applications exhibit different patterns of usage of these security-sensitive methods in these sensitive subgraphs, in comparison to benign applications – specifically, high usage of these methods (which is related to our concept of incidence), associated with low user interaction. Relying on the assumption that, in piggybacked applications, the malicious parts are detached from the original benign code, the authors of [17] employ Class-Dependency Graphs to identify disjoint dependency regions in the applications under analysis. They then employ call graph analysis to extract behavioral features of these parts, such as the usage of security-critical and user interaction APIs, to train a classifier. Our work, on the other hand, is not based on the assumption that malicious components are disconnected from the benign parts of the programs or components under analysis, and is able to analyze connections between security-critical methods that, compared to the sensitive subgraphs in Dapasa, are further apart. Furthermore, we employ an anomaly detection approach that, as already mentioned, we believe to be a more robust approach. Lastly, to the best of our knowledge, our work is the first in the context of Java SSC security, or Java security in general, to use the relations defined by Java packages to perform high-level topological analyses for anomaly detection.

7 Conclusion and Future Works

This work proposes a novel technique to detect malicious excerpts inserted into Java components through SSC attacks. We employ a binary static analysis approach that relies on the extraction of topological properties from the complete binaries of Java programs. By employing the concept of junction nodes, those properties are then used to analyze how security-sensitive methods are typically connected through the call graph of uncompromised programs. A simple Unsupervised Machine Learning model is then used to represent such typical connections, and to identify anomalies as malicious insertions. We used the XCorpus dataset to train and test our model, achieving high accuracy level.

As future work, we plan to perform further evaluations of the proposed method, implement improvements to it, and generalize it to other technologies. Specifically, we intend to test the method with more recent versions of the programs in the XCorpus dataset, as well as additional programs and libraries not in that dataset. We notice that such evaluation requires non-trivial modifications to the software components we developed so far. We also plan to generalize our proposal to other technologies, starting with those closely related to Java (e.g., Android, Scala, Kotlin and Groovy), and them moving to other prominent alternatives (e.g., Python and Node.js). The performance of the technique in these technologies constitutes an interesting research question in itself. Lastly, we plan to implement some of the robustness improvements discussed in Sect. 5.

References

1. Amusuo, P.C., et al.: Ztd$_{JAVA}$: mitigating software supply chain vulnerabilities via zero-trust dependencies (2024). https://arxiv.org/abs/2310.14117
2. Chen, M., Tu, T., Zhang, H., Wen, Q., Wang, W.: Jasmine: a static analysis framework for spring core technologies. In: Proceedings of the 37th IEEE/ACM International Conference on Automated Software Engineering. ACM, New York (2023)
3. Dietrich, J., Schole, H., Sui, L., Tempero, E.: XCorpus - an executable corpus of java programs. J. Object Technol. **16**(4) (2017)
4. Doll, J., McCarthy, C., McDougall, H., Bhunia̋, S.: Unraveling Log4Shell: analyzing the impact and response to the Log4j vulnerabil. Preprint: arXiv:2501.17760 (2025)
5. Falci, D.H.M., Gomes, O.A., Parreiras, F.S.: Complex networks analysis for software architecture: an hibernate call graph study. CoRR **abs/1706.09859** (2017)
6. Fan, M., Liu, J., Wang, W., Li, H., Tian, Z., Liu, T.: DAPASA: detecting android piggybacked apps through sensitive subgraph analysis. IEEE Trans. Inf. Forensics Secur. **12**(8), 1772–1785 (2017)
7. Herr, T., Lee, J., Loomis, W., Scott, S.: Breaking trust – shades of crisis across an insecure software supply chain. Tech. rep., Atlantic Council (2020). ISBN: 978-1-61977-112-3
8. Howard, M.: Sonatype statement: struts2 and equifax breach. Tech. rep., Sonatype Inc. (2017). https://www.sonatype.com/blog/sonatype-statement-struts2-and-equifax-breach
9. Ladisa, P., Plate, H., Martinez, M., Barais, O.: SoK: taxonomy of attacks on open-source software supply chains. In: IEEE S&P, pp. 1509–1526 (2023)
10. Ladisa, P., Plate, H., Martinez, M., Barais, O., Ponta, S.E.: Towards the detection of malicious java packages. In: ACM Works. on Software Supply Chain Offensive Research and Ecosystem Defenses (SCORED'22), pp. 63–72. ACM (2022)
11. n.a.: State of the software supply chain: a decade of data. Tech. rep., Sonatype Inc. (2024). https://www.sonatype.com/state-of-the-software-supply-chain/2024
12. Ohm, M., Plate, H., Sykosch, A., Meier, M.: Backstabber's knife collection: a review of open source software supply chain attacks. In: Detection of Intrusions and Malware. and Vulnerability Assessment, pp. 23–43. Springer, Cham (2020)
13. Ohm, M., Stuke, C.: SoK: practical detection of software supply chain attacks. In: 18th International Conference on Availability, Reliability and Security (ARES 23). ACM (2023)
14. Rahman, I., Paramitha, R., Plate, H., Wermke, D., Williams, L.: What's in a package? Getting visibility into dependencies using security-sensitive API calls (2025). https://arxiv.org/abs/2408.02846
15. Santos, J.C.S., Dolby, J.: Program analysis using WALA (tutorial). In: 30th ACM Joint European Software Engineering Conference and Symposium on the Foundations of Software Engineering (ESEC/FSE 2022), p. 1819. ACM (2022)

16. Schlumberger, J., Kruegel, C., Vigna, G.: Jarhead analysis and detection of malicious java applets. In: Proceedings of the 28th Annual Computer Security Applications Conference (ACSAC'12), pp. 249–257. ACM (2012)
17. Tian, K., Yao, D., Ryder, B.G., Tan, G., Peng, G.: Detection of repackaged android malware with code-heterogeneity features. IEEE Trans. Dependable Secure Comput. **17**(1), 64–77 (2020)
18. Zeng, Y., et al.: Wolf in sheep's clothing: shearing the camouflage of malicious Java components in Maven. IEEE Trans. Softw. Eng. **51**(10), 2847–2863 (2025)

Verification of Security Properties of Cryptographic Protocols Through Model Checking Execution Traces

Simona Bernardi(✉) and Joaquín Ezpeleta

Universidad de Zaragoza, Zaragoza, Spain
{simonab,ezpeleta}@unizar.es

Abstract. Developing cryptographic protocols that are free from vulnerabilities is very challenging, as evidenced by numerous design flaws and implementation bugs in widely used protocols. Recent work has leveraged model checking to generate *negative* test cases, that is test cases that include the behavior of Dolev-Yao attacker. In this paper, we address the problem of analyzing execution traces produced by testing protocol implementations with such negative test cases. We propose a method for detecting violations of security properties by model-checking execution traces. Our approach employs a model checker based on DLTL, a Linear Temporal Logic with freeze operators, and supports automation of the testing process through pre- and post-processing steps. These steps, respectively, bridge the gap between different levels of abstraction and synthesize the results produced by the model checker. We present experimental results as a proof of concept for the feasibility of the proposed method.

Keywords: Cryptographic protocols · Security properties · DLTL · Testing

1 Introduction

Developing cryptographic protocols free from vulnerabilities is very challenging. Many protocols have been repeatedly found to contain design flaws and implementation bugs [15]. Symbolic techniques have been successfully applied primarily to the verification of cryptographic protocols in their early stages enabling the identification of vulnerabilities exploitable by Dolev-Yao attacks [5], where the attacker has full control over the network. The main drawback of tools supporting symbolic analysis is their limited scalability [5], which can make the verification of real-world case studies extremely time-consuming.

At later stages of development, recent works have leveraged model checking to generate test cases for protocol implementations [11,16]. In particular, in [11] a method is proposed for generating *negative* test cases, i.e., test cases that

© The Author(s), under exclusive license to Springer Nature Switzerland AG 2027
K. Renaud et al. (Eds.): EICC 2026, CCIS 3012, pp. 234–244, 2027.
https://doi.org/10.1007/978-3-032-28957-5_14

incorporate the Dolev-Yao intruder. A negative test case is automatically derived from counterexamples generated by model checking protocol specifications against security goals, such as secrecy and authentication.

In this paper, we build on [11] and take a step further by addressing the problem of analyzing execution traces produced by testing protocol implementations with such negative test cases. Specifically, we propose a method to determine whether the implementation pass or not a test case by verifying the execution traces against a security goal defined in the original protocol specification. This technique can serve as an alternative when symbolic techniques are impractical, for instance due to time constraints.

The analysis is conducted using a model checker based on DLTL logic [9], an extension of LTL that allows the specification and evaluation of properties involving relations between attribute values across different states. Given a protocol execution trace and a DLTL formula encoding the violation of a security goal, the model checker verifies whether the formula is satisfied by the trace.

The proposed method facilitates automation of the testing process by including pre/post-processing steps. Pre-processing of execution traces establishes the mapping between the bitstrings in exchanged messages and the assets in the security goals. Besides, post-processing of the model checking results enables the test outcome to be interpreted in terms of the security goal and, when the goal is violated, allows for the identification of the specific traces that breach the property. As a proof of concept, we have developed a tool that has been used in the experiments with a set of protocols from the AnBx library [3].

The paper is organized as follows: the rest of this section reviews briefly the related work; Sect. 2 gives an overview of the approach; Sect. 3 defines the DLTL model and formulas for checking security properties; Sect. 4 describes trace checking workflow supported by the *goal_checker* tool; Sect. 5 reports the results from the experiments; and Sect. 6 provides conclusions and future work.

1.1 Related Work

Many tools are available to support the formal verification of cryptographic protocol specifications through symbolic analysis [5]. Given a protocol and a security property to be verified these tools typically return an attack trace as a counterexample if the property is violated. In this work, we use the OFMC [7] model checker in the experiments. One significant limitation we faced while using the OFMC is the verification time, which increases exponentially with the number of protocol sessions.

Besides, various frameworks have been developed to assist non-experts in formal methods by providing user-friendly languages for specifying protocols and security properties, while also integrating multiple analysis tools as backends [4,10]. In particular, AnBx [10] supports the specification of protocols and properties in "Alice&Bob" formal languages [8], and includes OFMC as a backend. Additionally, the AnBx framework provides model-to-code generation features [13] for rapid Java prototyping. This feature is leveraged in the approach proposed in this work to generate negative test cases.

Formal methods have been mainly used in the verification of cryptographic protocol specifications at design stage. Recent proposals leverage them for fuzzing or runtime verification [2,14]. Both works make use of formal logics, similar to the linear temporal logic used in the Tamarin tool [6], to express security properties: in [2], the logic is used to guide the fuzzer in identifying violations of security properties caused by Dolev-Yao attacks; in [14], it is used to monitor protocol execution and detect potential violations of security properties. Although our work differs in focus from [2,14], it shares a common challenge: coping with different levels of abstraction, particularly between the specification of security properties and the observation of concrete execution traces.

Linear Temporal Logic (LTL) is a formalism used to specify and reason about the behavior of systems over time, where properties are expressed along a single *trace* (a linear sequence of states or events). It introduces temporal operators such as X (neXt), F (Eventually), G (Globally), and U (Until) to describe how propositions evolve across future points in the trace [12].

Alur and Henzinger introduced Timed Propositional Temporal Logic (TPTL) [1], which augments LTL with freeze quantifiers, a mechanism that allows one to bind a variable to a state and subsequently compare these frozen state values with later states, within a temporal formula. This makes TPTL capable of expressing data-dependent temporal constraints, whereas standard LTL only reasons about Boolean propositions without memory of specific values. Thus, TPTL with freeze operators is strictly more expressive than classical LTL. *DLTL* logic [9] adopts the same perspective as TPTL, but adapting it to the case of finite traces, and also adding past operators.

2 Approach Overview

Figure 1 shows the overall approach for testing security properties of cryptographic protocols with checking of execution traces. Our approach relies upon the AnBx framework [13] and, in particular, the *Alice&Bob* protocol narration as input specification (AnB) and the automatic generation of a protocol Java-based implementation.

Given an initial AnB specification of a protocol P and a goal to be verified G, a set of negative test cases is considered. Each test case P_G_j is an AnB specification that, besides the original actors and goal, includes an intruder who may invalidate the property of interest. A test case can be semi-automatically generated (*Test cases generation*): indeed, the AnBx framework already provides support to automatically generate test cases [11] albeit only for a subset of attack scenarios, due to the limited expressivity of the *Alice&Bob* languages. Similarly, a Java-based implementation of a test case can be semi-automatically generated with the AnBx framework (*Model-2-code generation*).

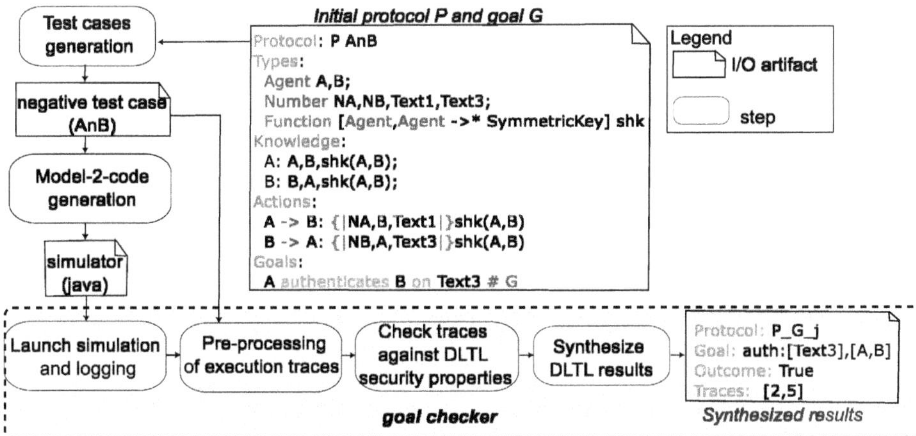

Fig. 1. Workflow and dataflow of the overall approach.

The *goal checker* workflow, emphasized by the dotted rounded rectangle, addresses the verification of the goal G through DLTL checking of the protocol execution traces and it is the focus of this paper.

The Java-based implementation of a negative test case P_G_j is used to launch a protocol simulation and log the execution traces: a trace represents the execution of a session of the protocol. Then, the execution traces are pre-processed to get a structured log and the goal as DLTL formula (i.e., goal violation), which are fed to the DLTL model checker. The latter produces outcomes for each trace in the log against the DLTL security property: the results are finally synthesized to summarize the outcome of the verification of the log against of the original goal. A list of traces in the log that violated the goal are listed in the summary.

3 Specification and Verification with DLTL

To verify that protocol executions comply with their intended specifications, we employ DLTL [9]. In our context, a trace is defined as a chronologically ordered sequence of events representing a protocol's execution. Each event encapsulates the metadata of a communication action, such as the participating agents, the action type, and the associated parameters, represented as typed attributes.

As we will demonstrate, protocol requirements often go beyond than simple sequences of actions, requiring the correlation of attribute values across different states in the trace. For instance, a property might specify that a nonce identifier captured during an initial event must be identical to the identifier appearing in a subsequent event.

By utilizing freeze operators, we can evaluate these cross-temporal properties without prior knowledge of the data domain (e.g., the set of all possible nonce IDs). While traditional LTL would require a distinct proposition for every possible value, leading to an unmanageable or even infinite state space, freeze operators allow us to bind an event to a variable and perform direct comparisons later in the trace. This makes this kind of logic to be well-suited for verifying protocols where the specific data values are not known in advance, but the consistency of those values is critical.

3.1 DLTL Model

The DLTL model $\sigma = e_1 e_2 \ldots e_n$ is the ordered sequence of events e_i, where each e_i corresponds to the ith event of the log and is a tuple of the form:

$$e_i \equiv \{a_i^{action}, n_i^{session}, n_i^{step}, s_i^{active}, s_i^{passive}, @_i^{mess}, s_i^{type}\}.$$

Each element of the tuple corresponds to an attribute as shown in Table 1: in particular, $a^{action} \in \mathcal{A}$ is the action ($\mathcal{A}$ is the set of actions, described in Table 2), $n^{session}$ and n^{step} are the session and step numbers respectively, s^{active} and $s^{passive}$ are the active and passive participants, $@^{mess}$ is the message content, and s^{type} is the type of the message.

Table 1. Event attributes.

Attribute	Description	DLTL Domain
session	Number of the session	integer
step	Number of the step	integer
active	Concrete agent who performs the action	string
passive	Concrete agent who is involved in the action	string
action	Type of action (cfr. Table 2)	atomic proposition
mess	Content of the message	set of strings
type	The type of the message	string

The possible actions are listed in Table 2. In addition to the send and receive actions, the remaining ones are related to message preparation on the sender side or post-processing on the receiver side.

3.2 DLTL Formulas

In this paper, we use DLTL formulas to represent goal violations. Depending on the goal a different DLTL formula is set as detailed in the following.

Authentication. Let us consider the goal `"A authenticates B on N"`, where `A` and `B` are two agents and `N` is an asset exchanged in the protocol. The authentication goal is violated when `B` does not agree on `N` or eventually `A` is not able to recover `N`. The DLTL formula (1) expresses such a violation:

$$\phi_{A,B,N} \equiv \mathrm{G}\ \mathrm{z}.\Big(\neg agrees(\mathrm{z})\ \vee \neg\ \mathrm{XF}\ \mathrm{y}.\big(can\_recover(\mathrm{z},\mathrm{y})\big)\Big) \quad (1)$$

Table 2. Type of actions in the events of a trace.

Action	Description
sent	A message is sent
received	A message is received
generateNumber*	The active participant generates a fresh value
AnBx_Params	The active participant prepares the message content to be sent
eq_check*	The active participant verifies the message content with an equality check
inv_check*	The active participant verifies the message content with a crypto-operation (e.g., decryption) or a projection operation followed by another check (e.g., extracts an element of the tuple and performs an equality check)

*It can be postfixed with the name of an *asset* (Agent or Number type) in the AnB specification.

where $agrees(z)$ and $can\_recover(z, y)$ are boolean functions: the former returns $true$ if and only if there is an event where B can compose N, the second returns $true$ if and only if there is an event where A can recover N and the agreed and recovered values are equal. Formally:

$$agrees(\mathrm{z}) \equiv \left(s_z^{active} == B \ \wedge \ can\_composeN\right) \tag{2}$$

$$recover(\mathrm{y}) \equiv \left(s_y^{active} == A \ \wedge \ can\_recover \ \wedge \ @_z^{mess} == @_y^{mess}\right) \tag{3}$$

The propositions $can\_composeN$ and $can\_recover$ are defined in Table 3. In particular, $can\_composeN$ is defined as the disjunction of three actions listed in Table 2, each postfixed with the name of the asset included in the goal, i.e., `N`. Depending on the protocol specification, one of these actions corresponds to the event in which `B` either generates or recovers the asset.

In case of a goal including several assets, i.e., `"A authenticates B on N_1,...,N_n"`, the DLTL formula is generalized as follows:

$$\phi_{A,B,\{N_1,\ldots,N_n\}} \equiv \bigvee_{\alpha=1}^{n} \phi_{A,B,N_\alpha} \tag{4}$$

Secrecy. Let us consider the goal `"N secret between A_1,...,A_m"`, where `A_1,...,A_m` is a set of m agents and `N` is an asset exchanged in the protocol. The secrecy goal is violated if one agent of the set generates `N` and someone different from the agents in the set can read `N`. The DLTL formula (5) expresses such a violation:

$$\psi_{N,\{A_1,\ldots,A_m\}} \equiv \mathrm{F}\ \mathrm{z}.\Big(generates(\mathrm{z}) \ \wedge \ \mathrm{X}\ \mathrm{F}\ \mathrm{y}.\big(leaked(\mathrm{z},\mathrm{y})\big)\Big) \tag{5}$$

Table 3. Component boolean functions

Function	Definition
can_composeN	$a^{action} == (generateNumberN \vee inv\_checkN \vee eq\_checkN)$
can_recover	$a^{action} == (inv\_check \vee eq\_check)$
generatesN	$a^{action} == generateNumberN$
can_decrypt	$a^{action} == inv\_check \wedge \neg(s^{type} ==$ "*javax.crypto.SealedObject.toString*"$)$

where *generates*(z) and *leaked*(z, y) are boolean functions: the former returns *true* if and only if there is an event where an agent from the set $\{A_\alpha\}_{\alpha=1,\dots,m}$ is generating N, and the latter returns *true* if and only if there is an event where an agent R, with $R \notin \{A_\alpha\}_{\alpha=1,\dots,m}$, can successfully decrypt N. Formally:

$$generates(\mathrm{z}) \equiv \Big(\bigvee_{\alpha=1}^{m} s_z^{active} == A_\alpha \ \wedge generatesN \Big) \tag{6}$$

$$leaked(\mathrm{z}, \mathrm{y}) \equiv \Big(\bigwedge_{\alpha=1}^{m} \neg(s_y^{active} == A_\alpha) \wedge can\_decrypt \wedge @_z^{mess} == @_y^{mess} \Big) \tag{7}$$

The propositions *generatesN* and *can_decrypt* are defined in Table 3. The former denotes the nonce-generation action listed in Table 2, postfixed with the name of the asset included in the goal, i.e., `N`. The latter identifies the event in which a decryption operation is successfully performed and the resulting message is not sealed.

4 The Goal Checker Workflow

In this section, we describe in detail the steps of the *goal checker* workflow of Fig. 1 together with the Python modules of the tool developed as a support.

4.1 Launch Simulation and Logging

The first step of the workflow aims to produce the execution traces of a negative test case protocol. Our approach relies upon the AnBx framework [13] to generate a Java-based implementation of the protocol. The *sim_launcher* module is in charge of compiling the protocol and launching it, by running a concurrent process for each protocol agent (i.e., role) and saving the execution traces of each process in a separate textual file. For each role R in the protocol, as many execution traces $\{t_i^R\}_{i=1,\dots,n}$ are generated as the number of sessions n specified as configuration parameter: each trace t_i^R includes a set of ordered events involving the concrete agent playing the role R in the protocol (i.e., instantiated value of the role R).

4.2 Pre-Processing of Execution Traces

The execution traces produced by the protocol need to be pre-processed to be fed to the DLTL model checker. This step produces the formatted traces and the DLTL formulas in two sub-steps as follows.

Extract Log and Map Agents to Protocol Roles. In the first sub-step, carried out by the *log_extractor* module, the execution traces of each role are filtered, considering the type of actions listed in Table 2. The filtered execution traces of the roles are then merged together to get a unique log, where the events are ordered according to the session number (i.e., trace identifier) and the number of the step. In particular, the latter corresponds to the number of the action in the AnB specification of the protocol. Then, the events in the log are formatted and unique identifiers are assigned to asset values as well as to encrypted messages.

Listing 1.1. Excerpt of the log produced by the *log_extractor*

```
1 #session,step,active,passive,action,mess,type
2 1,0,charlie,-,generateNumber,X0,anbxj.Crypto_ByteArray
3 1,0,charlie,-,generateNumber,X1,anbxj.Crypto_ByteArray
4 1,0,charlie,-,AnBx_Params,X0;bob;X1,anbxj.AnBx_Params
5 1,0,charlie,alice,sent,@3b6eb2ec,javax.crypto.SealedObject
6 1,0,alice,charlie,received,@3b6eb2ec,javax.crypto.SealedObject
7 1,0,alice,-,inv_check,@c2e1f26,javax.crypto.SealedObject.toString
```

The Listing 1.1 shows an excerpt of the log produced by the *log_extractor* from the execution traces of a negative test case. In particular, the logged events correspond to the exchange of a message encrypted with the secret key shared by roles A and B. The message is a triplet consisting of a nonce NA, the role name B and nonce $Text1$.

The first three events (lines 2–4) are actions performed by `charlie` (role A) to prepare the message, i.e., generating the two nonce values `X0` and `X1`, respectively, and setting the triplet {`X0,bob,X1`}, where `bob` plays the role B. The lines 5–6 correspond to the sending and the reception of the encrypted message `@3b6eb2ec`, where the receiver is `alice` —who plays the role of the intruder I. The last event (line 7) is the action performed by `alice`, who tries to decrypt the message without success (the `inv_check` operation returns an encrypted value, i.e., `@c2e1f26`).

The *log-extractor* module is also responsible for mapping the agents in the log to the roles defined in the AnB protocol and for generating the component boolean functions listed in Table 3, taking into account the goal to be verified.

Generate DLTL Model and Formula. In the second sub-step, the *dltl_generator* module generates both the DLTL model and the DLTL formula. The DLTL model closely resembles the log produced in the first step but conforms to the specific input format required by the DLTL tool. The DLTL formula is generated based on the goal specified in the negative test case.

4.3 Check Traces and Synthesize Results

The *dltl_model_checker* is in charge of checking the DLTL formula against each trace in the DLTL model. As a result, it produces a boolean value for each trace. In the last step, performed by the *res_synthesis* module, the results are synthesized in order to have a unique boolean outcome for all the traces in model. In particular, if at least a trace i satisfies the formula then the outcome is *true* and the trace identifier i is added to the list of traces violating the goal. Figure 1 shows the synthesized result produced for the negative test case $P6\_A2$ in Table 4, that represents a replay attack, considering a log with ten traces, where two out of ten traces violate the authentication goal.

5 Experiments

As a proof of concept for the feasibility of the proposed method, we have developed a tool[1] that has been validated with a set of protocols from the AnBx library [3]. Table 4, shows the protocols P_i used in the experiments with their original name (column 1) and the goals (column 2). The AnB specification of each protocol has been verified against the goals with OFMC [7], as a result all the protocols are vulnerable to attacks against authentication or secrecy.

For each protocol and goal, we have considered at least a negative test case (column 3) involving an intruder whose behavior is intended to violate the corresponding goal and follows an attack pattern: *passive*, reading the message in transit; *replay*, forwarding a received message multiple times across different protocol sessions; *reflection* receiving a message from a participant and sending it back; *mitm*, impersonating one or more legitimate parties in the protocol execution. Most of the test cases were generated in an semi-automatic manner with the support of the AnBx tool [11]. Depending on the type of goal, either a single session —for a secrecy goal— or multiple sessions —for an authentication goal— were simulated (column 4).

The last column shows the results, which are consistent with those produced by OFMC. Indeed, for all protocols, the *goal_checker* finds traces of at least one negative test case that satisfy the DLTL formula associated with the goal— that is, violate the goal. Negative test cases aimed at checking authentication goals have been run for several sessions, i.e., 10 sessions, to assess strong authentication: while checking strong authentication with OFMC is quite time-consuming —even for very few sessions, such as 2 or 3, the time to check is in the order of minutes or greater— the goal_checker takes few seconds to analyze 10 traces[2].

[1] Available at: https://github.com/simber72/goal-checker/.

[2] Experiments have been conducted using a MacOS laptop with 2,3 GHz Intel Core i7 with 4 kernels.

Table 4. Results from the experiments

Protocol (P)	Goal*	Test case (attack type)	Sessions	Results
AndrewSecureRPC (P_1)	A	P_1_A (replay)	10	✓(2, 3, 4, 5, 8, 10)
From_A_and_Back (P_2)	S	P_2_S (passive)	1	✓
GSM (P_3)	S	$P_3\_S_1$ (passive)	1	✓
		$P_3\_S_2$ (mitm)	1	✓
NSPK (P_4)	S	$P_4\_S_1$ (passive)	1	✗
		$P_4\_S_2$ (mitm)	1	✓
ISOPK2PMAP (P_5)	$A1$	P_5_A1 (reflection)	10	✓(all)
	$A2$	P_5_A2 (reflection)	10	✓(2, 4, 6, 8, 10)
ISOSK2PMAP (P_6)	$A1$	P_6_A1 (replay)	10	✓(2, 4, 9, 10)
	$A2$	P_6_A2 (replay)	10	✓(2, 5)
Kerberos (P_7)	A	P_7_A (mitm)	10	✓(all)
SSO (P_8)	$A1$	P_8_A1 (mitm)	10	✓(all)
	$A2$	P_8_A2 (replay)	10	✓(3, 5)
	S	P_8_S (mitm)	1	✓

* S is a secrecy goal and A, $A1$ and $A2$ are authentication goals.

6 Conclusions

We have presented a method for testing the security properties of cryptographic protocols by model-checking execution traces. The model-checker takes an execution trace and a DLTL formula expressing the violation of a security goal and verifies whether the formula is satisfied by the trace. Pre-processing of execution traces and post-processing of the results facilitate the automation of the testing process.

The experiments were conducted using simple protocols from the AnBx library; as future work, we are planning to apply the method to a real-world case study in the critical infrastructure domain.

Achieving full automation in the generation of negative test cases remains an open challenge, which we are currently working to address. We recognize limitations in the expressiveness of *Alice&Bob* languages which prevents the specification of negative test cases with attacks that involve multiple protocol sessions (such as replay and reflection attacks), and we aim to overcome them.

Acknowledgments. The authors were supported by the Spanish Ministry of Science and Innovation [ref. PID2020-113969RB-I00, PID2024-159323NB-I00] and by the European Union-Spanish "Agencia Estatal de Investigación" [ref. PID2023-148202OB-C22].

Disclosure of Interests. The authors have no competing interests to declare that are relevant to the content of this article.

References

1. Alur, R., Henzinger, T.A.: A really temporal logic. J. ACM **41**(1), 181–203 (1994). https://doi.org/10.1145/174644.174651
2. Ammann, M., Hirschi, L., Kremer, S.: DY fuzzing: formal Dolev-Yao models meet cryptographic protocol fuzz testing. In: Proceedings of the 45th Symposium on Security and Privacy (S&P'24). IEEE, San Francisco, CA, USA (2024). https://doi.org/10.1109/SP54263.2024.00096
3. AnBx Compiler and Java Code Generator. https://paolo.science/anbx/
4. Armando, A., et al.: The AVISPA tool for the automated validation of internet security protocols and applications. In: Etessami, K., Rajamani, S.K. (eds.) CAV 2005. LNCS, vol. 3576, pp. 281–285. Springer, Heidelberg (2005). https://doi.org/10.1007/11513988_27
5. Barbosa, M., et al.: Sok: computer-aided cryptography. In: Proceedings of the 2021 Symposium on Security and Privacy (S&P'21), pp. 777–795. IEEE CS (2021). https://doi.org/10.1109/SP40001.2021.00008
6. Basin, D., Cremers, C., Dreier, J., Sasse, R.: Symbolically analyzing security protocols using Tamarin. ACM SIGLOG News **4**(4), 19–30 (2017). https://doi.org/10.1145/3157831.3157835
7. Basin, D., Mödersheim, S., Viganò, L.: OFMC: a symbolic model checker for security protocols. Int. J. Inf. Secur. **4**(3), 181–208 (2005). https://doi.org/10.1007/s10207-004-0055-7
8. Bugliesi, M., Calzavara, S., Mödersheim, S., Modesti, P.: Security protocol specification and verification with AnBx. J. Inf. Secur. Appl. **30**, 46–63 (2016). https://doi.org/10.1016/j.jisa.2016.05.004
9. Couvreur, J.M., Ezpeleta, J.: A linear temporal logic model checking method over finite words with correlated transition attributes. In: Ceravolo, P., van Keulen, M., Stoffel, K. (eds.) Data-Driven Process Discovery and Analysis. SIMPDA 2017. LNBIP, vol. 340, pp. 89–104. Springer, Cham (2019). https://doi.org/10.1007/978-3-030-11638-5_5
10. Garcia, R., Modesti, P.: A practical approach to formal methods: an eclipse integrated development environment (IDE) for security protocols. Electronics **13**(23) (2024). https://doi.org/10.3390/electronics13234660
11. Garcia, R., Modesti, P.: Automatic generation of security protocols attacks specifications and implementations. Cyber Secur. Appl. **2** (2024). https://doi.org/10.1016/j.csa.2024.100038
12. Manna, Z., Pnueli, A.: The Temporal Logic of Reactive and Concurrent Systems. Springer-Verlag New York, Inc., New York, NY, USA (1992). https://doi.org/10.1007/978-1-4612-0931-7
13. Modesti, P.: AnBx: automatic generation and verification of security protocols implementations. In: Garcia-Alfaro, J., Kranakis, E., Bonfante, G. (eds.) FPS 2015. LNCS, vol. 9482, pp. 156–173. Springer, Cham (2016). https://doi.org/10.1007/978-3-319-30303-1_10
14. Morio, K., Künnemann, R.: SpecMon: modular black-box runtime monitoring of security protocols. In: Proceedings of the 2024 on ACM SIGSAC Conference on Computer and Communications Security, pp. 2741–2755. ACM, New York, NY, USA (2024). https://doi.org/10.1145/3658644.3690197
15. NIST: National Vulnerabilities Database. https://nvd.nist.gov/
16. Yao, J.: A model checking-based framework for testing security properties of protocols under development. Comput. Netw. **265** (2025). https://doi.org/10.1016/j.comnet.2025.111259

Monitoring and Forensics

Seeing Threats Without Looking: An Encrypted Anomaly Detection Framework for Privacy-Aware Intrusion Detection

Jero Schäfer(✉) and Lena Wiese

Institute of Computer Science, Goethe University, Robert-Mayer-Str. 10, 60325 Frankfurt am Main, Germany
{jeschaef,lwiese}@cs.uni-frankfurt.de

Abstract. As the Web increasingly serves as the backbone of critical infrastructures, safeguarding its security while respecting user privacy has become a central challenge. Automation, large-scale data flows, and AI-driven attacks amplify risks and undermine public confidence. Traditional Intrusion Detection Systems (IDS) require deep inspection of network traffic, creating inherent tensions between effective threat detection and the protection of sensitive personal or organizational data. This socio-technical dilemma makes privacy-preserving security mechanisms essential for the future web. Fully Homomorphic Encryption (FHE) offers a compelling solution by enabling computation directly on encrypted data, thereby allowing collaborative analysis and anomaly detection (AD) without exposing underlying content. Our work investigates the practicality and potential of implementing IDS within an FHE environment. We enhance the detection quality of a state-of-the-art unsupervised AD approach for intrusion detection and provide an encrypted prototype.

Keywords: IDS · FHE · Anomaly Detection · Clustering

1 Introduction

We live in a time when the web has become indispensable. It is a central component of our everyday lives and opens up almost unlimited possibilities. The anonymity of the internet yields individuals freedom of expression but also facilitates illegal activities that offer lucrative returns and cause immense financial damage [29]. Ensuring cyber security remains a permanent challenge as cyber threats evolve and exploit vulnerabilities in complicated web infrastructures.

Intrusion detection systems (IDS) serve as the first line of defense against cyber threats. They aim to analyze network behavior and identify abnormal events such as unauthorized access attempts and malicious activities within computer networks [9]. Continuous monitoring, identification, and containment

© The Author(s), under exclusive license to Springer Nature Switzerland AG 2027
K. Renaud et al. (Eds.): EICC 2026, CCIS 3012, pp. 247–266, 2027.
https://doi.org/10.1007/978-3-032-28957-5_15

of potential intrusions into network systems is essential to protect critical data and maintain the integrity of web infrastructures. As a manual monitoring of data flows can not compete with the speed, complexity and volume of incoming traffic, automated solutions are required for this task. Hence, modern IDS usually use machine learning (ML) techniques to train sophisticated detection models [13,38]. Despite their advantages, IDS implicate data privacy challenges as they deeply analyze various types of data or sometimes cooperate with other IDS [4,25]. This complicates the relationship between analyzing potentially sensitive data, e.g., user information or behavior, and intruder protection. Clearly, preventing attackers from gaining unauthorized access provides security but simultaneously puts the disclosure of private information at risk.

(Fully) Homomorphic Encryption (FHE) techniques [1] yield a promising outlook on the mitigation of this tradeoff. In theory, they enable safe computation and analysis on encrypted data on different levels of operational complexity. As the computations do not require decryption, no sensitive information is revealed to third parties. The final results can only be decrypted by the client (data owner) if necessary. This constitutes a potential for user privacy and simultaneously open up an attractive market for offering privacy-aware computing services of any kind, e.g., cyber security services.

We envision a privacy-preserving IDS that preserves the privacy of personal or internal data. While being blind for sensitive information, the IDS is trained on encrypted, real-world data allowing data sharing between multiple companies or organizations without fearing information disclosure. Our contribution in this paper is the evaluation of the technical practicality of applying ML techniques for IDS in an FHE environment in terms of efficiency and performance. We introduce an encrypted version of the Pearson correlation coefficient and DBSCAN clustering forming the core of our unsupervised anomaly detection (AD) model prototype. The proof-of-concept implementation highlights the significant potential of utilizing HE in IDS for enhanced privacy while maintaining precision in analysis. The rest of this article is outlined as follows: Sect. 2 gives an overview on related work in the fields of FHE and privacy-aware IDS. Our proof-of-concept algorithm is described in Sect. 3 and the encrypted implementation of a prototype within an FHE environment in Sect. 4. The experimental setup and results are presented in Sect. 5, discussed in Sect. 6 and summarized in Sect. 7.

2 Related Work

The state-of-the-art models for IDS usually use supervised learning techniques relying on deep learning [5,18], hybrid models [33] or ensembles of models [14, 20]. The training requires ground-truth labels, large training datasets, and/or powerful hardware for intense computations. Furthermore, supervised models often lack the capability of predicting zero-day attacks [2] as no training samples are present beforehand. Unsupervised prediction models, in contrast, provide sophisticated AD for IDS that have the capability to handle zero-day attacks by treating them as anomalous traffic [2].

Verkerken et al. [36] demonstrate four unsupervised models (autoencoder, OCSVM, isolation forest and principal component analysis) in their research for identifying cyber attacks. They found the OCSVM models to be the most performant on average in their experiments. The combination with subspace selection techniques that enable analysis on a reduced amount of features [8,27] facilitates the recognition of certain characteristics as a subspace can be sufficient to detect these. The SSC-EA [8] algorithm uses a technique called subspace clustering to obtain multiple partitions of the data by applying a density-based clustering in different subspaces. For each partition, they use evidence accumulation (EA) to rank all outliers with a dissimilarity vector that accumulates outliers distances. The distances are computed between the outliers and the biggest cluster of a subspace. This approach is extended by the SSC-OCSVM [27] algorithm. It uses OCSVM models that are trained on a subspace to learn to distinguish normal and anomalous points. We improve these methods by reducing the amount of subspaces to analyze as we integrate a correlation measure to extend subspaces by a specific attribute instead of all possible combinations. We also train OCSVM models on the subspaces but accumulate their scores for all points and not just outliers found in the DBSCAN clustering. With our approach, we outperform the SSC-EA and SSC-OCSVM in terms of a smaller execution time by a linear subspace complexity. Simultaneously, we maintain a similar or better performance on the detection of anomalous network traffic.

Privacy and security have also become more prominent topics in ML and privacy-aware models are common nowadays [21,34]. The PFMLP framework [12] uses partial HE and federated learning to compute privacy-preserving multi-layer perceptrons for multiple clients. The clients encrypt their locally extracted gradients with the Paillier scheme and share them to a server which integrates the gradients and transmits back the results. The "Concrete ML" framework uses the TFHE scheme [11] to provide encrypted versions of commonly used ML models. The framework is limited to 16-bit integer computations and requires a quantization of floating point numbers, which might yield inaccurate results depending on the size of the gap between consecutive representable values in the data and the quantization parameters. The Cheon-Kim-Kim-Song (CKKS) scheme [26] supports the application of certain operators on floating point numbers instead and omits the loss in precision due to quantization. However, practical deployment remains constrained by noise growth, limited supported operations, and high computational cost.

There have also been advances towards privacy-aware solutions in IDS using different techniques [25], e.g., federated learning [16], public key encryption with keyword search (PEKS) [7] or differential privacy [30]. Current privacy-preserving approaches, such as differential privacy or federated learning, still expose metadata or require partial trust in aggregation servers, leaving fundamental attack surfaces unaddressed. FHE presents a cryptographically robust alternative by enabling arbitrary computation on ciphertexts and presents itself as a promising enabler of data protection and preserver of the secrecy of sensitive information [1,32]. The approach in [32] proposes a collaborative IDS that clusters intrusion alerts to identify large-scale attacks on multiple targets. A third party processes the shared alert data and clusters them according to precomputed distance matrices. The clustering reveals similar alert events on inter-organizational level to affected collaborators. The requirement for clients to also perform "unfeasible" computations [32] in the PHE scheme is a limitation. Due to the shared key-pair among all clients, limited data leakage can occur when information are revealed to other organizations. To overcome this, we use CKKS and avoid computations by the client wherever possible.

3 Unsupervised Anomaly Detection

In this section, we present our proof-of-concept, a novel unsupervised AD method. This detector is trained to identify network-based attacks but does not implement any privacy protection yet. Instead, it serves as algorithmic basis which will be encrypted in the CKKS scheme to render it privacy-aware. We summarize all notational symbols in Table 1. Given $\mathcal{D} \subset \mathbb{R}^m$, the goal is to learn a mapping $f : \mathbb{R}^m \mapsto [0, 1]$ assuming a preprocessing of $\mathcal{D}$ into a numerical representation with dimension m. The learned mapping f distinguishes normal ($\hat{y} = 0$) and anomalous data points ($\hat{y} = 1$) by a score $\hat{y} = f(x)$ between 0 and 1 which is binarized by a threshold λ. The approximation $\hat{y} \approx y$ determines the performance. Furthermore, $\mathcal{D}[A]$ denotes the projection of the dataset on a single attribute or a set of attributes.

Table 1. Notation Symbols

Symbol	Description	Symbol	Description
x	Data point from $\mathcal{R}^m$	$y, \hat{y}$	Label & predicted label
$\mathcal{D}$	Dataset of N data points	$\mathcal{A}$	Attribute set of $\mathcal{D}$
$\mathcal{D}[A]$	Projection to attribute(s) A	$\mathcal{C}$	Clustering of $\mathcal{D}$
ϵ	Neighborhood radius	s_{min}	Neighborhood density
$N_\epsilon(x)$	Neighborhood of point x	o_{max}	Max. number of outliers
d_{min}, d_{max}	Min./max. search depth	$\mathcal{S}$	Set of subspaces
D	Score vector of $\mathcal{D}$	S_i	Candidate subspace
ν	Regulariz. parameter		

Correlation. The dependence of a feature A_i onto another feature A_j is referred to as correlation and correlation measures $Corr(A_i, A_j)$ describe this dependency mathematically. In most cases, the correlation expresses the linear relationship between two features A_i and A_j. A common method for correlation calculation is the Pearson correlation coefficient $Corr(A_i, A_j) = \frac{Cov(X,Y)}{\sigma_X \sigma_Y}$. It divides the covariance $Cov(X, Y)$ of the random variables X and Y that correspond to the features A_i and A_j, respectively, by the product of their standard deviations σ_X and σ_Y.

DBSCAN. A density-based clustering relies on the idea of finding regions in a dataset with a high density of points that are separated from other high-density regions by areas with low density. Any points, that fall into the same dense region, are considered a cluster whereas points in low-density regions are interpreted as outliers or noise without cluster membership. The well-known DBSCAN algorithm computes such a clustering of a dataset $\mathcal{D}$ such that every point $x \in \mathcal{D}$ belongs to at most one cluster or is considered an outlier. Based on a distance function d, it uses ϵ-neighborhoods $N_\epsilon(x) = \{y \in \mathcal{D} \mid d(x,y) \leq \epsilon\}$ to define core points if $|N_\epsilon(x)| \geq s_{min}$ for some density threshold $s_{min} \in \mathbb{N}$. If core points are found, the clusters around them are expanded across dense regions by connecting reachable points, i.e., neighbors $y \in N_\epsilon(x)$ and chains of neighbors $y_1 \in N_\epsilon(y_2), \ldots, y_{i-1} \in N_\epsilon(y_i), y_i \in N_\epsilon(x)$.

OCSVM. The task of distinguishing data points that should belong together and foreign data points can be interpreted as one-class classification problem. An algorithm solves this by learning a function to separate points of the one class from the others. A specific category of these algorithms are One-Class Support Vector Machines (OCSVMs) that try to maximize the margin between a group of points and points that are no group members such that a decision boundary is found. To this end, a hypersphere with center a and radius r is constructed to enclose as many of the training data points $z \in Z_i$ that are members of the group as possible. Meanwhile, a small volume of the hypersphere is maintained by minimizing r^2. The data points closest to the boundary are support vectors.

As the data are usually not linearly separable, the kernel method maps them into a higher dimensional space. Ideally, the fitting of the hypersphere to the data is facilitated there. Thus, a more complex boundary in the original space can be calculated linearly in the kernel space. The data points to be tested are then compared to whether they are lying inside or outside of the hypersphere in kernel space. All points inside are considered normal and abnormal points lie outside. The regularization parameter ν controls the fraction of data to be separated and balances the tradeoff between the fraction of falsely classified data points during training and the fraction of support vectors.

Our proposed Correlation-Based Subspace (CBS) OCSVM adopts the SSC-OCSVM algorithm [27]. The idea is to divide the problem into smaller pieces by training an ensemble of classifiers on carefully selected subspaces and integrate their responses into a prediction. First, we generate meaningful subspaces $\mathcal{S}$ bottom-up (cf. Algorithm 1). The search starts with single feature sets $S_i = \{A_i\}$, for $i = 1, \ldots, m$, to $\mathcal{S}$ in the initial set of subspaces $\mathcal{S}^0$. The subspaces $S_i \in \mathcal{S}^d$ are then extended in each iteration by one additional attribute $A \notin S_i$. The chosen attribute A^* is the one with the least mean absolute correlation $C(S_i, A) = \frac{1}{|S_i|} \sum_{A_j \in S_i} |Corr(A_j, A)|$ to the other features in the subspace S_i. This ensures growth along dimensions with as little redundancy as possible but more variety.

Algorithm 1. Correlation-based subspace search

Require: Dataset $\mathcal{D}$ over attributes $\{A_1, \ldots, A_m\}$; DBSCAN parameters ϵ, s_{min}; Search parameters o_{max}, d_{min}, d_{max}
Ensure: Set of subspaces $\mathcal{S}$

```
S^0 ← {{A_1}, ..., {A_m}}
d ← 0
while True do
    S^{d+1} ← {}
    for S_i ∈ S^d do
        A* = argmin_{A ∉ S_i} C(S_i, A)
        C, O ← DBSCAN(D[S_i ∪ {A*}], ε, s_min)
        if |O| ≥ o_max and d ≥ d_min then
            S^∞ ← S^∞ ∪ {S_i}
        else
            S^{d+1} ← S^{d+1} ∪ {S_i ∪ {A*}}
        end if
    end for
    d ← d + 1
    if d ≥ d_max then
        return S^∞ ∪ S^d
    end if
end while
```

The subspace extension is constrained by two factors: the minimal and maximal depth of the search ($d_{min} \leq |S_i| \leq d_{max}$), and a maximal number of outliers o_{max}. For each selected candidate subspace $S_i \cup \{A^*\}$, a DBSCAN clustering is computed on the projection of the data to the subspace only (denoted as $\mathcal{D}[S_i \cup \{A^*\}]$) using the usual parameters ϵ and the minimal number of samples s_{min}. The clusters $\mathcal{C}$ consist of the grouped border and core points and a set of outlier points O. If $|O| \geq o_{max}$ and $d \geq d_{min}$, the candidate subspace $S_i \cup \{A^*\}$ is discarded. Instead, the original subspace S_i is added to the final set of subspaces $\mathcal{S}^\infty$. Otherwise, if at least one of the two conditions breaks, the candidate subspace is added to $\mathcal{S}^{d+1}$ for further exploration in the next iteration. As soon as the maximal depth is reached, $\mathcal{S}^\infty$ is returned. The algorithm produces at maximum m subspaces (no duplicates).

An OCSVM is then trained on the core points $Z_i \subseteq \mathcal{D}[S_i]$ as computed by DBSCAN on each $S_i \in \mathcal{S}$. As a result, a decision function $f_i : \mathbb{R}^{|S_i|} \mapsto \mathbb{R}$ is learned to predict scores $f_i(\mathcal{D}[S_i])$ of data projected to the according subspaces S_i. The vector $D = [D_1, \ldots, D_N] \in \mathbb{R}^n$ accumulates the OCSVM scores. In contrast to SSC-OCSVM, the scores are accumulated for inliers, too, to improve the scoring as the gap between inliers and outliers in the accumulated scores should diverge more. The score vector D is then min-max normalized and subtracted from a vector of only ones before being returned. A decision threshold λ can then be used to separate normal and anomalous points. This transformation allows for a convenient interpretation of the scores in the range $[0, 1]$ where values close to 0 indicate normal data points and values close to 1 represent potential attacks. The score vector D is then cut off by a threshold $\lambda \in [0, 1]$. The obtained binary labels $\hat{y}$ are then used to further process the traffic flows. Normal traffic flows are maintained as before but anomalous traffic can be investigated deeply or handled immediately, e.g., malicious traffic is reported and blocked.

4 Privacy-Aware IDS

It is crucial to adopt powerful machine learning algorithms to the FHE environment to train sophisticated, privacy-aware detection models. Hence, we verify that our method is capable of being processed in encrypted form while still maintaining an appropriate detection rate. FHE provides significantly better opportunities for configuration to achieve an optimal trade-off between data protection and the detection of attacks than other techniques. Especially, the CKKS scheme supports operations on floating point numbers allowing for an adequate precision and our privacy-aware IDS prototype was implemented in Pyfhel [15].

CKKS. The CKKS scheme is an approximate HE scheme with predefined precision [26] and becomes fully homomorphic by applying the bootstrapping technique. It is currently one of the most efficient schemes to perform approximate homomorphic computations over real and complex numbers [19]. The CKKS scheme relies on the hardness of the Ring Learning With Errors (RLWE) assumption, which is a variant of the Learning With Error (LWE) problem [1,23]. LWE involves a secret vector $s \in \mathbb{Z}_q^n$, a number of pairs of the form (a_i, b_i) where $a_i, b_i \in \mathbb{Z}_q^n$ and the different a_i are chosen independently from the uniform distributed $\mathbb{Z}_q^n$ and a small error $e_i, e_i \in \mathbb{Z}_q^n$ [28]. The goal is to find the secret vector s so that $a_i \cdot s$ equals b_i with the small errors e_i. Instead of vectors in $\mathbb{Z}_q^n$, RLWE uses uses polynomials in $\mathbb{Z}_q[X] \ /(X^n + 1)$ where n is a power of two [23]. This results in smaller keys since all operations are performed between polynomials. It also allows for faster operations, e.g., multiplication can be executed in loglinear time complexity (instead of polynomial) using Fast Fourier Transform [23,28].

Smart and Vercauteren [31] developed a ciphertext packing strategy that enables up to $\frac{n}{2}$ numbers to be encoded into one plaintext object and encrypted into a single packed ciphertext object. This enhances the homomorphic execution of Single-Instruction Multiple-Data (SIMD) operations on the ciphertext data that can operate on all elements simultaneously. These executions support operations such as addition or multiplication of two ciphertexts that add or multiply the elements component-wise.

Limitations of CKKS. FHE is very complex and still faces some limitations and challenges. These include the fact that not all mathematical operations are supported yet. Two of these operations are the square root and the ciphertext-ciphertext division which our CBS-OCSVM requires. Since that is not currently possible, we needed to combine homomorphic and plaintext operations. In particular, we calculate the square root on plaintext and replace the ciphertext-ciphertext division with a ciphertext-plaintext division.

Since the size of ciphertexts is fixed by the chosen CKKS parameter n, the last object is usually not entirely filled with values as $N \bmod \frac{n}{2} \not\equiv 0$. The remaining spaces are internally filled with zeros and theoretically remain zero when using operations like element-wise addition or multiplication. In practice, the remaining values are only approximately zero due to the tiny bit of noise added in CKKS and grow under certain mathematical operations to such an extent that results render unusable. This particularly affects the outcome of the Pyfhel function cumul_add(), that cumulatively adds all values in a ciphertext. We reduce the noise accumulation by using shifting operations and element-wise addition to calculate cumulative sums and provide an auxiliary function cumulatively adding only up to N values. This improved the precision of our results tremendously.

Pearson Correlation. Two previous works already use the Pearson correlation in their FHE frameworks [22,37]. In [37], the Pearson correlation coefficient is used to measure the similarity between a centrally trained IDS model and other client-trained IDS models in order to obtain a more robust model. However, the Pearson correlation coefficients are not calculated on the encrypted data but on the plain text. Liu et al. [22] developed a secure KNN classification scheme for cloud servers that uses the Pearson correlation coefficient to determine the similarity between data vectors. Our implementation performs the necessary calculations on the ciphertexts in CKKS as far as possible and utilizes packing for more efficient computation. This also simplifies the interaction with the client.

Algorithm 2. Pearson correlation in Pyfhel

Require: Lists of ciphertexts cx, cy of features A_x, A_y; number of data points N
Ensure: Pearson correlation coefficient $Corr(A_x, A_y)$
$cx_{cov} \leftarrow$ _cov(cx, N), $cy_{cov} \leftarrow$ _cov(cy, N)
$cxy_{cov} \leftarrow \sim (cx_{cov} \cdot cy_{cov})$
$cxy_{sum} \leftarrow$ sum_array(cxy_{cov}, N)
$cx_{std} \leftarrow \sim (cx^2_{cov})$, $cy_{std} \leftarrow \sim (cy^2_{cov})$
$cx_{stdsum} \leftarrow$ sum_array(cx_std, N)
$cy_{stdsum} \leftarrow$ sum_array(cy_std, N)
rescale_to_next(cx_{stdsum}), rescale_to_next(cy_{stdsum})
$cxy_{std} \leftarrow \sim (cx_{stdsum} * cy_{stdsum})$
$cxy_{sqr} =$ inv_square_root(cxy_{std})
return $cxy_{sum} \cdot cxy_{sqr}$

Algorithm 2 shows the implementation in Pyfhel [15]. The input parameters are two lists $cx = [cx_1, ..., cx_\mu]$ and $cy = [cy_1, ..., cy_\mu]$ of ciphertexts, the number N of data points, and the CKKS contexts (client and server). For the sake of simplicity, the CKKS contexts required for computation and the strict separation of client and server context, which would be absolutely necessary in a real setup, were deliberately omitted for the time being. The ciphertext lists represent the encrypted feature vectors of the two attributes A_x and A_y packed into μ ciphertext objects (cf. Packed DBSCAN).

First, the covariance cxy_{sum} (numerator) is calculated. All individual values are accumulated in the _cov() function (l. 1) by first adding all cx_i element-wise and then cumulatively adding the result using the Pyfhel function cumul_add(). The element-wise division by N yields the arithmetic mean to be subtracted from all feature values. This is also performed for cy and both results are multiplied. Due to the increasing number of polynomials caused by multiplications, a relinearization (l. 2) is necessary to reduce the dimension. Finally, a separate auxiliary function sum_array() is used to form the cumulative sum of the first N elements of the intermediate values (individual products) stored as a list of ciphertexts (l. 3). The result is a single ciphertext cxy_{sum} with the value of the entire sum in the first position. This is necessary because the Pyfhel library function also adds up the whole elements including noise leading to an excessively distorted result due to the previous operations. Instead, the auxiliary function first sums $\mu - 1$ ciphertexts element-wise, adds the resulting ciphertext cumulatively and finally adds the ciphertext with the cumulative sum of the values of the elements from the μ-th ciphertext. The last cumulative sum only includes the remaining values until a total of N values have been calculated using clever rotations and element-wise additions.

The product of the standard deviations cxy_{std} (square of denominator) is formed by squaring the previously calculated intermediate covariance values (l. 4) and cumulatively adding them (l. 5–6). After rescaling with the Pyfhel function rescale_to_next (l. 7), the ciphertexts cx_{stdsum} and cy_{stdsum} are multiplied (l. 8). The reciprocal of the root of cxy_{std} must then be calculated from this product by the client, as this operation is not polynomial and therefore cannot be performed natively in the CKKS scheme. The final result is then determined by multiplying the covariance of the two attributes with the inverse root cxy_{sqr}. The return value is a ciphertext with the correlation value $Corr(A_x, A_y)$ of the two attributes in the first position.

Packed DBSCAN. The crucial part of the DBSCAN algorithm are $\mathcal{O}(N^2)$ pairwise distance computations. Our approach uses a packing scheme to significantly improve it utilizing SIMD operations. In contrast to the vertical encryption for the Pearson correlation, the $x_i \in \mathcal{D}$ are now encrypted horizontally. As comparisons are not supported in CKKS, the distance matrix has to be decrypted by the client. The remaining steps are then carried out on the plaintext distance matrix and use the default sklearn DBSCAN() class.

Figure 1 shows this packing strategy graphically on the three levels. Under the assumption $m \ll \frac{n}{2}$, all instances are padded with zeros and multiple padded instances are batch-wise packed into a single ciphertext. The padding ρ is computed as $\min\{z \in \mathbb{N} \mid \frac{n}{2} \bmod z \equiv 0 \wedge z \geq m\} - m$ based on the dimension m and the number of available slots per ciphertext $\frac{n}{2}$ as shown on the lowest level of Fig. 1. This ensures an equal spacing to keep the operations in the algorithm

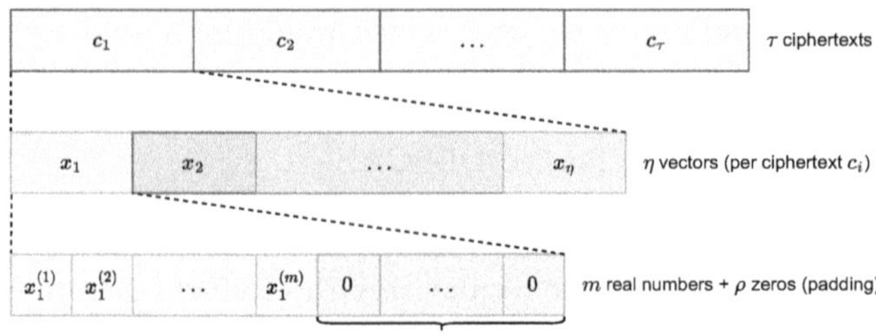

Fig. 1. Packing strategy for efficient encryption.

such as shifts and additions simple. The smallest divisor is chosen to tightly fit (i.e., without random data between padded instances) exactly $\eta = \frac{n/2}{m+\rho} \in \mathbb{N}$ padded x_i of size $m + \rho$ into one ciphertext (cf. middle level of Fig. 1). The i-th ciphertext is $c_i = [x_{(i-1)\cdot\eta+1}, x_{(i-1)\cdot\eta+2}, \ldots, x_{i\cdot\eta}]$ whereas the last ciphertext $c_\tau = [x_{(\tau-1)\cdot\eta+1}, \ldots, x_N, r]$ contains some random encrypted data r at the end filling the remaining $(\eta - (N \ mod \ \eta)) \cdot (m + \rho)$ slots. In total, $\tau = \lceil \frac{N}{\eta} \rceil$ ciphertexts (cf. top level of Fig. 1) encrypt all data instances $x_i \in \mathcal{D}$ in order with this packing strategy.

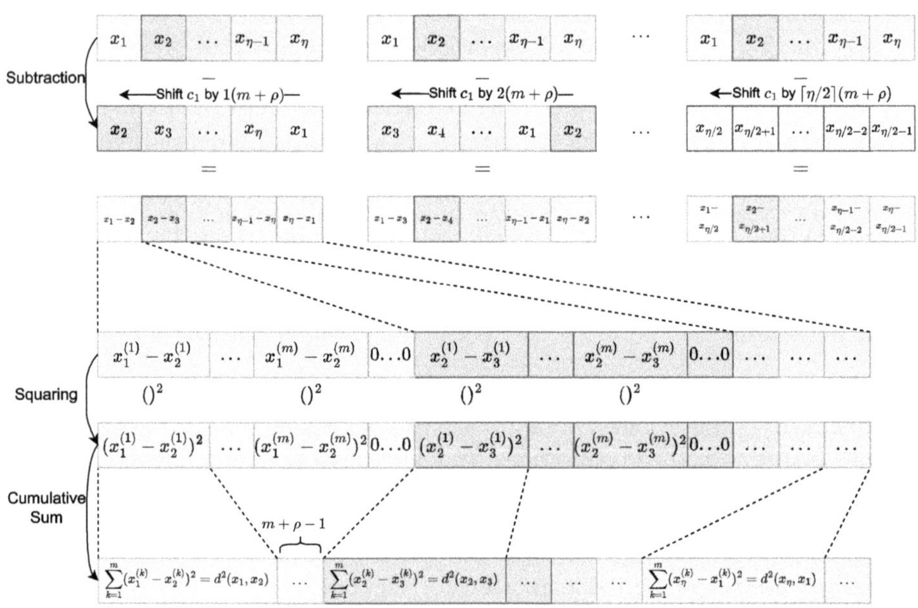

Fig. 2. Scheme of the pairwise distances with packing (first algorithm cycle).

To compute the distance matrix, pairs of ciphertexts are shifted and subtracted such that the resulting ciphertext can be squared component-wise and summed cumulatively. The algorithm runs three nested loops that pair each ciphertext with all the other ciphertexts for each necessary shifting offset to compute $\mathcal{O}(\tau^2)$ ciphertexts $c_{i,j}$ each containing $\mathcal{O}(\eta)$ pairwise distances. The indices $i = 1, \ldots, \tau$ and $j = 1, \ldots, \tau - i + 1$ determine the ciphertext pair c_j and c_{j+i} for calculation. The ciphertext c_{j+i-1} is then rotated with offset $k(m+\rho)$ for $k = 1, \ldots, \eta$ or $k = 2, \ldots, \lceil \frac{\eta}{2} \rceil$ depending on whether $i > 1$ or $i = 1$, respectively. In case of $i = 1$, each ciphertext is paired with itself ($c_j = c_{j+i-1}$) and only half of the rotation offsets are required due to symmetry.

Figure 2 exemplifies this process for the first inner iteration ($j = 1, i = 1, k = 1$) that pairs c_1 with itself. Here, c_1 is rotated by $k = 1$ vector length $m + \rho$ and subtracted from itself resulting in $x_1 - x_2, x_2 - x_3, \ldots, x_\eta - x_1$. The resulting ciphertext holding these differences is self-multiplied, i.e., squared component-wise, and then summed up cumulatively such that the m relevant slots of the data instances (without padding) are accumulated into the first of these slots each. The final result is the ciphertext $c_{1,1,1}$ which stores $d^2(x_1, x_2), d^2(x_2, x_3), \ldots,$ $d^2(x_\eta, x_1)$ at the slots $(l-1) \cdot (d+\rho)$ for $l = 1, \ldots, \eta$. The computation of $c_{i,j,k}$ requires $d - 1$ additions, one subtraction, d shifts and one multiplication (squaring) for η pairwise squared Euclidean distances.

OCSVM Classification. Preliminary works [3,6,10] on encrypted (OC)SVM classification only refer to the implementation of inference on encrypted data, but the model itself is trained on unencrypted data and the relevant parameters such as the support vectors are subsequently encrypted. We let the client train multiple OCSVM models with linear kernels on the CBS subspaces obtained from the preceeding encrypted subspace search. The location of analyzed data points is inspected wrt. the learned hyperspheres in each of the subspaces. The scores for each individual subspace are then accumulated for a final decision. The learned parameters of each model in the ensemble are then encrypted as described in the previous works and used for inference on the encrypted data on the server. First, the learned coefficients and intercepts are padded according to the DBSCAN packing and encrypted. All coefficients are then multiplied with the data. The cumulative sums are computed on the corresponding subspaces and the intercepts are added at the end. The result is a list of lists of ciphertexts where each (except the last one) contains the scores of η data points to be predicted.

5 Results

In this section, we provide our experimental evaluation of the plain and encrypted version of our approach. All the experiments were run on an Intel Xeon Silver 4314 (16 cores, 2.4 GHz). First, we compare the predictive performance of our CBS-OCSVM method (proof-of-concept) with two simple unsupervised baseline methods based on k-Means or DBSCAN clustering $\mathcal{C}$, which could also be implemented in CKKS. For the baselines, the Mahalanobis distances between the data points and the centroid z_{max} of the maximal size clustering $C_{max} \in \mathcal{C}$ are used to establish a ranking for the baselines as it improves the outlier distinction [8]. The ranking is cut off by a threshold λ to separate inliers and outliers. Furthermore, we validate the correctness of the encrypted IDS prototype by comparing (intermediate) results to the results of plain computation. As CKKS is an approximate FHE scheme, a minor deviation of the actual value of computational results has to be expected. This deviation has to be as small as possible to not distort the interpretation or analysis of the results. Additionally, we provide measurements on the efficiency of our DBSCAN packing strategy.

5.1 Datasets and Preprocessing

We used the NSL-KDD dataset [35] and the UNSW-NB15 dataset [24] in our experiments. Both comprise normal and malicious network data. The NSL-KDD dataset provides a training (125972 rows) and a testing version (22543 rows) which both contain 41 mixed features and ground truth labels (only used for evaluation) in binary (0 = normal, 1 = attack) or categorical form ("normal", 22 attack categories). The attack categories are grouped into four broader categories DoS, R2L (root to local attack), U2R (user to root attack) and Probe (probing attacks) [17]. From the UNSW-NB15 dataset, we just present the experiments on the training data (175341 rows, $m = 42$, labels) as the test data showed similar results. The categories contain the attack types analysis, backdoor, DoS, exploits, fuzzers, generic, reconnaissance, shellcode and worms.

We preprocess both datasets independently. Categorical features are one-hot encoded and then all features are scaled into $[0, 1]$ by min-max normalization. We select only specific features based on the Pearson correlation coefficient and we remove $A_i \in \mathcal{A}$ sharing a low absolute correlation with the binary label ($|PCC(y, A_i)| < 0.2$). We also identify redundant pairs $A_i, A_j \in \mathcal{A}$ correlating highly with each other ($|PCC(A_i, A_j)| > 0.9$) and remove one attribute.

Evaluation Subsets. Following [27], we sample different subsets from the datasets. One subset for each kind of attack type is generated and a subset containing all attack types mixed from each dataset. The single attack subsets of the NSL-KDD datasets contain 2000 normal samples and 50 (R2L, U2R) or 100 attack samples (DoS, Probe) and the mixed subsets contain 9000 normal and $2 \cdot 100 + 2 \cdot 50 = 300$ attack samples. The single attack UNSW-NB15 subsets are sampled with the same 2000:100 ratio except for the worm attacks from the testing subset (2000:25). The results are also compared to the SSC-OCSVM algorithm [27] on the NSL-KDD dataset. However, we could not reimplement this algorithm due to lacking information or execute an existing implementation. The mixed UNSW-NB15 subsets also contain 9000 normal samples and $9 \cdot 50 = 450$ attack samples. In the following, we refer to them as "$\langle$Dataset$\rangle$ $\langle$Version$\rangle$ $\langle$Category/Mixed$\rangle$", e.g., NSLTrainDoS for the DoS subset of the training NSL-KDD dataset.

For the sake of simplicity, we fix the parameters throughout all experiments on the same dataset. On the NSL-KDD training and testing subset we use the parameters values $o_{max} = 50, d_{min} = 1, d_{max} = 5, \epsilon = 0.5, s_{min} = 30, \nu = 0.5$ and a linear kernel function for the OCSVM models. The k-Means clustering is run for $k = 5$ and the DBSCAN clustering with the same ϵ and s_{min} as the CBS-OCSVM. For the UNSW-NB15 training subsets, we reuse the same values for the parameters o_{max} and d_{min} as before. The other parameters are adapted: $d_{max} = 4, \epsilon = 0.1, s_{min} = 50, \nu = 0.1$ and the kernel is a radial basis function. The baseline algorithms reuse the same parameter values as before.

5.2 Proof-Of-Concept

Table 2. NSL-KDD subsets best TPR & FPR

Approach	Training data subsets				
	Probe	DoS	R2L	U2R	Mixed
CBS-OCSVM	0.95, 0.19	0.91, 0.03	0.96, 0.47	1.00, 0.26	0.92, 0.24
k-Means	1.00, 0.33	0.97, 0.34	1.00, 0.43	1.00, 0.45	1.00, 0.44
DBSCAN	1.00, 0.23	1.00, 0.23	1.00, 0.48	1.00, 0.50	1.00, 0.49
	Testing data subsets				
	Probe	DoS	R2L	U2R	Mixed
CBS-OCSVM	1.00, 0.10	0.98, 0.08	1.00, 0.12	1.00, 0.06	0.91, 0.17
k-Means	0.96, 0.21	0.75, 0.03	0.98, 0.23	0.96, 0.21	1.00, 0.32
DBSCAN	0.99, 0.21	0.99, 0.20	0.98, 0.23	1.00, 0.20	1.00, 0.32

Table 3. UNSW-NB15 training subsets best TPR & FPR

Approach	Analysis	Backdoor	DoS	Exploits	Fuzzers	Generic
CBS-OCSVM	0.43, 0.05	0.86, 0.29	0.83, 0.25	0.93, 0.28	0.84, 0.41	0.80, 0.16
k-Means	0.67, 0.06	0.77, 0.05	0.75, 0.06	1.00, 0.49	0.99, 0.44	0.99, 0.00
DBSCAN	1.00, 0.51	0.79, 0.19	0.76, 0.19	1.00, 0.49	0.99, 0.46	0.99, 0.00
	Reconn.	Shellcode	Worms	Mixed		
CBS-OCSVM	0.91, 0.38	0.90, 0.34	0.91, 0.33	0.76, 0.23		
k-Means	0.98, 0.41	1.00, 0.36	0.96, 0.44	0.57, 0.06		
DBSCAN	0.95, 0.40	0.98, 0.42	0.99, 0.48	1.00, 0.50		

Different pairs of FPR and TPR are obtained for different λ. We select the best λ^* maximizing $TPR(D_\lambda, y) - FPR(D_\lambda, y)$. This threshold yields a reasonable balancing of TPR and FPR for comparison where the TPR should me maximized and the FPR should be minimized. The resulting (TPR, FPR)-pairs are reported in Table 2 for all NSL-KDD subsets and in Tables 3 & 4 for the UNSW-NB15 subsets. The CBS-OCSVM clearly outperforms the baselines on certain NSL-KDD subsets with a similar or better TPR value but a much lower FPR (Train: U2R; Test: Probe, DoS, R2L, U2R). The baseline model performances

Table 4. UNSW-NB15 testing subsets best TPR & FPR

Approach	Analysis	Backdoor	DoS	Exploits	Fuzzers	Generic
CBS-OCSVM	0.97, 0.62	0.53, 0.06	0.55, 0.17	0.65, 0.41	0.93, 0.73	0.98, 0.18
k-Means	0.91, 0.08	0.87, 0.16	0.74, 0.10	1.00, 0.71	1.00, 0.62	0.96, 0.00
DBSCAN	0.91, 0.16	0.87, 0.16	0.75, 0.17	1.00, 0.71	1.00, 0.62	0.96, 0.00
	Reconn.	Shellcode	Worms	Mixed		
CBS-OCSVM	0.73, 0.40	0.80, 0.38	0.84, 0.34	0.76, 0.37		
k-Means	0.98, 0.55	0.95, 0.50	0.88, 0.35	0.62, 0.18		
DBSCAN	0.98, 0.55	0.95, 0.55	0.88, 0.40	0.62, 0.18		

are quite similar throughout all subsets with one exception on the DoS testing subset. They often achieve a max. TPR in trade for a suboptimal FPR (0.2–0.5) whereas our method provides only slightly smaller TPRs but in relation much better FPRs, e.g., on both mixed subsets our approach attains rate differences of 0.68 (train) and 0.74 (test). The baselines, instead, have only a maximal rate difference of 0.56 (train) and 0.68 (test). The trade-off by missing some of the attacks is worth the decreased rate of false alarms as it is also assumed that in reality regular traffic is much more frequent and, thus, a high FPR potentially overwhelms the IDS.

The baseline algorithms have a very similar performance on the UNSW-NB15 training subsets with only a few exceptions. Optimizing again for the equally weighted tradeoff between detections and false alarms, our approach shows the best performance on the mixed dataset (0.76, 0.23) whereas k-Means has only a TPR of 0.57 and a low FPR. The DBSCAN, in contrast, reaches a perfect detection rate but also a 50% false alarm rate. However, the performance on single attack subsets is more divergent and the CBS-OCSVM method is not always on top. At least one baseline performs better on the analysis, fuzzers, reconnaissance and shellcode subset with a similar FPR but a better TPR. The k-Means baseline provides a better balance of TPR and FPR on the backdoor and DoS attack subset than our method. Additionally, the CBS-OCSVM has a slightly better AUC score (0.83) than the baseline approaches (k-Means 0.82, DBSCAN 0.78) on the mixed training dataset.

The execution times of CBS-OCSVM range between 65 s and 105 s on the mixed subsets. We could not reproduce the execution times of the SSC-OCSVM and no hardware information was given [8,27]. Based on theoretical considerations, we assume CBS-OCSVM to be much quicker. The SSC-OCSVM identifies $\mathcal{O}(m^2)$ subspaces whereas our approach only investigates $\mathcal{O}(m)$ subspaces which significantly affects the run time when upscaling the dataset size. Additionally, our approach can be customized more flexibly by adjusting the parameters such as d_{min} or d_{max} to constrain the search space. Under the assumption that the SSC-OCSVM threshold was optimized analogously, the rate differences in their experiments indicate similar performance on the NSL-KDD subsets to our approach.

5.3 Encrypted Performance

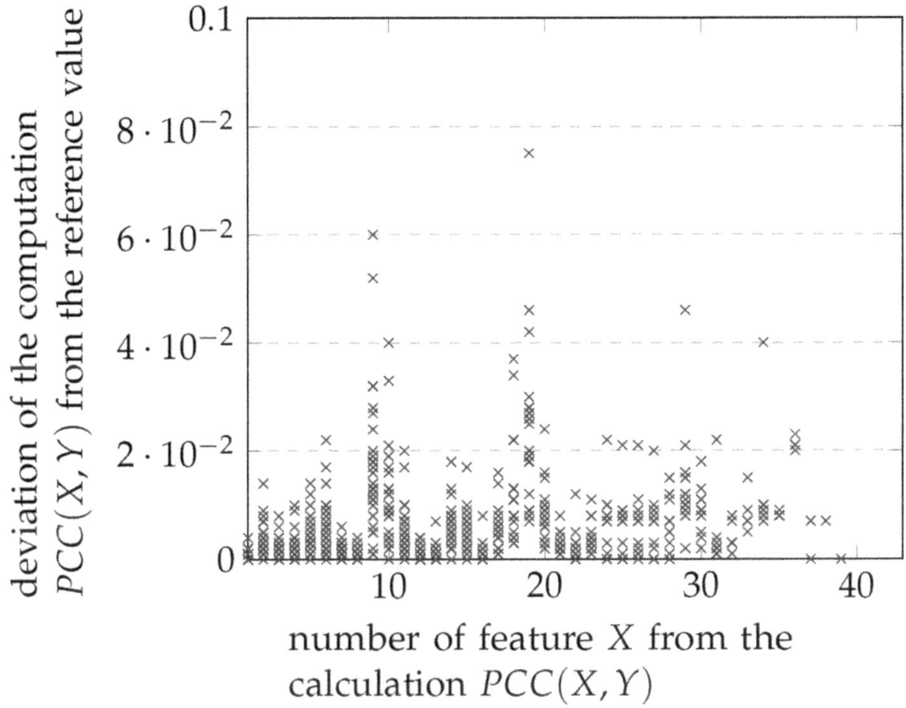

Fig. 3. Results of the Pearson correlation in encrypted computation for each attribute of the NSL-KDD dataset. Absolute deviation values for pairs of features are plotted on the y-axis against the fixed features as represented on the x-axis.

We investigated the deviation of the Pearson correlation in encrypted computation for each attribute of both IDS dataset. Figure 3 shows the absolute deviation values (y-axis) of $PCC(X, Y)$ for a feature X (x-axis) against all other features Y for the NSL-KDD dataset. The maximum deviation to the actual value across all features is $\approx 8 \cdot 10^{-2}$. However, the deviation of the majority of the correlation values smaller than $1 \cdot 10^{-2}$ and sometimes even so small that it is not even visible from the plot. The plot for the NSL-KDD values is omitted due to similarity but the maximal deviation of $\approx 5 \cdot 10^{-3}$ is even one magnitude less. Even the maximum deviation of $\approx 8 \cdot 10^{-2}$ does not affect the interpretation of correlation values in CBS-OCSVM negatively and, thus, the precision of encrypted Pearson computation is sufficient for our approach.

For visualization purposes, our encrypted DBSCAN implementation was first benchmarked using six toy datasets ($N = 500, m = 2$) for clustering from the

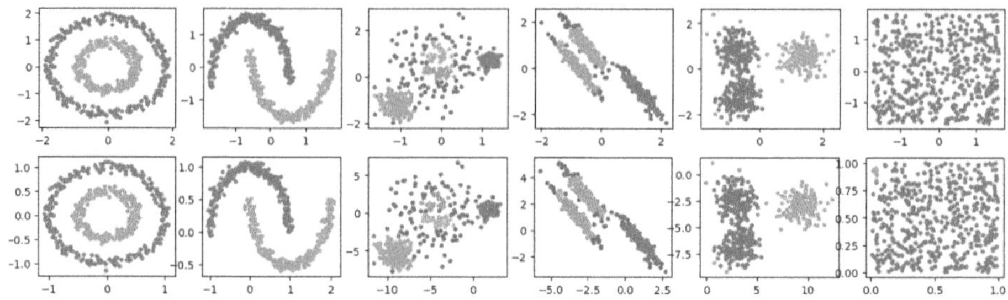

Fig. 4. Toy dataset benchmark with six datasets (columns 1–6) for clustering with plaintext DBSCAN (top row) and our encrypted DBSCAN (bottom row). The cluster membership is represented by the point colors.

sklearn library[1]. They consist of two nested circles, two interleaving half circles (moons), three blobs with varying variance, three anisotropic blobs, three isotropic blobs, and unstructured data as shown in Fig. 4 (col. 1–6). The top row shows the plaintext DBSCAN result compared to our encrypted DBSCAN at the bottom row for the same parameter values ϵ and s_{min}. The point colors indicate the cluster membership and outliers are brown (col. 3&4) or green (col. 5). Visually, there are almost no differences between any pair of clustering result except a minor deviation in the last column reveals (second small cluster in top left corner) - most probably due to noise incurred in the CKKS scheme. We verified the equality of clustering results with standard external cluster validation indices (adjusted Rand index & normalized mutual information). As the graphics already showed, the index values confirm perfect or almost perfect ($\approx$1) matching indicating the correctness of our implementation.

We also measured execution time of two encrypted DBSCAN variants (with and without packing) and compared it to the plaintext version for the six toy datasets. The slowest computation of 1662.62 s on average was measured for the DBSCAN without packing and the packing approach significantly reduced this to 15.58 s on average. For reference, the plaintext DBSCAN ($7.5 \cdot 10^{-3}$s) is on average more than 2000 and 200000 times faster than the encrypted DBSCAN with and without packing, respectively, in trade for privacy. Our packing strategy yields a tremendous speedup of being around 100 times faster than the native implementation. Benchmarking of the data protection-compliant prototype was limited to evaluation using the same partial data sets as for evaluation of the proof-of-concept algorithm. For all subsets of the UNSW-NB15 and NSL-KDD datasets, the same subspaces and the same results were achieved. The noise added by the encryption and calculations in CKKS did neither affect the correlation-based subspace search nor the inference or calculation of the anomaly scores with the OCSVM ensemble. This also results in the exact same TPR and FPR as for the proof-of-concept evaluation on plaintext data (cf. Sect. 5.2).

[1] https://scikit-learn.org/stable/auto_examples/cluster/plot_cluster_comparison.html.

6 Discussion

Our experiments show the decent results of our algorithm applied to the two IDS datasets. Our improvement does not limit the performance and even boosts the run time by a factor of $\mathcal{O}(m)$ in comparison to other subspace AD approaches [8,27]. The clever parallelization of instructions by SIMD operations in our packing strategy improves the run time in comparison to a native implementation significantly. This represents an enormous time advantage in terms of application for larger IDS datasets and the fact that CBS-OCSVM requires multiple DBSCAN clusterings to be calculated in each of the iteration steps depending on the parameters for search depth, maximum sub-space size, and maximum number of outliers. Nevertheless, the gain in privacy and confidentiality of potentially sensitive data by using FHE comes at the expense of extremely increased processing time. The significantly increased model training time is acceptable, as this process only needs to take place once or occasionally. The inference time for given potential attacks, however, must be as fast as possible in order to trigger a timely response that prevents attempted intrusion and wards off damage. This can still pose quite a challenge to the requirements of modern IDS having to handle huge numbers of messages continuously especially at higher scales.

The stopping criteria of the CBS and the parameters enable a customization of the AD making it adaptable. Depending on the dataset structure, the parameter selection might be sensitive to changes and some parameters also counteract each other producing unexpected results (s_{min} & o_{max}). The toy data sets (cf. Fig. 4) clearly show the structures suiting DBSCAN. The nested circles (col. 1) or the interleaving moons (col. 1&2) can be correctly separated into two clusters (e.g., this would be impossible for k-means). The diagonally scattered regions and the isotropic blobs (col. 4&5) are also grouped relatively well. In these cases, changing the clustering parameters again would possibly yield better results with the correct number of clusters recognized. However, this is irrelevant for the purpose of verifying the accuracy of the encrypted DSCAN.

A balance of ν and the clustering based on ϵ and s_{min} is also crucial for an effective computation. The limit for training errors and the fraction of support vectors are balanced by the parameter ν of the OCSVM model. It influences how sharp the boundary is that encloses the training points considered normal. If the clusterings in the subspaces match the ground-truth normal data points well, sharper boundaries can be chosen. If the clusterings do not reflect them very well, ν can be adjusted to learn weaker boundaries which account for more errors and provide a better generalization capability. We experienced the best result on the experiments on the NSL-KDD for a balanced choice of $\nu = 0.5$ yielding an equal tradeoff between the training errors and fraction of support vectors. The more complex UNSW-NB15 subsets required a weaker boundary to be learned ($\nu = 0.1$) in the OCSVM models to better mitigate the matching error between the clusterings and the normal data points.

7 Conclusion

We analyzed the feasibility of FHE techniques in IDS to improve privacy and protect sensitive data while identifying malicious activities in the web. CBS-OCSVM is a proof-of-concept for a flexible, unsupervised AD model that can be trained for the purpose of an IDS. The parameters of the algorithm provide a well configurable setup that can be adjusted and fine-tuned according to the scenario. Our results prove a similar performance to a state-of-the-art method and outperforming baselines. We adopted the proposed CBS-OCSVM algorithm in an FHE scheme which required the encryption of different parts of operations like the DBSCAN clustering, the Pearson correlation measure and OCSVM models. The runtime improvement through reducing the number of subspace candidates is a necessary step towards executing the method in the computation-heavy FHE setting. The identification of network-based attacks through encrypted computations with our anomaly detector contributes a concrete step towards a cyber-resilient and privacy-aware web.

For future work, it would be technically interesting to incorporate methods like polynomial inverse square root approximation, homomorphic comparison, or OCSVM training in CKKS to fully protect privacy by avoiding the need for the decryption of intermediate results and eliminate client interaction. A deep investigation of the legal conformity of the approach like the compliance with GDPR regulations is also highly recommended. This benefits digital forensics and legal prosecution of identified breaches, too, as found evidence can be stored encrypted and tamper-proof and only the evidential data has to be revealed (data minimization). Furthermore, our encrypted anomaly detector is also applicable in different application domains, e.g., to detect outliers in medical data to distinguish patterns like electrocardiograms of potentially ill patients.

Acknowledgments. This study received funding by the state of Hessen (Distr@l framework - Machbarkeitsstudien Förderlinie 1) through the CryptScan project (Project Number 71674641).

Disclosure of Interests. The authors have no competing interests to declare that are relevant to the content of this article.

References

1. Acar, A., Aksu, H., Uluagac, A.S., Conti, M.: A survey on homomorphic encryption schemes. ACM Comput. Surv. **51**(4), 1–35 (2018). https://doi.org/10.1145/3214303
2. Ahmad, R., Alsmadi, I., Alhamdani, W., Tawalbeh, L.: Zero-day attack detection: a systematic literature review. Artif. Intell. Rev. 1–79 (2023)
3. Al Badawi, A., Chen, L., Vig, S.: Fast homomorphic svm inference on encrypted data. Neural Comput. Appl. **34**(18), 15555–15573 (2022)
4. Arshad, J., Azad, M.A., Amad, R., Salah, K., Alazab, M., Iqbal, R.: A review of performance, energy and privacy of intrusion detection systems for IoT. Electronics **9**(4), 629 (2020)

5. Awajan, A.: A novel deep learning-based intrusion detection system for iot networks. Computers **12**(2), 34 (2023)
6. Brand, M., Pradel, G.: Practical privacy-preserving machine learning using fully homomorphic encryption. Cryptology ePrint Archive (2023)
7. Canard, S., Diop, A., Kheir, N., Paindavoine, M., Sabt, M.: Blindids: market-compliant and privacy-friendly intrusion detection system over encrypted traffic. In: Proceedings of the 2017 ACM on Asia Conference on Computer and Communications Security, pp. 561–574 (2017)
8. Casas, P., Mazel, J., Owezarski, P.: Unsupervised network intrusion detection systems: detecting the unknown without knowledge. Comput. Commun. **35**(7), 772–783 (2012)
9. Chen, P., Li, F., Wu, C.: Research on intrusion detection method based on pearson correlation coefficient feature selection algorithm. In: Journal of Physics: Conference Series, vol. 1757, p. 012054 (2021). https://doi.org/10.1088/1742-6596/1757/1/012054
10. Chen, Y., Mao, Q., Wang, B., Duan, P., Zhang, B., Hong, Z.: Privacy-preserving multi-class support vector machine model on medical diagnosis. IEEE J. Biomed. Health Inform. **26**(7), 3342–3353 (2022)
11. Chillotti, I., Gama, N., Georgieva, M., Izabachène, M.: Tfhe: fast fully homomorphic encryption over the torus. J. Cryptol. **33**(1), 34–91 (2020)
12. Fang, H., Qian, Q.: Privacy preserving machine learning with homomorphic encryption and federated learning. Future Internet **13**(4), 94 (2021)
13. He, K., Kim, D.D., Asghar, M.R.: Adversarial machine learning for network intrusion detection systems: a comprehensive survey. IEEE Commun. Surv. Tutor. (2023)
14. Hossain, M.A., Islam, M.S.: Ensuring network security with a robust intrusion detection system using ensemble-based machine learning. Array **19**, 100306 (2023)
15. Ibarrondo, A., Viand, A.: Pyfhel: python for homomorphic encryption libraries. In: Proceedings of the 9th on Workshop on Encrypted Computing & Applied Homomorphic Cryptography, pp. 11–16 (2021)
16. Idrissi, M.J., et al.: Fed-anids: federated learning for anomaly-based network intrusion detection systems. Expert Syst. Appl. **234**, 121000 (2023)
17. Ieracitano, C., et al.: Statistical analysis driven optimized deep learning system for intrusion detection. In: Ren, J., et al. (eds.) BICS 2018. LNCS (LNAI), vol. 10989, pp. 759–769. Springer, Cham (2018). https://doi.org/10.1007/978-3-030-00563-4_74
18. Kasongo, S.M.: A deep learning technique for intrusion detection system using a recurrent neural networks based framework. Comput. Commun. **199**, 113–125 (2023)
19. Kim, A., Papadimitriou, A., Polyakov, Y.: Approximate homomorphic encryption with reduced approximation error. In: Galbraith, S.D. (ed.) CT-RSA 2022. LNCS, vol. 13161, pp. 120–144. Springer, Cham (2022). https://doi.org/10.1007/978-3-030-95312-6_6
20. Lazzarini, R., Tianfield, H., Charissis, V.: A stacking ensemble of deep learning models for iot intrusion detection. Knowl.-Based Syst. **279**, 110941 (2023)
21. Liu, B., Ding, M., Shaham, S., Rahayu, W., Farokhi, F., Lin, Z.: When machine learning meets privacy: a survey and outlook. ACM Comput. Surv. (CSUR) **54**(2), 1–36 (2021)
22. Liu, J., Wang, C., Tu, Z., Wang, X.A., Lin, C., Li, Z.: Secure knn classification scheme based on homomorphic encryption for cyberspace. Secur. Commun. Netw. **2021**, 1–12 (2021). https://doi.org/10.1155/2021/8759922

23. Lyubashevsky, V., Peikert, C., Regev, O.: On ideal lattices and learning with errors over rings. J. ACM **60**(6) (2013). https://doi.org/10.1145/2535925
24. Moustafa, N., Slay, J.: Unsw-nb15: a comprehensive data set for network intrusion detection systems (unsw-nb15 network data set). In: 2015 military communications and information systems conference (MilCIS), pp. 1–6. IEEE (2015)
25. Niksefat, S., Kaghazgaran, P., Sadeghiyan, B.: Privacy issues in intrusion detection systems: a taxonomy, survey and future directions. Comput. Sci. Rev. **25**, 69–78 (2017)
26. Panda, S.: Principal component analysis using CKKS homomorphic encryption scheme. In: Dolev, S., Margalit, O., Pinkas, B., Schwarzmann, A. (eds.) Cyber Security Cryptography and Machine Learning. CSCML 2021. LNCS, vol. 12716, pp. 52–70. Springer, Cham (2021). https://doi.org/10.1007/978-3-030-78086-9_4
27. Pu, G., Wang, L., Shen, J., Dong, F.: A hybrid unsupervised clustering-based anomaly detection method. Tsinghua Sci. Technol. **26**(2), 146–153 (2020)
28. Regev, O.: On lattices, learning with errors, random linear codes, and cryptography. J. ACM **56**(6) (2009). https://doi.org/10.1145/1568318.1568324
29. Sharif, M.H.U., Mohammed, M.A.: A literature review of financial losses statistics for cyber security and future trend. World J. Adv. Res. Rev. **15**(1), 138–156 (2022)
30. Siachos, I., Kaltakis, K., Papachristopoulou, K., Giannoulakis, I., Kafetzakis, E.: Comparison of machine learning algorithms trained under differential privacy for intrusion detection systems. In: 2023 IEEE International Conference on Cyber Security and Resilience (CSR), pp. 654–658. IEEE, Venice, Italy (2023). https://doi.org/10.1109/CSR57506.2023.10225010
31. Smart, N.P., Vercauteren, F.: Fully homomorphic simd operations. Des. Codes Crypt. **71**, 57–81 (2014)
32. Spathoulas, G., Theodoridis, G., Damiris, G.-P.: Using homomorphic encryption for privacy-preserving clustering of intrusion detection alerts. Int. J. Inf. Secur. **20**(3), 347–370 (2020). https://doi.org/10.1007/s10207-020-00506-7
33. Talukder, M.A., et al.: A dependable hybrid machine learning model for network intrusion detection. J. Inf. Secur. Appl. **72**, 103405 (2023)
34. Tanuwidjaja, H.C., Choi, R., Baek, S., Kim, K.: Privacy-preserving deep learning on machine learning as a service–a comprehensive survey. IEEE Access **8**, 167425–167447 (2020)
35. Tavallaee, M., Bagheri, E., Lu, W., Ghorbani, A.A.: A detailed analysis of the kdd cup 99 data set. In: IEEE Symposium on Computational Intelligence for Security and Defense Applications, pp. 1–6. IEEE (2009)
36. Verkerken, M., D'hooge, L., Wauters, T., Volckaert, B., De Turck, F.: Towards model generalization for intrusion: Unsupervised machine learning techniques. J. Netw. Syst. Manag. **30**, 1–25 (2022)
37. Wang, J., Xia, Z., Chen, Y., Hu, C., Yu, F.: Intrusion detection framework based on homomorphic encryption in ami network. Front. Phys. **10**, 1102892 (2022). https://doi.org/10.3389/fphy.2022.1102892
38. Zhang, C., Jia, D., Wang, L., Wang, W., Liu, F., Yang, A.: Comparative research on network intrusion detection methods based on machine learning. Comput. Secur. 102861 (2022)

OSINT Monitoring in German Organizations: A Benchmark Study

Thomas Fritzler(✉), Toni Le(✉), and Michael Massoth

Department of Computer Science, Hochschule Darmstadt (h_da) - University of Applied Sciences, Member of European University of Technology (EUt+), Darmstadt, Germany
{thomas.fritzler,toni.le}@stud.h-da.de, michael.massoth@h-da.de

Abstract. Open-source intelligence (OSINT) monitoring is widely used in cyber threat intelligence, yet there is limited empirical evidence on how organizations operationalize OSINT monitoring across use cases and how such capabilities can be benchmarked in a reproducible manner. This study employs a cross-sectional survey design and develops a standardized measurement approach to capture organizational OSINT monitoring in terms of monitoring breadth and monitoring intensity. We contribute (i) a content-validated, reusable survey instrument for organizational OSINT monitoring, (ii) a transparent 0–100 maturity index based on explicit aggregation assumptions, and (iii) an exploratory empirical benchmark from a German organizational sample. The results show that OSINT monitoring adoption concentrates on technically proximate use cases such as domain security, vulnerability and exploit intelligence, and external attack surface management, while counterfeit/fake-shop and disinformation monitoring remain limited; overall maturity is skewed toward early stages and is significantly higher in larger organizations. The proposed instrument and maturity index provide a practical and reproducible basis for benchmarking organizational OSINT monitoring and support future research on its determinants, governance, and effectiveness.

Keywords: OSINT monitoring · Cyber threat intelligence · Survey instrument development

1 Introduction

Open-source intelligence (OSINT)—intelligence derived from publicly accessible sources—is widely discussed as an enabler for cyber threat intelligence (CTI) and proactive security operations [4,30]. In security practice, OSINT monitoring is expected to support early warning, prioritization, and decision-making in the presence of uncertainty. Relevant signals may include newly registered domains or DNS changes that indicate brand impersonation, exposure of services and

T. Fritzler and T. Le—These authors contributed equally to this work.

© The Author(s), under exclusive license to Springer Nature Switzerland AG 2027
K. Renaud et al. (Eds.): EICC 2026, CCIS 3012, pp. 267–286, 2027.
https://doi.org/10.1007/978-3-032-28957-5_16

assets on the public Internet, leaked credentials and documents, or early exploit discussions and proof-of-concepts [22,26,27].

However, a recurring challenge lies in the distinction between the existence of technical OSINT methods and the institutionalization of *organizational* monitoring capability. Threat landscape reporting by governmental agencies and supranational bodies describes attack categories and trends, yet typically does not quantify how organizations implement OSINT monitoring as a routine capability [5,9]. At the same time, practice-oriented guidance (e.g., security frameworks) provides normative structure for cyber security outcomes but leaves open the question of how OSINT monitoring is operationalized and measured across organizations [20]. This gap is particularly salient for small and medium-sized organizations, for which comprehensive frameworks can be perceived as complex and resource-intensive [2].

1.1 Problem Statement and Motivation

Two observations motivate the work.

First, the external attack surface of organizations produces signals that are, at least partly, observable in open sources. Domain-related abuse provides a concrete example. The domain name life cycle constitutes an attack surface in which typosquatting, expired-domain abuse, and exploitation of dangling records are repeatedly documented and mapped to preventive controls [12]. Several of these control classes are monitoring-centric: detection of new domain registrations that mimic protected identifiers, monitoring of DNS records, and monitoring of domain expiry and renewal processes. Such monitoring can be performed without access to internal networks, yet it requires systematic governance and integration into response workflows to be effective.

Second, organizations face constraints and boundary conditions that complicate systematic monitoring. Constraints include limited personnel, fragmented tooling, and the difficulty of translating heterogeneous open-source signals into actionable CTI or incident response decisions. In addition, OSINT monitoring is subject to legal and ethical constraints: publicly accessible does not necessarily imply unrestricted processing, and organizational policies must align monitoring activities with lawful bases, proportionality, and accountability requirements [4]. These constraints are particularly pronounced in public-sector and governmental settings, where the protected objects may include official designations and sovereign digital identities rather than trademarks, and where transparency and proportionality constraints may be stronger.

Given these conditions, a systematic empirical picture of current OSINT monitoring practice is needed. Such a picture requires (i) a measurement instrument with defensible validity and reliability properties, and (ii) an operational definition of OSINT monitoring maturity that allows comparison and benchmarking.

1.2 Research Questions and Contributions

The study is guided by three research questions (**RQ**):

1. **Patterns and benchmarking:** Which patterns of OSINT monitoring (use-case coverage and monitoring frequency) are observable across organizations, and how can the results be operationalized as a benchmark?
2. **Organizational determinants:** How do OSINT monitoring maturity levels differ by organizational size and industry sector?
3. **Measurement quality:** How can content validity and reliability be established for a questionnaire intended to measure organizational OSINT monitoring?

Three contributions are provided.

1. **Instrument contribution:** A standardized survey instrument is developed using established questionnaire engineering procedures and expert validation (CVR/CVI) [13,15,16,18,24].
2. **Model contribution:** A transparent 0–100 maturity index is defined, combining breadth (use-case coverage) and intensity (monitoring frequency). The index enables reproducible benchmarking and explicit discussion of assumptions [21].
3. **Empirical contribution:** Exploratory empirical results from a German cross-sectional survey ($n = 33$ organizations) are reported, including (i) adoption patterns across use cases, (ii) maturity distribution and benchmark tables, (iii) a robust size effect on maturity, and (iv) measurement-quality evidence (internal consistency and exploratory factor structure).

2 Background and Related Work

This section positions organizational OSINT monitoring conceptually, defines the use-case lens adopted in the instrument, and clarifies why maturity is interpreted here as a benchmark of monitoring activity rather than as a full governance audit.

2.1 OSINT and CTI: Conceptual Positioning

OSINT is commonly defined as intelligence derived from publicly accessible sources [4]. In practice, OSINT is often embedded within threat intelligence processes, where open sources are combined with commercial feeds, closed communities, and internal telemetry to form CTI products [30]. This positioning implies two conceptual distinctions relevant for measurement.

First, OSINT as an *information source class* must be distinguished from OSINT as an *organizational capability.* While the former concerns where information can be obtained, the latter concerns how systematically information is collected, validated, interpreted, disseminated, and acted upon. Second, OSINT monitoring must be distinguished from ad hoc information gathering. Monitoring is interpreted here as repeated, institutionalized acquisition and analysis of open-source signals aligned with defined monitoring objectives (use cases), which are embedded in decision-making processes (e.g., triage, incident response, governance).

The literature reports numerous technical applications of OSINT for security operations, such as extraction of infrastructure and user information from artifacts [27], identification of threat indicators and campaigns, and discovery of exposure on the public Internet [26]. At the same time, OSINT research is described as relatively young and characterized by methodological gaps, particularly regarding evaluation and organizational adoption [22]. Several recent works focus on OSINT practices and challenges, providing conceptual overviews or practitioner-oriented syntheses [1,3,25,28]. These contributions underscore the need for measurement instruments that can capture organizational implementation beyond individual techniques.

2.2 Use-Case Perspective: Making OSINT Monitoring Measurable

To make OSINT monitoring measurable, the present study decomposes monitoring capability into a set of recurring monitoring objectives (use cases). This decomposition follows the logic that organizational maturity is multidimensional: monitoring is rarely uniform across all external threats, and resource allocation typically prioritizes a subset of objectives.

The instrument operationalizes eleven OSINT monitoring use cases: (i) brand protection; (ii) domain security; (iii) social media monitoring; (iv) public visibility/media presence; (v) external attack surface management; (vi) vulnerability/exploit intelligence; (vii) credential and document leaks; (viii) supply-chain security; (ix) counterfeit/fake-shop monitoring; (x) disinformation monitoring; and (xi) dark-web monitoring. This selection is consistent with prior taxonomies and discussions of OSINT and CTI use cases [22,26,30].

These use cases were not proposed by respondents during scoring; rather, the questionnaire presented the same fixed battery to all participants, which supports direct comparison across organizations. For interpretive clarity, the use cases can be summarized as follows: brand protection captures monitoring for misuse of brands or official identifiers; domain security covers suspicious registrations, expiry, and DNS-related abuse; social media monitoring targets impersonation or abuse on major platforms; public visibility/media presence captures public mentions and media narratives; external attack surface management covers Internet-facing assets and exposures; vulnerability/exploit intelligence captures disclosures, exploit discussions, and vulnerability prioritization; credential and document leaks address exposed credentials or documents; supply-chain security targets partner or third-party exposures; counterfeit/fake-shop monitoring addresses fraudulent shops and product abuse; disinformation monitoring captures manipulative narratives affecting the organization; and dark-web monitoring addresses criminal forums and markets relevant to the organization.

The use-case lens also supports interdisciplinary integration. Brand protection and domain security intersect with trademark and identifier protection, but also with technical security outcomes (phishing, impersonation, DNS abuse). Conversely, disinformation monitoring intersects with information security and organizational risk, but also with communication strategy and legal constraints. By measuring adoption and intensity at the use-case level, the approach avoids collapsing OSINT monitoring into a single binary "uses OSINT" variable, which would obscure such heterogeneity.

2.3 Domain Security and Identifier Protection

Domain security monitoring is of particular relevance because domain names serve as user-facing identifiers and as anchors for digital trust. The domain name life cycle creates an attack surface that includes pre-registration targeting (e.g., typosquatting), active registration abuse (e.g., phishing sites), expiry-related takeover, and malicious re-registration [12]. Monitoring can therefore include both *asset protection* (ensuring ownership and correct DNS configuration) and *threat detection* (identifying newly registered domains that target a brand or public institution).

This duality connects technical and legal perspectives. In private-sector contexts, domain abuse and brand impersonation intersect with trademark rights and enforcement processes. In public-sector contexts, the protected objects include official designations and institutional identifiers, which may be protected through administrative law and policy rather than trademark law, but still require operational detection of impersonation and misrepresentation. The empirical instrument developed in this study treats domain security as a use case but also includes detail items that reflect governance-relevant practices (expiry monitoring, DNS monitoring, detection of newly registered domains).

2.4 Maturity Models and Explicit Scoring

Maturity models are used to represent staged capability development and to enable comparison across organizations. The Capability Maturity Model (CMM) is a canonical example that distinguishes five levels from "Initial" to "Optimizing" [23]. In security governance, maturity models are often used qualitatively, and benchmarks are sometimes derived from self-assessments. For OSINT monitoring, the challenge is to define maturity in a way that is (i) transparent, (ii) reproducible, and (iii) aligned with observable organizational practices.

The present study operationalizes maturity as a composite index of monitoring activity on a 0–100 scale, not as a full measure of overall OSINT program maturity. The index combines breadth (coverage across use cases) and intensity (monitoring frequency). Conceptually, the approach is consistent with standard index construction methods such as min–max normalization and arithmetic aggregation [21]. Governance quality, legal-compliance quality, staffing depth, and process integration are conceptually important, but they are not directly included in the score. The explicit score nonetheless enables methodological transparency: assumptions about weights, handling of missing values, and mapping to maturity categories become visible and can be challenged or improved in follow-up work. Alternative aggregation approaches include use-case-specific weighting, sector- or risk-adjusted weighting, non-linear penalties for imbalance between breadth and intensity, or the parallel reporting of separate sub-scores instead of a single scalar benchmark.

2.5 Legal and Ethical Boundaries as Part of Maturity

OSINT monitoring relies on publicly accessible information, yet public accessibility does not automatically imply unrestricted processing or storage. Constraints may arise from data protection requirements, contractual terms of service, and organizational ethics [4]. Moreover, in public-sector settings, additional constraints related to proportionality, transparency, and accountability apply. Legal and ethical considerations are therefore treated as part of the maturity problem rather than as an external constraint: mature monitoring programs require governance that aligns monitoring scope, retention, and dissemination practices with legal and ethical boundaries.

3 Methodology

This section summarizes the survey design, the development and validation of the instrument, the operationalization of the maturity index, and the statistical procedures used for the exploratory analyses.

3.1 Research Design

A cross-sectional survey design is employed to measure OSINT monitoring adoption across organizations. The methodological objective is twofold: (i) instrument construction and validation, and (ii) exploratory benchmarking and group comparisons.

The survey targets organizations operating in Germany, spanning multiple sectors. Recruitment occurred via direct invitations (e-mail and professional network outreach) and amplification through professional social-network posts. Because open recruitment channels were used, the sample is best interpreted as an exploratory convenience sample. The analysis focuses on descriptive patterns, benchmark construction, and hypothesis-generating group comparisons.

3.2 Instrument Development and Validation

Instrument development followed established questionnaire engineering principles: construct definition, item generation, expert review, iterative refinement, and attention to completion time to reduce dropout [13,15,16]. A target completion time of approximately 10 min was set to reduce survey abandonment.

Expert-Based Content Validity (CVR and CVI). Content validity was assessed via an expert panel ($N = 7$). Two complementary indices were used.

Content Validity Ratio (CVR). Following Lawshe, CVR for item j is defined as $\mathrm{CVR}_j = \frac{n_e - N/2}{N/2}$, where n_e denotes the number of experts rating the item as "essential" and N is the number of panelists [18]. For $N = 7$, the critical value is $\mathrm{CVR} \geq 0.99$; consequently, only items unanimously rated essential pass the strict threshold.

Content Validity Index (CVI). The item-level CVI is operationalized as the proportion of experts rating an item as sufficiently relevant (e.g., upper categories on a relevance scale). A commonly used threshold is I-CVI ≥ 0.78 [24]. Scale-level indices were computed to summarize section-wise content validity.

At section level, S-CVI/Ave values ranged from 0.50 to 1.00 across sections, with the use-case section achieving 1.00 and several sections exceeding 0.80 (acceptable) [24]. S-CVI/UA values were reported as strict reference points but were treated as conservative due to their stringency [24]. The expert review was conducted on a broader prototype instrument comprising 81 items across sections A–K. Under the strict Lawshe threshold for $N = 7$, 14 items met the unanimity criterion (CVR = 1.00); at the item level, 54 prototype items reached the I-CVI threshold. Sections F (budget) and G (usage outcomes) showed the weakest section-level content validity (S-CVI/Ave = 0.50 and 0.68, respectively) and were therefore not retained as part of the final core instrument used for the paper's analyses. Table 1 summarizes the section-level CVI results.

Table 1. Section-level content validity indices (CVI) from the expert panel.

Section	S-CVI/Ave	S-CVI/UA
A	0.81	0.33
B	0.84	0.33
C (use cases)	1.00	1.00
D	0.83	0.00
E	0.78	0.00
F	0.50	0.00
G	0.68	0.00
H	0.85	0.42
I	0.85	0.13
J	0.83	0.00
K	0.83	0.14

Notes. Common benchmarks: S-CVI/Ave $\geq$ 0.90 (excellent) and $\geq$ 0.80 (acceptable); S-CVI/UA is conservative [24].

Based on the combined CVI/CVR evidence, expert comments, and the 10-minute completion target, the broader prototype was reduced to a focused final instrument. The analyses reported in this paper therefore rely on the validated core components: screening items, a fixed use-case battery for adoption and relevance, a parallel frequency battery for monitored use cases, and three domain-security detail items. No pilot test with target respondents or formal cognitive pretest was conducted before fielding; accordingly, the evidence reported here should be interpreted as expert-validated content evidence combined with exploratory field results rather than as a fully pretested survey instrument.

3.3 Measures

The final analyzed instrument operationalizes OSINT monitoring as a fixed, respondent-independent battery of use cases. All participants received the same use-case set, which supports direct benchmarking across organizations. The original questionnaire wording is archived in a public repository snapshot [11].

1. **Use-case adoption (breadth).** For each of the 11 use cases, respondents indicate whether the use case is monitored (binary yes/no).
2. **Perceived relevance.** For each use case, perceived relevance is rated on a 5-point Likert scale from "not relevant at all" (1) to "very relevant" (5).
3. **Monitoring frequency (intensity).** For use cases that are monitored, frequency is rated on an ordinal 6-point scale from daily/continuous (1) to yearly (6).

The questionnaire logic was sequential: respondents first indicated adoption for each predefined use case, then rated its perceived relevance, and provided a frequency rating only when the use case was monitored. In addition, three dichotomous domain-security detail items captured concrete practices: (i) monitoring for newly registered domains with organizational relevance, (ii) monitoring expiry dates of registered domains, and (iii) monitoring public DNS entries. Optional open-ended questions in the broader survey were used only to capture qualitative comments or missing cases and were not included in the maturity index.

3.4 Construction of the 0–100 OSINT Monitoring Maturity Index

The maturity index $\mathcal{M}$ is designed to capture two orthogonal dimensions: breadth of coverage and intensity of monitoring. Let $K = 11$ denote the number of use cases and let $u_{ik} \in \{0, 1\}$ indicate whether organization i monitors use case k. Define the breadth component as: $B_i = \frac{1}{K} \sum_{k=1}^{K} u_{ik}$, mapped to a 0–100 scale as $B_i^{(100)} = 100 \cdot B_i$.

For monitoring intensity, let $f_{ik} \in \{1, \dots, 6\}$ denote the reported frequency category (defined only for monitored use cases). The scale is reversed such that higher values correspond to higher monitoring intensity: $f_{ik}^{\text{rev}} = 7 - f_{ik}$. Let $K_i = \{k \mid u_{ik} = 1 \wedge f_{ik} \text{ observed}\}$ denote the set of monitored use cases with a valid frequency rating. The intensity component is computed as the mean of reversed values: $I_i = \frac{1}{|K_i|} \sum_{k \in K_i} f_{ik}^{\text{rev}}$. $I_i \in [1, 6]$ is min–max normalized to $I_i^{(100)} = 100 \cdot (I_i - 1)/(6 - 1)$ [21].

Finally, the maturity index is defined as an equal-weight aggregation:

$$\mathcal{M}_i = \frac{1}{2} B_i^{(100)} + \frac{1}{2} I_i^{(100)}. \tag{1}$$

Missing frequency values are handled by computing I_i only over observed monitored use cases, and breadth uses valid adoption responses. The index is mapped to four descriptive categories: *Low* (0–24.99), *Emerging* (25–49.99), *Advanced* (50–74.99), and *Mature* (75–100).

3.5 Analysis Procedures

Analyses are descriptive and exploratory. Adoption rates are reported as percentages; relevance scores are summarized using means and standard deviations; and frequency is summarized via medians. Internal consistency of relevance ratings is assessed using Cronbach's α.

Group comparisons for maturity by organization size use the Mann–Whitney U test due to non-normality and ordinal characteristics of underlying components. Effect size is reported using Cliff's δ [8,19]. Industry differences are tested using one-way ANOVA for maturity (as an index) and Kruskal–Wallis tests for item-level relevance where applicable; effect sizes include η^2 for ANOVA.

Exploratory factor analysis of relevance ratings uses principal axis factoring and oblimin rotation. Sampling adequacy is assessed using the Kaiser–Meyer–Olkin (KMO) measure and Bartlett's test of sphericity [7].

Sensitivity Analysis. A G*Power sensitivity analysis (two-sided, $\alpha = 0.05$, power $= 0.80$; $n_1 = 18$, $n_2 = 14$) yields a minimal detectable effect of $d = 1.03$ (Cohen), i.e., a large effect, underscoring the exploratory nature of group comparisons [6,10,17].

4 Results

This section reports sample characteristics, use-case patterns, benchmark distributions, group comparisons, and the exploratory measurement-quality results.

4.1 Sample Characteristics

The final dataset includes $n = 33$ organizational responses. Table 2 summarizes organization size, industry sector, and respondent roles; the sample is heterogeneous but not sector-balanced.

4.2 Use-Case Adoption, Relevance, and Monitoring Frequency

Table 3 summarizes adoption rates, perceived relevance, and typical monitoring frequency for the eleven OSINT use cases.

Table 2. Sample characteristics (valid responses).

Characteristic	n	%
Organization size ($n = 32$)		
≤ 10,000 employees	18	56.3
> 10,000 employees	14	43.8
Industry sector ($n = 30$)		
IT/Cyber/Tech	12	40.0
Industry/Production/Energy	9	30.0
Services/Trade/Public sector	9	30.0
Respondent role ($n = 31$)		
Security leadership/management	10	32.3
Security specialist/operations	9	29.0
Other management/business/tech roles	12	38.7

Percentages are within each block (valid responses; missing excluded).

Interpretive Summary. The pattern in Table 3 suggests a technically centered monitoring portfolio. The strongest adoption clusters around domain security, vulnerability/exploit intelligence, external attack surface management, and credential/document leaks, i.e., use cases that are closely coupled to established cyber defense operations. By contrast, counterfeit/fake-shop and disinformation monitoring remain marginal, despite their potential relevance for fraud, communication, and reputational risk. Among adopters, the median frequencies indicate that monitored use cases are usually handled as recurring routines rather than occasional checks, so the main benchmark gap appears to concern portfolio breadth more than day-to-day cadence.

Table 3. Adoption, perceived relevance, and monitoring frequency of OSINT use cases.

Use case	Adoption (% yes)	n	Relevance M (SD)	Freq. median
Brand protection	51.5	33	3.30 (1.43)	1
Domain security	74.2	31	3.71 (1.19)	1
Social media monitoring	35.7	28	2.68 (1.31)	1
Public visibility/media presence	25.0	28	3.14 (1.21)	2
External attack surface management	66.7	27	3.81 (1.18)	1
Vulnerability/exploit intelligence	70.4	27	3.85 (1.17)	1
Credential & document leaks	63.0	27	3.70 (1.27)	1
Supply-chain security	48.1	27	3.37 (1.39)	1
Counterfeit/fake-shop monitoring	22.2	27	2.22 (1.25)	1
Disinformation monitoring	18.5	27	2.56 (1.22)	1
Dark-web monitoring	51.9	27	3.15 (1.32)	1

Notes. Relevance: 1–5 Likert (1 = not relevant at all, 5 = very relevant). Frequency: 1 = daily/continuous, 2 = weekly, 3 = monthly, 4 = quarterly, 5 = semiannual, 6 = yearly (median for adopters). Missing values are excluded at item level.

4.3 Domain Security Detail Items

Domain security adoption is complemented by three governance-relevant detail indicators (Table 4). Expiry monitoring is most common (91.3% yes), while monitoring newly registered domains with organizational relevance and monitoring public DNS entries are each reported by 73.9% of valid responses ($n = 23$). This suggests that domain monitoring is often anchored in asset governance (expiry control) and, in many cases, extended toward external-signal monitoring [12].

Table 4. Domain security monitoring practices (detail items; valid responses $n = 23$).

Detail item (yes/no)	Yes (%)	No (%)
Early detection of newly registered domains with organizational relevance	73.9	26.1
Monitoring expiry dates of registered domains	91.3	8.7
Monitoring public subdomain/domain/DNS entries	73.9	26.1

Percentages are within item (valid responses; missing excluded).

4.4 Maturity Distribution and Benchmarking

Applying Eq. (1) yields the maturity distribution in Table 5. Most organizations fall into *Low* or *Emerging*, indicating limited breadth and/or intensity of OSINT monitoring.

Table 5. Distribution of OSINT monitoring maturity levels ($\mathcal{M}$).

Maturity level (0–100)	n	%
Low (0–24.99)	12	36.4
Emerging (25–49.99)	11	33.3
Advanced (50–74.99)	5	15.2
Mature (75–100)	5	15.2

Because $\mathcal{M}$ aggregates breadth and intensity, it distinguishes narrow-but-frequent monitoring from broader-but-less-frequent monitoring, which is operationally relevant for identifying coverage gaps.

4.5 Benchmark Tables by Organization Size and Industry

Benchmark tables disaggregate maturity levels by organization size and industry sector (Tables 6 and 7). The benchmark is intended for descriptive self-positioning rather than for normative certification. In practical terms, *Low* usually indicates either narrow coverage or infrequent monitoring; *Emerging* indicates that recurrent routines exist for selected use cases but portfolio coverage remains partial; *Advanced* reflects broader coverage combined with regular monitoring across several use cases; and *Mature* indicates both comparatively wide coverage and high monitoring intensity within the limits of the activity-based index. Because $\mathcal{M}$ does not directly capture governance quality, two organizations with similar scores may still differ substantially in formalization, legal review, or response integration.

Table 6. Benchmark: maturity levels by organization size (valid $n = 32$).

Size group	Low	Emerging	Advanced	Mature
$\leq$ 10,000 employees ($n = 18$)	10 (55.6%)	5 (27.8%)	2 (11.1%)	1 (5.6%)
$>$ 10,000 employees ($n = 14$)	1 (7.1%)	6 (42.9%)	3 (21.4%)	4 (28.6%)

Row percentages within group (valid responses; missing excluded).

Table 7. Benchmark: maturity levels by industry sector (valid $n = 30$).

Industry sector	Low	Emerging	Advanced	Mature
IT/Cyber/Tech ($n = 12$)	5 (41.7%)	4 (33.3%)	2 (16.7%)	1 (8.3%)
Industry/Production/Energy ($n = 9$)	3 (33.3%)	2 (22.2%)	2 (22.2%)	2 (22.2%)
Services/Trade/Public sector ($n = 9$)	3 (33.3%)	5 (55.6%)	0 (0.0%)	1 (11.1%)

Row percentages within sector (valid responses; missing excluded).

4.6 Group Comparisons: Organization Size and Industry

Organization Size. Organizations with $>$ 10,000 employees exhibit higher maturity than those with $\leq$ 10,000 employees (Table 8). The Mann–Whitney U test is significant ($U = 195.0$, $p = 0.008$) with a large effect (Cliff's $\delta = -0.55$); median maturity increases from 13.18 to 50.00 (valid $N = 32$) [8,19].

Industry Sector. Industry differences in maturity are not statistically supported in the present sample. An ANOVA comparing the three sector groups yields $F(2, 27) = 0.58$ with $p = 0.567$ and an effect size of $\eta^2 = 0.041$, suggesting that sector membership explains only a small portion of variance in maturity [6,29]. Given the limited sample size and coarse sector categories, these results should be interpreted as absence of evidence rather than evidence of absence.

Industry Differences in Perceived Relevance. Kruskal–Wallis tests of relevance ratings by sector do not indicate statistically significant differences for any use case in this sample (Holm-adjusted post-hoc comparisons not applicable due to non-significance). The largest exploratory effects occur for selected use cases (e.g., domain security), but effect sizes remain moderate and should be treated as hypotheses for larger follow-up studies.

Table 8. Maturity by organization size: descriptive statistics and Mann–Whitney test.

Size group	n	Median [IQR]	Mean (SD)
$\leq$ 10,000 employees	18	13.18 [38.18]	27.02 (27.76)
$>$ 10,000 employees	14	50.00 [48.18]	55.00 (26.98)

Notes. Mann–Whitney $U = 195.0$, $p = 0.008$; Cliff's $\delta = -0.55$. Valid $N = 32$ (one missing size).

4.7 Measurement Quality: Internal Consistency and Exploratory Factor Structure

Internal Consistency. Perceived relevance ratings across the eleven use cases exhibit high internal consistency ($\alpha \approx 0.88$; $k = 11$ items), indicating that organizations that rate one use case as relevant tend to rate others as relevant as well [15]. This reliability evidence supports the use of relevance ratings as a consistent descriptive measure, but does not imply unidimensionality.

Exploratory Factor Analysis. An exploratory factor analysis of relevance ratings indicates adequate sampling adequacy (KMO = 0.717; Bartlett $p < 0.001$) [7]. A two-factor solution with oblimin rotation yields the pattern matrix in Table 9. Several items load strongly on a first factor (counterfeit/fake-shop monitoring, disinformation, supply-chain security, public visibility, social media), while a second factor is dominated by vulnerability/exploit intelligence and external attack surface management. Domain security and brand protection exhibit cross-loadings, consistent with their boundary position between technical and external-perception monitoring.

Given the sample size, these results are interpreted as exploratory structure hints rather than as confirmatory evidence [7,14]. Nonetheless, the factor pattern is substantively coherent and provides a useful lens for interpreting monitoring portfolios.

5 Discussion and Implications

This section interprets the benchmark results, clarifies the conceptual boundaries of the activity-based maturity index, and derives implications for organizational and public-sector monitoring practice.

Table 9. Exploratory factor analysis of relevance ratings (oblimin rotation; pattern matrix).

Use case	Factor 1	Factor 2	Uniqueness
Counterfeit/fake-shop monitoring	0.995	–	0.071
Disinformation monitoring	0.801	–	0.339
Supply-chain security	0.734	–	0.351
Public visibility/media presence	0.724	–	0.471
Social media monitoring	0.672	–	0.427
Domain security	0.499	0.549	0.313
Brand protection	0.490	0.459	0.437
Credential & document leaks	0.398	0.422	0.579
Vulnerability/exploit intelligence	–	0.874	0.280
External attack surface management	–	0.864	0.168
Dark-web monitoring	–	0.368	0.725

Notes. KMO = 0.717; Bartlett $p < 0.001$. Exploratory due to sample size [7]. Uniqueness is unexplained variance [14].

5.1 Empirical Picture: Relevance Acknowledged, Implementation Uneven

The empirical findings characterize OSINT monitoring as a capability that is widely perceived as relevant but unevenly implemented. Several use cases are rated as highly relevant (especially vulnerability/exploit intelligence, external attack surface management, and domain security), and adopters typically perform monitoring frequently. Yet the maturity distribution indicates that breadth is often limited, and that systematic coverage across a portfolio of monitoring objectives is lacking for a substantial fraction of organizations.

Importantly, the reported maturity values should be interpreted as activity-based monitoring maturity, i.e., as a benchmark of what organizations monitor and how regularly they do so. The index does not directly measure whether monitoring is embedded in formal governance, legal review, escalation playbooks, or performance management. Accordingly, a higher score should be read as evidence of broader and/or more frequent monitoring activity, not as proof of a fully mature OSINT program in the broader organizational sense.

This gap between perceived relevance and measured maturity suggests that barriers likely reside in operationalization rather than awareness. Such barriers may include limited personnel, lack of standardized monitoring processes, fragmentation between tooling and governance functions, and insufficient integration of open-source signals into response processes. While the present survey does not directly measure these causal factors, the strong size effect provides indirect evidence consistent with resource and specialization differences between large and smaller organizations.

5.2 Domain Security as a Boundary Object: Technical and Governance Dimensions

Domain security monitoring illustrates the interdisciplinary nature of OSINT monitoring. On the technical side, domain-related indicators can support early detection of impersonation campaigns, suspicious registrations, and DNS misconfigurations. On the governance side, monitoring expiry and ownership is a prerequisite for preventing loss of control and for enforcing brand or institutional identifiers. The high prevalence of expiry monitoring (91.3 %) indicates that domain asset governance is commonly recognized. The substantial prevalence of monitoring newly registered domains and public DNS entries (73.9 % each) further suggests that a majority of organizations have begun to operationalize domain security beyond purely administrative controls.

However, the maturity model emphasizes breadth: domain security monitoring alone does not constitute a mature OSINT monitoring program. Domain abuse and phishing threats often co-occur with credential leaks, exploit activity, and broader attack-surface issues [12]. The maturity index therefore supports the interpretation that OSINT monitoring programs should combine domain-centric governance with complementary technical monitoring objectives (attack-surface and vulnerability intelligence) and with external-situation monitoring where relevant.

5.3 Blind Spots and Externalities

The lowest adoption and relevance are observed for counterfeit/fake-shop monitoring and disinformation monitoring. These objectives sit at the intersection of security, fraud, and communication. Low prioritization may reflect limited internal ownership, unclear response processes, or the perceived complexity of monitoring and attribution. The absence of significant industry differences suggests that these objectives are not exclusively relevant to a narrow set of sectors in this sample; rather, they may represent systematic blind spots in OSINT monitoring portfolios.

From a risk perspective, such blind spots can create externalities. For example, disinformation campaigns may target organizations indirectly through reputational damage or manipulation of stakeholder perceptions. Counterfeit/fake-shop activity may create customer harm and undermine trust while also being used as a vehicle for payment fraud and credential harvesting. These threats are partially technical (e.g., malicious infrastructure) and partially socio-technical (e.g., narrative manipulation), and therefore require cross-functional governance that is often absent in early maturity stages.

5.4 Interpreting the Two-Factor Structure

The exploratory factor analysis suggests a two-dimensional conceptualization of perceived relevance: (i) an *external situation awareness* dimension (disinformation, counterfeit/fake shops, public visibility, social media, supply chain), and

(ii) a *technical attack-surface and vulnerability management* dimension (vulnerability/exploit intelligence and external attack surface management). Domain security and brand protection load on both dimensions, reflecting their dual nature as technical controls and identifier-protection activities.

This structure can guide the design of monitoring programs. Organizations that focus exclusively on one dimension (e.g., only technical scanning) may neglect socio-technical threats and reputational harm, while organizations that focus exclusively on external perception may neglect technical exposure signals. Mature monitoring should cover both dimensions in a manner consistent with organizational risk models and legal constraints.

5.5 Implications for Public-Sector Organizations

Although the sample is not designed as a public-sector study, the instrument and maturity model are applicable to public-sector and governmental settings. In such contexts, protected objects include official designations, institutional identifiers, and sovereign digital identities. Several measured use cases translate directly: domain security and impersonation monitoring are essential for citizen-facing services; vulnerability and attack-surface monitoring supports protection of public infrastructure; and disinformation monitoring is relevant for the integrity of public communication.

Public-sector constraints imply that maturity cannot be interpreted solely as "more monitoring". Rather, maturity requires lawful, proportionate monitoring with robust governance, transparency, and accountability [4]. The maturity index presented here captures operational breadth and intensity but does not directly capture governance quality; accordingly, public-sector extensions should incorporate explicit governance indicators.

6 Limitations and Future Work

Sampling and Generalizability. The sample size ($n = 33$), voluntary recruitment, and convenience sampling introduce selection bias and limit generalizability. In addition, IT/cyber/tech organizations account for 40% of valid sector responses, so the sample should not be read as sector-balanced. The sensitivity analysis indicates that only large effects can be detected reliably [6,10,17]. Accordingly, non-significant sector differences should not be interpreted as proof of absence; rather, they motivate larger, stratified follow-up studies with finer-grained sector categories.

Self-report Measurement. Measures are self-reported and may not perfectly correspond to actual operational practice. Social desirability bias may inflate reported monitoring; conversely, respondents in non-security roles may underreport monitoring performed elsewhere. Future work should triangulate survey responses with objective indicators (e.g., existence of monitoring playbooks, ticketing-system evidence, domain portfolio governance documentation).

Modeling Assumptions in the Maturity Index. The maturity index $\mathcal{M}$ reflects explicit modeling choices: equal weighting of breadth and intensity, and equal weighting of use cases in breadth. These decisions enable transparency and reproducibility, but they may not reflect incident impact or sector-specific risk. Moreover, the score captures monitoring activity rather than the full maturity of governance, legal review, staffing, or response integration. Alternative approaches include outcome-weighted aggregation (weights derived from incident cost or threat prevalence), non-linear aggregation (penalizing extreme imbalance between breadth and intensity), or multi-dimensional maturity reporting rather than a single scalar score. The present work should therefore be interpreted as an initial operationalization that enables benchmarking and can be refined as empirical evidence accumulates [21].

Exploratory Factor Analysis and Group Comparisons. The factor structure is exploratory and derived from a small sample. With roughly 27 valid cases for the relevance battery across 11 items, the exploratory factor analysis should be read as a provisional structure hint rather than as robust evidence for a stable latent model. The same caution applies to group comparisons based on small cell sizes and coarse sector categories. Confirmatory factor analysis, measurement-invariance testing, and more robust sector comparisons require substantially larger samples [7,14].

Outcome Linkage. The study measures adoption and perceptions but does not link maturity to security outcomes such as incident rates, time-to-detection, or reduction of domain-abuse exposure. Establishing such links is essential for moving from descriptive benchmarking to causal claims.

6.1 Future Work

Future work should therefore (i) conduct larger, sector-stratified surveys with improved representativeness; (ii) add formal cognitive pretesting and pilot testing before field deployment; (iii) incorporate governance and legal-compliance indicators into the maturity model; (iv) combine survey measures with objective telemetry (e.g., monitoring events, case management, domain-portfolio controls); and (v) evaluate associations between activity-based maturity and operational security outcomes. In addition, the domain-security sub-instrument can be extended toward *risk-based prioritization* of newly observed domains. This includes the integration of explicit relevance scoring models (e.g., 0–100) for domain names in relation to brands or protected identifiers, enabling targeted monitoring and early detection of targeted attacks. Such extensions would align domain-security monitoring with broader CTI prioritization workflows and support both private-sector and governmental identifier protection.

7 Conclusion

This paper presents a validated questionnaire and a transparent 0–100 maturity index for benchmarking organizational OSINT monitoring. Exploratory evidence from a German cross-sectional sample indicates that OSINT monitoring is perceived as relevant—particularly for vulnerability/exploit intelligence, external attack surface management, and domain security—yet maturity remains uneven and is strongly associated with organizational size. The instrument and maturity index provide a reproducible baseline for future research on determinants and effectiveness of OSINT monitoring and support practical benchmarking of monitoring portfolios, including domain-security monitoring at the intersection of technical security and identifier governance.

Language Assistance Statement. The original manuscript text was drafted in German and subsequently translated into American English with the assistance of ChatGPT 5.2 for language editing only.

References

1. Adjibi, B.V., et al.: The guardians of name street: studying defensive registration practices of the fortune 500. In: Proceedings 2025 Network and Distributed System Security Symposium. San Diego, CA, USA: Internet Society (2025). isbn: 979-8-9894372-8-3. https://doi.org/10.14722/ndss.2025.241202
2. Benz, M., Chatterjee, D.: Calculated risk? A cybersecurity evaluation tool for SMEs. Bus. Horiz. **63**(4), 531–540 (2020). https://doi.org/10.1016/j.bushor.2020.03.010. ISSN: 00076813
3. Bizouarn, K.M., Abdulnabi, M., Tan, J.: OSINT and AI: a powerful combination for company vulnerability detection. In: IEEE 21st Student Conference on Research and Development (SCOReD). Kuala Lumpur, Malaysia, pp. 246–250. IEEE (2023). isbn: 979-8-3503-1882-1. https://doi.org/10.1109/SCOReD60679.2023.10563226
4. Böhm, I., Lolagar, S.: Open source intelligence: introduction, legal, and ethical considerations. Int. Cybersecur. Law Rev. **2**(2), 317–337 (2021). https://doi.org/10.1365/s43439-021-00042-7. issn: 2662-9720, 2662-9739
5. Bundesamt für Sicherheit in der Informationstechnik (BSI). Die Lage der IT-Sicherheit in Deutschland 2024. de. https://www.bsi.bund.de/SharedDocs/Downloads/DE/BSI/Publikationen/Lageberichte/Lagebericht2024.pdf (visited on 09/09/2025)
6. Cohen, J.: Statistical Power Analysis for the Behavioral Sciences. 2. ed., reprint. Psychology Press, New York (2009). isbn: 978-0-8058-0283-2
7. Dalawi, I., Isa, M.R., Aimran, N.: Exploratory factor analysis on the development and validation of the understanding, attitude, practice and health literacy questionnaire on COVID-19 in Malay language. Sci. Rep. **15**(1), 19654 (2025). https://doi.org/10.1038/s41598-025-04517-z. issn: 2045-2322
8. Daniel, W.W., Cross, C.L.: Biostatistics: A Foundation for Analysis in the Health Sciences. https://onlinelibrary.wiley.com/doi/epdf/10.1002/1097-0258%2820010130%2920%3A2%3C324%3A%3AAID-SIM635%3E3.0.CO%3B2-O. Accessed 10 Aug 2025
9. European Union Agency for Cybersecurity. ENISA threat landscape 2024: July 2023 to June 2024. en. LU: Publications Office (2024). https://data.europa.eu/doi/10.2824/0710888. Accessed 09 Sept 2025
10. Faul, F., et al.: Statistical power analyses using G*Power 3.1: tests for correlation and regression analyses. Behav. Res. Methods **41**(4), 1149–1160 (2009). https://doi.org/10.3758/BRM.41.4.1149. issn: 1554-351X, 1554-3528

11. Fritzler, T., Le, T., Massoth, M.: EICC2026 OSINT Questionnaire. GitLab repository snapshot, commit e32fc9dc (2026). https://code.fbi.h-da.de/thomas.fritzler/EICC2026_OSINTQuestionnaire/commit/e32fc9dc7000cb05c6ba00f2ceab0c50d93c9702
12. Fritzler, T., Massoth, M.: The Domain Name Life Cycle as an Attack Surface: Systematic Threat Mapping and Defense Recommendations. https://www.thinkmind.org/library/CYBER/CYBER_2025/cyber_2025_1_10_80054.html. Accessed 16 Sept 2025
13. Gupta, S., et al.: Item generation and establishing face and content validity of a rating scale: a primer. Indian J. Psychiatry **67**(8), 816–822 (2025). https://doi.org/10.4103/indianjpsychiatry_750_25. issn: 0019-5545, 1998-3794
14. Hair, J.F., et al.: Multivariate Data Analysis. 8th ed. Andover, Hampshire: Cengage Learning EMEA (2018). isbn: 9781473756540
15. Karimian, Z., Chahartangi, F.: Development and validation of a questionnaire to measure educational agility: a psychometric assessment using exploratory factor analysis. BMC Med. Educ. **24**(1), 1284 (2024). https://doi.org/10.1186/s12909-024-06307-z. issn: 1472-6920
16. Kishore, K., et al.: Practical guidelines to develop and evaluate a questionnaire. Indian Dermatol. Online J. **12**, 266–275 (2021). https://doi.org/10.4103/idoj.IDOJ_674_20
17. Lakens, D.: Sample size justification. Collabra: Psychol. **8**(1) (2022). https://doi.org/10.1525/collabra.33267. issn: 2474-7394
18. Lawshe, C.H.: A quantitative approach to content validity. Pers. Psychol. **28**(4), 563–575 (1975). https://doi.org/10.1111/j.1744-6570.1975.tb01393.x. issn: 0031-5826, 1744-6570
19. Meissel, K., Yao, E.S.: Using Cliff's Delta as a Non-Parametric Effect Size Measure: An Accessible Web App and R Tutorial. https://doi.org/10.7275/pare.1977
20. National Institute of Standards and Technology. The NIST Cybersecurity Framework (CSF) 2.0. Tech. rep. NIST CSWP 29. Gaithersburg, MD (2024). https://doi.org/10.6028/NIST.CSWP.29
21. OECD et al.: Handbook on Constructing Composite Indicators: Methodology and User Guide. In: (2008). https://doi.org/10.1787/9789264043466-en
22. Pastor-Galindo, J., et al.: The not yet exploited goldmine of OSINT: opportunities, open challenges and future trends. IEEE Access **8**, 10282–10304 (2020). https://doi.org/10.1109/ACCESS.2020.2965257. issn: 2169-3536
23. Paulk, M.C., et al.: Capability Maturity Model for Software, Version 1.1: In: (1993). https://doi.org/10.21236/ADA263403
24. Polit, D.F., Beck, C.T.: The content validity index: are you sure you know what's being reported? Critique and recommendations. Res. Nurs. Health **29**(5), 489–497 (2006). https://doi.org/10.1002/nur.20147. issn: 0160-6891, 1098-240X
25. Romanova, N., Gryzunov, V.: OSINT model of privacy violator. In: International Russian Smart Industry Conference (SmartIndustryCon). Sochi, Russian Federation, pp. 929–933. IEEE (2025). isbn: 979-8-3315-1124-1. https://doi.org/10.1109/SmartIndustryCon65166.2025.10985976
26. Szymoniak, S., Foks, K., Pyrkosz-Dziubczyk, A.: Application of OSINT Methods in Ensuring Cybersecurity. https://doi.org/10.58245/ipsi.tir.2502.05
27. Urban, T., Große-Kampmann, M., Tatang, D., Holz, T., Pohlmann, N.: Plenty of Phish in the Sea: analyzing potential pre-attack surfaces. In: Chen, L., Li, N., Liang, K., Schneider, S. (eds.) ESORICS 2020. LNCS, vol. 12309, pp. 272–291. Springer, Cham (2020). https://doi.org/10.1007/978-3-030-59013-0_14

28. Vrhovec, S., Markelj, B.: We need to aim at the top: factors associated with cybersecurity awareness of cyber and information security decision-makers. PLoS ONE **19**(10), e0312266 (2024). https://doi.org/10.1371/journal.pone.0312266. issn: 1932-6203
29. Zhou, Y., Zhu, Y., Wong, W.K.: Statistical tests for homogeneity of variance for clinical trials and recommendations. Contemp. Clin. Trials Commun. **33**, 101119 (2023). https://doi.org/10.1016/j.conctc.2023.101119. issn: 24518654
30. Zibak, A., Sauerwein, C., Simpson, A.C.: Threat intelligence quality dimensions for research and practice. Digital Threats: Res. Pract. **3**(4), 1–22 (2022). https://doi.org/10.1145/3484202. issn: 2692-1626, 2576-5337

Forensic Examination of the Withings Body Smart Weight Scale Ecosystem

Felix Fischer(✉), Patrick Stjepanovic, and Dirk Labudde

Faculty Applied Computer Sciences and Biosciences, Hochschule Mittweida University of Applied Sciences, Mittweida, Germany
{fischel1,stjepano,labudde}@hs-mittweida.de

Abstract. This paper reports an exploratory forensic examination of the Withings Body smart weight scale in the Internet of Things (IoT) context. Evidence was identified, preserved, and analyzed in device storage, companion smartphone app, Bluetooth communication, WiFi traffic to vendor cloud and the vendor cloud itself. The methods included reverse engineering of the device's key-value store, access via the onboard Universal Asynchronous Receiver/Transmitter (UART) interface, analysis of Android app artifacts, Bluetooth Host Controller Interface (HCI) snoop capture, and inspection of app-cloud Hyper Text Transport Protocol Secure (HTTPS) traffic. Actionable information was recovered from most endpoints, whereas the scale-to-cloud link was protected by Transport Layer Security (TLS) and did not allow payload inspection. The artifacts included personal data such as name, age, weight, and height, as well as device and account metadata. The access to UART proved especially valuable, allowing the extraction of personal data, plain text credentials, certificates, and other security-relevant material. The findings support digital investigations and IoT security assessments and argue for collecting UART data and applying targeted reverse engineering as early priorities in the initial assessment phase.

Keywords: smart home · withings · firmware · data extraction · IoT

1 Introduction

Due to ongoing technological advances, the Internet of Things (IoT) has become an integral part of many people's daily life. A representative survey conducted in Germany with 2,518 respondents found that about half of all households use more than four internet-connected devices [2]. According to Statista, this development reflects a broader global trend, as the number of IoT devices worldwide grew from 13.8 billion in 2022 to 19.8 billion in 2025 and is expected to double again by 2034 [18].

The wide range of applications has reached the health and fitness sector, making bathroom scales "smart" as well. In addition to body weight, these smart bathroom scales record other biometric data and additional information, such as

© The Author(s), under exclusive license to Springer Nature Switzerland AG 2027
K. Renaud et al. (Eds.): EICC 2026, CCIS 3012, pp. 287–303, 2027.
https://doi.org/10.1007/978-3-032-28957-5_17

the Body Mass Index (BMI), and automatically transmit the results to smartphone apps or cloud services. For example, the Withings Body transmits data via Bluetooth to mobile devices or via a Wireless Local Area Network (WLAN) to cloud services [19,21]. The sale and distribution of smart bathroom scales is growing steadily [14]. Thus, data collected and produced by IoT devices can provide valuable digital traces that may prove pivotal to criminal investigations, inquiries into data-protection infringements, or civil proceedings.

An illustrative example is the criminal case against Richard Dabate in the United States. In the 2022 murder trial, movement data from his deceased wife's Fitbit activity tracker contributed significantly to refuting his account of events and his alibi [8,15]. The recorded step data further indicated that the victim was still moving inside the house after the time at which Dabate claimed she had been killed and also suggested the approximate time of death, as the Fitbit stopped registering steps at approximately the time of the shooting [15]. In 2025, the Connecticut Supreme Court affirmed the admissibility of Fitbit data as evidence [15].

However, IoT devices also present new challenges for digital forensic examiners, as such devices tend to be heterogeneous and proprietary due to the large number of manufacturers and application domains. Against this background, this work presents an exploratory forensic case study of the Withings Body ecosystem.

The study is guided by the following research questions:

RQ1: Which forensically relevant artifacts can be recovered from the Withings Body ecosystem, including the scale, the companion Android app, Bluetooth communication, WiFi traffic, and cloud exports?

RQ2: Which acquisition and analysis methods, including UART-based access and network-based collection, are effective for recovering these artifacts across the different evidence carriers?

This study deliberately excludes invasive hardware access. Furthermore, we restrict app analysis to a rooted Android device, while the iOS app is out of scope. We also do not conduct deep cryptographic attacks against TLS or Bluetooth, and the analysis of the paid Withings+ services is likewise out of scope.

To answer the questions, we develop and apply a reproducible forensic acquisition and analysis workflow for the Withings Body ecosystem, covering the scale, the companion Android app, Bluetooth communication, WiFi traffic, and cloud exports. Further, we demonstrate direct device access via the on-board UART interface and reverse engineer the proprietary Withings DBlib key–value store, enabling the recovery and categorization of personal, health-related, device, account, and configuration artifacts. We complete our research by assessing the forensic relevance of the recovered artifacts across the examined evidence carriers, including cross-layer correlations and firmware-related findings relevant to IoT investigations.

The remainder of this paper is organized as follows. Section 2 reviews related work, while Sect. 3 outlines the methodological workflow and resources used in the examination. Section 4 describes device functionality and hardware I/O, and Sect. 5 details UART-based acquisition, including interface identification, backup, data preparation, and reverse engineering of the DBlib users field. Section 6 analyzes rooting, acquisition, and artifacts from the Withings companion app, while Sect. 7 reports vendor cloud exports and Sect. 8 examines network communication including Bluetooth, the scale-to-cloud WiFi link, and app-to-cloud HTTPS. Section 9 lists findings which are discussed in Sect. 10. Finally, Sect. 11 concludes and describes future work.

2 Related Work

A useful theoretical foundation for the forensic examination of IoT ecosystems is the distinction between end-device forensics, network forensics, and cloud forensics [23]. This perspective is particularly suitable for smart-scale ecosystems, where potentially relevant artifacts are not confined to the device itself but may also reside in companion applications, communication channels, and cloud services. The present study follows this view and therefore structures the examination across these three layers.

The work most closely related to this paper is the exploratory case study by Grispos et al., who investigated the iHealth HS6 smart-scale ecosystem [7]. Their study shows that a smart scale ecosystem can contain forensically relevant residual data across multiple components, including scale, smartphone applications, and cloud-related artifacts. In contrast, the present study additionally includes the communication channels of the ecosystem by examining both Bluetooth communication and WiFi-based traffic alongside device and cloud artifacts.

A complementary perspective is provided by Wu et al., who analyzed the communication channels of 32 consumer IoT devices with regard to encryption, traffic metadata, and the feasibility of intercepting app-to-cloud communication via a proxy [22]. Their dataset includes a Withings smart scale and demonstrates that IoT network traffic can provide evidential value even when the investigation is limited to the network layer.

Against this background, the contribution of the present work to IoT forensics lies in combining these perspectives for the Withings Body ecosystem. More specifically, the paper integrates end-device analysis, network-oriented examination of Bluetooth and WiFi communication, and cloud-related artifact analysis in a single exploratory study. In this way, it extends prior smart-scale work beyond device, app, and cloud artifacts and contributes an ecosystem-wide view that is directly aligned with the layered structure commonly used to describe IoT forensics.

3 Methodology

Figure 1 summarizes the methodological workflow of the investigation and links the individual acquisition and analysis steps to Sects. 5–8. The workflow consists of four components: acquisition from the scale via UART, artifact extraction from the rooted Android app, analysis of Bluetooth and WiFi communication, and examination of cloud-exported data. The recovered artifacts are then consolidated across evidence carriers in Sect. 9.

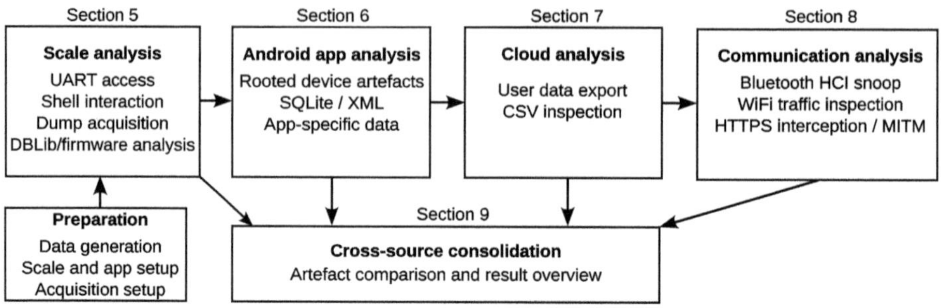

Fig. 1. Methodological workflow of the forensic examination

Table 1 centralizes the main hardware and software resources used throughout the investigation.

Table 1. Main resources used in the examination

Resource	Used for
Withings Body WBS06	Target device
Xiaomi Redmi 4a with Android 15 installed	Withings App Version 7.5.1 analysis, Bluetooth capture, HTTPS interception
Raspberry Pi 5	UART-based access
minicom	Interactive UART communication
ADB	Android access and HCI snoop activation
Wireshark	Bluetooth traffic capture and analysis
Burp Suite	HTTPS interception, inspection, parameter tampering
Magisk/Cert-Fixer	System-level certificate installation
Binwalk	Flash and SRAM image analysis
ImHex	Analysis of proprietary binary structures
DB Browser for SQLite	SQLite database content analysis
Microsoft Excel	CSV content analysis

4 Device Function Overview

The Withings Body offers a wide range of features. The scale provides body weight measurement and calculates the Body Mass Index (BMI) using the measured weight and the height stored in the app. After weight measurement, the display shows the weight history for the last eight measurements and a local weather forecast is provided. The scale supports multiple users and automatically recognizes up to eight people, with all data synced to the Withings app via Bluetooth or WiFi. There is also a baby mode for young families [21].

If enabled in the app, the scale displays the step count recorded by a smartphone or pedometer and subsequently synchronizes with the scale via the Withings cloud. The weight sensors not only measure weight but, by means of Position Control Technology^{TM}, also detect standing position on the scale. If several users have similar weights, the correct user can be selected by slightly shifting the body to the right or left. The connection button serves to initiate Bluetooth pairing with the smartphone [21]. Located directly above this button is the UART interface. UART is a hardware-based serial interface for communication between two devices [10]. On the Withings Body the UART interface, though not intended for end users, is physically accessible through three small openings in the housing.

5 Data Extraction via UART

UART is a widely used asynchronous serial communication interface which manufacturers commonly use for debugging and production testing [10]. Nevertheless, it is best practice to disable this communication channel after production for security reasons.

In a PCB image provided by the Federal Communications Commission (FCC), the UART contacts were clearly visible and labeled Tx, Rx, and GND [4]. In addition, the designation of the Micro Controller Unit (MCU) could be identified, which was relevant for later analysis of the firmware. The required data rate of 1 MBaud and the UART configuration of 8N1 were determined from a Reddit post [16]. A connection was established by using "minicom" via the GPIO interfaces of a Raspberry Pi. Connecting to UART, the scale provides a readable shell and offers numerous commands, which can be listed using the command `help` or `ls`.

All available shell commands were systematically executed to identify relevant forensic artifacts in the output. The shell provides commands to extract data from the SPI flash, MCU flash and SRAM in the form of hex dumps.

In addition, the locally stored key–value store used by Withings, under the name DBLib, can also be backed up via a dedicated command. The dumped DBLib entries include an ID, a size specification, and data which are provided in hexadecimal format. The following commands were used to acquire and preserve the memory contents and the DBlib users field: `spiflash dump 0x0 0x400000` for SPI flash, `mcuflash dump 0x0 0x80000` for MCU flash, `mcuflash dump 0x1FFF0000 0x20000` for SRAM and `dblib dump 1 5` for DBLib users field.

During data acquisition, it was observed that the scale no longer performs measurements as long as a UART connection is active and requires a reboot to resume normal functionality. Conversely, a UART connection cannot be established once a measurement has taken place. This significant behavior prevents immediate backup of the SRAM contents following a measurement.

The binary images extracted from the SPI flash, MCU flash, and SRAM were subsequently analyzed using the tool Binwalk. For the SPI flash and MCU flash, signature-based analysis yielded few meaningful results and produced several false positives, such as the misidentification of an ESP32 image, although the scale does not use an ESP32 microcontroller. The SRAM image did not yield any relevant results.

The Binwalk output suggests that the scale firmware follows a monolithic software architecture without an underlying operating system, appears to be precompiled, and is stored on both the MCU flash and the SPI flash.

The shell boot sequence indicates the presence of two firmware versions on the scale, suggesting an A/B update scheme with alternating firmware partitions. The entropy graph of Binwalk confirmed this partitioning of the images.

By executing the scale's shell commands and examining their output, numerous forensically relevant details were obtained. However, none of the commands directly yielded personal data such as user, weight, age, height, or synchronized step count. Access to this data required further analysis of the DBlib key–value store through reverse engineering. The analysis builds on a prior reverse engineering study presented by Michael Coppola at Sommercon in 2013 while he was a student at Northeastern University. Coppola examined a predecessor of the Withings Body, the Withings WS-30, and reconstructed parts of the fundamental structure of DBlib [3].

The DBlib field containing user information was examined in greater detail to access personal data. The following example exhibits the structure of DBlib dump.

```
ID(d.) SZ(d.) DATA(hex.)
5      1600   000fe0036452f00f00e000050...
```

The Hex bytes can be copied directly from the captured protocol log and converted into a binary file using the command `xxd -r -p > dump.bin`. Next, this file was analyzed with the ImHex hex editor. On first inspection of the Users data field, it became apparent that it contains the users' initials, each consisting of three letters. Starting from the first letter of the initials, the user entries are spaced 200 bytes apart. This appears plausible because the scale supports up to eight users, and the 1,600 bytes of the field divide exactly into eight blocks of 200 bytes each. Beginning at the first letter of each set of initials, the 200-byte user fields were examined with ImHex's Data Inspector, and each byte was checked for known or otherwise interpretable values.

The analysis shows that information such as weight, age, and height can be read in 32-bit floating-point format. The step count, user ID, data creation date, and date of birth (both Unix timestamp) are stored as 32-bit integers. Further, the first three letters of the initials followed by a Null terminating byte is encoded in ASCII. Table 2 shows the identified offsets in the user entries.

Table 2. Offsets in DBlib entries.

Offset	Length	Type	Content
0 Byte	4 Bytes	integer	UserID
4 Bytes	4 Bytes	ASCII	initial letters
8 Bytes	4 Bytes	unix time	data creation date
16 Bytes	4 Bytes	float	height in m
20 Bytes	4 Bytes	float	age in years
24 Bytes	4 Bytes	float	weight in kg
160 Bytes	4 Bytes	integer	step count
180 Bytes	4 Bytes	unix time	date of birth

6 Smartphone App Data

The Withings app is available for iOS and Android and supports the setup, use, and management of the Withings Body over Bluetooth or WiFi [21]. It can also be used independently, without a Withings device, as a health and fitness app for recording and viewing health-related data. In this mode, it supports physical activity via the smartphone's accelerometer, weight and body fat percentage via manual entry, heart rate via the smartphone camera, and blood pressure via manual entry [20]. It also supports synchronization with connected Withings devices, such as the Withings Body [19], and can be linked to third-party platforms to receive health data from those sources [21].

The app data analysis was conducted exclusively on a Xiaomi Redmi Note 4a running a rooted crDroid 11.4 (Android 15). The smartphone was connected to the examination workstation via USB for the acquisition of the app data at `/data_mirror/data_ce/null/0/com.withings.wiscale2` via ADB.

In the preserved copy of the Withings app directory the following folders can be observed: `app_`, `app_textures`, `app_webview`, `cache`, `code_cache`, `databases`, `files`, `no_backup`, `shared_prefs` and `trace`. The directory contents were examined sequentially for forensically relevant information. Most traces were located in the folders `databases` and `shared_prefs`.

The databases directory contained multiple SQLite database files, which were analyzed using DB Browser for SQLite. The databases stored personal data derived from measurements and user inputs, as well as non-personal information such as the scale's firmware version and synchronized third-party applications. Nearly all tables contained timestamps.

The `shared_prefs` directory contained Extensible Markup Language (XML) files that predominantly stored technical information related to device settings, application configuration, and the user.

7 Cloud Storage Data

User data was acquired from the Withings cloud through the export functionality available through the Withings mobile application and the Withings web portal (app.withings.com). In the mobile application, exports could be initiated for individual user profiles, and the resulting download link was sent to the email address associated with the account. In both cases, the exported data was provided as a ZIP archive containing comma-separated values (CSV) files. In the web portal, export access was limited to the primary account holder and required two-factor authentication. The exported files contained health and personal data recorded by the scale and the smartphone, while non-personal identifiers such as user IDs were absent. The CSV files were subsequently analyzed using Microsoft Excel.

8 Data in the Network Communication

The Withings Body supports WiFi as well as Bluetooth Low Energy (BLE) and Bluetooth Classic. Its WiFi interface supports IEEE 802.11 b/g/n in the 2.4 GHz band and is compatible with WEP, WPA, and WPA2. Once WiFi has been configured during the initial setup, the scale automatically connects to the access point and synchronizes measurements with the Withings cloud. Firmware updates can also be installed via WiFi without user intervention [21].

Bluetooth is used for initial provisioning, changing WiFi credentials, performing manual firmware updates, and operating the scale in Bluetooth-only mode through the Withings app. When WiFi is unavailable, the scale can continue to operate in this Bluetooth-only mode and synchronize data with the Withings cloud via BLE through the smartphone app. In this mode, however, weather and step-count features are unavailable [21].

To capture the Bluetooth traffic between the scale and the smartphone, Android's built-in Bluetooth Host Controller Interface (HCI) snoop logging facility was used. This facility records all packets processed by the HCI and stores them in a log file at `data/misc/bluetooth/logs`. The BTSnoop format resembles the Snoop format described in RFC 1761 and is supported by Wireshark.

To record Bluetooth communication, Android's Bluetooth HCI snoop log was enabled either through the Developer options or via ADB using `adb shell settings put secure bluetooth_hci_log 1` [6]. On rooted smartphones, the HCI snoop log could then be monitored in real time and captured with Wireshark. In this configuration, the log can be exposed through a BTSnoop Net interface by redirecting it to TCP port 8872 using the following ADB command: `adb shell su -c "nc -s 127.0.0.1 -p 8872 -L system/bin/tail -f -c +0 /data/misc/bluetooth/logs/btsnoop_hci.log"` [5]. After refreshing the available interfaces in Wireshark, the BTSnoop interface became available for packet capture.

During capture, all Bluetooth functions of the scale were exercised and subsequently analyzed in Wireshark. Captured frames were reviewed step by step, and the Packet Bytes pane was inspected for readable strings. The plaintext information was carried over RFCOMM, a Bluetooth transport protocol that emulates RS-232 serial communication over L2CAP [1], with most content transmitted in JSON structures.

During Bluetooth transmission, both health-related and personal data, such as first name, weight, age, and date of birth, as well as non-personal data, such as WiFi credentials, MAC addresses, user IDs, and synchronization timestamps, are transmitted.

Wireshark was used to capture a man-in-the-middle attack on the connection between the Withings Body and the Withings cloud. For this purpose, the Withings Body was connected via WiFi to the PC's WiFi hotspot.

This way, all traffic could be routed through the PC and recorded and analyzed with Wireshark. However, the WiFi traffic was encrypted using TLS 1.3, which meant that the only metadata that could be meaningfully interpreted in plaintext was the endpoint `wbs06-ws.withings.net` and the corresponding IP address `89.39.121.170`.

Burp Proxy, part of the Burp Suite, was used to perform a man-in-the-middle (MITM) attack on the connection between the Withings app and the Withings cloud. The proxy enables the interception, inspection, and modification of HTTPS traffic in both directions [12].

The smartphone was connected via WiFi to the PC's hotspot, which routed all HTTPS traffic through the PC for capture and analysis with Burp Proxy.

For the MITM attack, Burp Suite and smartphone were prepared for HTTPS traffic interception in accordance with official documentation [11,13]. The smartphone was connected to a desktop PC via USB with USB debugging enabled. On the desktop system, Burp Suite was configured as an intercepting proxy listening on port 8080 on all interfaces. A Burp Suite CA certificate was generated and installed on the smartphone to allow interception of TLS-protected traffic. Because user-level certificate installation was insufficient in this setting, the certificate was additionally deployed as a system certificate using the Magisk module Cert-Fixer [9]. Finally, the smartphone's WiFi proxy settings were configured manually to forward network traffic to the desktop PC via its Ethernet interface on port 8080.

All HTTPS traffic transmitted via the smartphone's WiFi connection was captured through the MITM attack and stored as an XML file for subsequent analysis with the Burp HTTP History Browser [17]. During acquisition, all functionalities of the Withings app were exercised, and additional traffic was recorded while the scale operated in Bluetooth-only mode. The captured traffic was then examined for readable content. This analysis revealed extensive information contained in JSON structures, including health-related and personal data such as first name, email address, weight, and date of birth, as well as non-personal data such as the scale's MAC address, user IDs, various timestamps, and battery status.

To obtain the firmware from the Withings cloud, parameter tampering was performed. For this purpose, the smartphone was connected to the scale via Bluetooth to trigger a firmware-update query. The corresponding HTTP request contained the parameters `currentfw=1181` and `httpslink=3`. Since firmware version 1181 was already the latest version, the HTTPS response did not return a firmware update. The request was therefore transferred from Burp Proxy to Burp Repeater, which allows HTTPS requests to be modified and resent. The parameter `currentfw` was changed to the older firmware version 981, and `httpslink` was set to 1 to obtain the firmware download URL. Resending the modified request caused the server to return a URL for downloading the firmware in binary format. To verify that the firmware acquired via UART matched the firmware downloaded from the cloud, SHA-256 hashes were computed for both binaries. The identical hashes confirmed that the two firmware files were the same.

9 Results

The Tables 7, 8, 9, 10, 11 and 12 show which evidence carriers (*Scale, Smartphone, Bluetooth, WiFi, Cloud*) hold which information type. In the *Source references* column, the corresponding references are given as source numbers referring to entries in Tables 3, 4, 5, and 6. These tables list the path, file name, function, or actions on network data to gain the information.

Table 3. Sources of the scale functions

Num.	Function	Num.	Function
1	Start-up process	12	perso mfgid
2	account_secret Dump	13	perso get_factory_fw
3	Connection_manager WiFi	14	perso get_battery_cal
4	connection_manager bt	15	perso get_csr
5	dblib dump 1 5	16	perso get_crt
6	device_setup btconnect firmware_update	17	probe
7	get_battery	18	spiflash dump 0x0 0x400000
8	locale get	19	settz get
9	perso get_mac	20	wifi init
10	perso get_pubkey	21	wifi get_rssi
11	perso get_secret		

Table 4. Sources on the smartphone

Num.	Path
22	/com.withings.wiscale2/databases/Withings-WiScale
23	/databases/room-healthmate.db
24	/app_webview/Default/Cookies
25	/databases/com.google.android.datatransport.events
26	/databases/rl_persistence.db
27	/cache/WebView/Default/HTTP Cache/Cache_Data /00da7e0136f4745f_0
28	/shared_prefs/rl_prefs.xml
29	/shared_prefs/account.xml

Table 5. Sources in the Bluetooth and Wi-Fi traffic

Num.	Bluetooth Action	Num.	WiFi Action
30	Initial setup of the scale	33	Initial setup of the scale
31	Firmware update retrieval	34	Measurements using the scale
32	Bluetooth-only mode	35	Interaction in the Withings app
		36	Synchronisation with Cloud

Table 6. Sources in the cloud storage

Num.	File	Num.	File
37	activities.csv	47	aggregates_distance.csv
38	raw_location_altitude.csv	48	aggregates_steps.csv
39	raw_location_gps-speed.csv	49	body_temperature.csv
40	raw_location_horizontal-radius.csv	50	bp.csv
41	raw_location_latitude.csv	51	devices.csv
42	raw_location_longitude.csv	52	height.csv
43	raw_location_vertical-radius.csv	53	other.csv
44	account.csv	54	user.csv
45	aggregates_calories_earned.csv	55	weight.csv
46	aggregates_calories_passive.csv		

Table 7. Accessible locations for personal information

Information	Scale	Phone	Bluetooth	WiFi	Cloud	Source references
Name abbreviation	X	X	X	X	-	3, 5, 22, 30, 31, 32, 33, 36
First name	X	X	X	X	X	3, 18, 22, 23, 30, 31, 32, 33, 36, 54
Last name	-	X	-	X	X	22, 23, 35, 36, 54
Age	X	-	X	X	-	3, 5, 18, 30, 31, 32, 33
Gender	X	X	X	X	X	3, 22, 30, 31, 32, 33, 36, 54
Date of birth	X	X	X	X	X	3, 5, 22, 30, 31, 32, 33, 36, 54
Height	X	X	X	X	X	3, 5, 18, 22, 30, 31, 32, 33, 34, 36, 52
E-Mail	-	X	-	X	X	22, 34, 36, 44

Table 8. Accessible locations for time-related information

Information	Scale	Phone	Bluetooth	WiFi	Cloud	Source references
Time zone information	X	X	X	X	X	3, 19, 22, 23, 26, 30, 31, 32, 33, 34, 35, 36, 37, 51
Creation date user data	X	X	X	X	-	3, 5, 22, 30, 31, 32, 33, 34, 35, 36
Last sync with cloud	X	X	X	X	X	3, 22, 30, 31, 32, 33, 36, 51
Account Creation Time	-	X	-	-	X	22, 44
Last Use of the Scale	-	X	-	-	X	22, 51

Table 9. Accessible locations for IDs, token and secrets

Information	Scale	Phone	Bluetooth	WiFi	Cloud	Source references
WiFi SSID & Password	X	-	X	-	-	3, 18, 21, 30
User ID	X	X	X	X	-	3, 5, 22, 23, 24, 25, 26, 28, 30, 31, 32, 33, 34, 35, 36
Account ID	X	X	-	X	-	2, 3, 23, 26, 28, 29, 33, 34, 35, 36
Device ID Scale	-	X	-	X	-	23, 33, 34, 35, 36
Device Token Phone	-	X	-	X	-	22, 26, 28, 29, 33, 34, 35, 36
Session IDs	X	X	X	X	-	3, 23, 25, 30, 31, 32, 33, 34
Secrets / Keys	X	X	X	X	-	2, 10, 11, 17, 22, 30, 31, 32, 35
CSR certificate request	X	-	-	-	-	15
CRT certificate	X	-	-	-	-	16

Table 10. Accessible locations for device, account and network-related information

Information	Scale	Phone	Bluetooth	WiFi	Cloud	Source references
MAC addr. scale	X	X	X	X	X	3, 4, 6, 9, 17, 18, 20, 22, 30, 31, 32, 33, 35, 51
Bluetooth MAC addr. scale	X	-	X	-	-	4, 6, 18, 30, 31, 32
Firmware version scale	X	X	-	X	-	13, 17, 18, 22, 33, 34, 35
Battery status of scale	X	X	-	X	X	1, 7, 14, 22, 25, 33, 35, 51
Local IP addr. scale	X	-	-	-	-	3
Local IP addr. access point	X	X	-	-	-	3, 27
External IP addr.	-	-	-	X	X	33, 44
Manufacturer ID	X	-	X	X	-	12, 17, 30, 31, 32, 35
Phone model	-	X	-	X	-	22, 25, 26, 33, 35, 36
Phones Android version	-	X	-	X	-	26, 33, 35
Scale model	X	X	-	X	-	3, 18, 22, 25, 36

Table 11. Accessible locations for health and activity data

Information	Scale	Phone	Bluetooth	WiFi	Cloud	Source references
Body weight	X	X	X	X	X	3, 5, 18, 22, 30, 31, 32, 33, 34, 35, 36, 55
Body temperature	-	X	-	X	X	22, 36, 49
Calories	-	X	-	X	X	22, 23, 36, 37, 45, 46
Activities performed	-	-	-	-	X	37
Distance covered	-	X	-	X	X	22, 23, 36, 37, 47
Heart rate	-	X	-	X	X	22, 36, 50
Step count	X	X	-	X	X	5, 18, 22, 23, 36, 37, 48
Speed	-	X	-	-	X	23, 37, 39
Blood pressure	-	X	-	X	X	22, 36, 50
Menstruation data	-	X	-	-	X	22, 53

Table 12. Accessible locations for location-based data

Information	Scale	Phone	Bluetooth	WiFi	Cloud	Source references
Country code	X	X	X	X	-	1, 3, 8, 18, 22, 24, 30, 31, 32, 33, 34, 35, 36
Geographical coordinates Activities	-	X	-	-	X	23, 37, 38, 39, 41, 42
Geographical coordinates Scale	-	X	-	X	X	22, 33, 35, 51
Horizontal & vertical radius data	-	-	-	-	X	37, 40, 43
Location of ISP	-	-	-	X	-	33

10 Discussion

This exploratory examination of the Withings Body ecosystem shows that forensically relevant artifacts can be recovered across four layers: local data on the scale, local data within the companion app, Bluetooth traffic between scale and smartphone, and app-cloud HTTP(S)-traffic over WiFi. By contrast, the direct scale-cloud connection did not yield inspectable payloads because transport is protected by TLS. Taken together, the accessible layers provide health and personal data such as name, age, weight, height and device/account metadata, enabling cross-source corroboration and timeline reconstruction.

The on-board UART interface emerged as the most productive source. Console access allowed retrieval of configuration information and, with an active

WiFi configuration, the return of health and personal data via a single command. In addition, full images of SPI flash, MCU flash, and SRAM were acquired. Reverse engineering of the DBlib users field revealed a stable layout keyed by three-letter initials, with fixed offsets for health and personal data, for example weight, height, age, and step count as well as non-personal data. like User IDs or time stamps. This structure supports direct parsing of memory images and improves the reliability of cross-layer correlation.

Several practical constraints limit transfer to routine casework. Root access was required to copy the app's private storage, to read Bluetooth HCI snoop logs directly, and to install a system CA for inspecting app-cloud HTTPS traffic. The required bootloader unlocking wipes all data, which would preclude this acquisition route in many investigations. A second Bluetooth capture method can operate without root, but typically requires physical access to an unlocked phone. Over-the-air capture of BLE and WiFi is feasible, yet decryption of protected payloads is generally not possible. These constraints place the emphasis on UART as a realistic first-line source when invasive methods like chip-off or soldering for JTAG are not pursued.

Attribution and provenance remain challenging. Memory images on the device expose initials rather than full names, and UART output includes only first names. App databases mix manual entries with device measurements, and the users field in DBlib as well as the JSON transfer reflect the latest weight regardless of origin. While the presence of device identifiers alongside measurements can help, a general rule for distinguishing manual inputs from sensor-derived values was not established. The parameter *att* in the JSON data structure might be able to contribute to differentiation, since it has been determined that the value is 0 for a measurement and 2 for an input. However, this observation must be confirmed by extensive research.

The security implications mirror the forensic opportunities. The same UART pathway that enables evidence collection can expose sensitive material under adversarial access, including WiFi credentials, certificates, and other secrets. Dumpster-diving scenarios for discarded devices and chip-off recovery of intact SPI flash are plausible threat vectors for this device. These observations underline that open debug interfaces on consumer IoT devices are double-edged: they facilitate lawful analysis yet enlarge the attack surface if left unprotected.

11 Conclusion and Future Work

This work analyzed sensitive data storage and transmission in a representative smart-scale ecosystem and provides a reproducible end-to-end workflow prioritizing UART as an accessible source for IoT forensic examination. UART delivered a high evidential yield, including health and personal data, plaintext credentials, certificates, configuration stores, and complete memory images. The presented acquisition and analysis procedures enable timeline reconstruction and offer practical guidance, while structured records in application storage and JSON traffic support robust cross-layer reasoning for health and personal data. At the same time, the results highlight security weaknesses of open debug interfaces and the operational challenge of modern mobile platforms that increasingly restrict rooting.

Future work should pursue three directions:

1. investigation of options for bootloader unlocking without mandatory data wipes,
2. examination of acquisition and analysis on iOS,
3. design of automated parsers and specialized machine-learning tools for firmware and artifact discovery that can detect anomalies, patterns, and hidden structures in proprietary formats.

In the near term, hardening of debug access and better segregation and protection of credentials are immediate steps that can improve the security posture of consumer IoT while preserving pathways for lawful forensic access.

Acknowledgments. This paper was funded by the European Union and the Free State of Saxony (Germany).

Kofinanziert von der Europäischen Union

Freistaat SACHSEN

Diese Maßnahme wird mitfinanziert durch Steuermittel auf der Grundlage des vom Sächsischen Landtag beschlossenen Haushaltes.

References

1. Bluetooth SIG, I.: RFCOMM with TS 07.10 Serial Port Emulation. https://www.bluetooth.com/specifications/specs/html/
2. Civey GmbH: Smart Home: Jeder zweite Haushalt in Deutschland nutzt bereits mehr als vier Geräte mit Onlinezugang (engl. Every second house hold already uses more than four devices with internet access.). https://www.eco.de/presse/smart-home-jeder-zweite-haushalt-in-deutschland-nutzt-bereits-mehr-als-vier-geraete-mit-onlinezugang/
3. Coppola, M.: Weighing in on Issues with "Cloud Scale". https://poppopret.org/wp-content/uploads/2013/06/weighing-in-on-issues-with-cloud-scale.pdf
4. Federal Communications Commission: FCC ID XNAWBS06 – Withings WBS06. https://fcc.report/FCC-ID/XNAWBS06/3414812
5. Gadgetbridge: Inspect Bluetooth packets. https://gadgetbridge.org/internals/development/bluetooth/
6. Google: Verify and debug. Android Open Source Project. https://source.android.com/docs/core/connect/bluetooth/verifying_debugging
7. Grispos, G., Tursi, F., Choo, K.K.R., Mahoney, W., Glisson, W.B.: A digital forensics investigation of a smart scale IoT ecosystem. In: 2021 IEEE 20th International Conference on Trust, Security and Privacy in Computing and Communications (TrustCom). IEEE (2021). https://doi.org/10.1109/trustcom53373.2021.00104
8. Office of the Chief State's Attorney, Connecticut: Jury Convicts Richard Dabate of the December 2015 Murder of His Wife, Connie Dabate. Press Release, Superior Court, Rockville, Connecticut. https://portal.ct.gov/dcj/press-room/press-releases/051022dabate
9. Pavithran, J.: pwnlogs/cert-fixer. GitHub. https://github.com/pwnlogs/cert-fixer

10. Peña, E., Legaspi, M.G.: UART: a hardware communication protocol understanding universal asynchronous receiver/transmitter. https://www.analog.com/media/en/analog-dialogue/volume-54/number-4/uart-a-hardware-communication-protocol.pdf
11. PortSwigger Ltd.: Configuring an Android device to work with Burp Suite. https://portswigger.net/burp/documentation/desktop/mobile/config-android-device
12. PortSwingger Ltd.: Burp Proxy. https://portswigger.net/burp/documentation/desktop/tools/proxy
13. Singh, P.: Install Burpsuite's or any CA certificate to system store in Android 15. https://pswalia2u.medium.com/install-burpsuites-or-any-ca-certificate-to-system-store-in-android-15-0aecd2979659
14. Statista: Smarte Waagen - Deutschland (engl. Smart scales - Germany). https://de.statista.com/outlook/hmo/digital-health/digital-fitness-well-being/fitness-tracker/smarte-waagen/deutschland
15. Supreme Court of Connecticut: State of Connecticut v. Richard G. Dabate. SC 20749. https://law.justia.com/cases/connecticut/supreme-court/2025/sc20749.html
16. SureshotM6 (pseudonym): Reddit comment regarding: are Withings scales hackable? Reddit. https://www.reddit.com/r/withings/comments/18vuckz/comment/m1h0gwu
17. Telange, A.: Burp http history browser (BHHB). GitHub. https://github.com/adityatelange/bhhb
18. Transforma Insights, Exploding Topics: Number of Internet of Things (IoT) connections worldwide from 2022 to 2023, with forecasts from 2024 to 2034. https://www.statista.com/statistics/1183457/iot-connected-devices-worldwide/
19. Withings: Device sync. https://developer.withings.com/sdk/v2/tree/overview/device-sync/
20. Withings: Withings App - Android - What is the Withings App and what does it do?. https://support.withings.com/hc/en-us/articles/202719068-Withings-App-Android-What-is-the-Withings-App-and-what-does-it-do
21. Withings: Withings Body Installation and operating instructions. https://support.withings.com/hc/de/article_attachments/360012313517
22. Wu, T., Breitinger, F., Niemann, S.: IoT network traffic analysis: opportunities and challenges for forensic investigators? Forensic Sci. Int.: Digit. Invest. **38**, 301123 (2021). https://doi.org/10.1016/j.fsidi.2021.301123
23. Zawoad, S., Hasan, R.: FAIoT: towards building a forensics aware eco system for the internet of things. In: 2015 IEEE International Conference on Services Computing, pp. 279–284. IEEE (2015). https://doi.org/10.1109/scc.2015.46

CyFRP – Special Session

PERUN: Protecting Sensitive Cyber Ecosystems from Upcoming Next Generation and AI-Generated Malware Threats

Marek Pawlicki[1,2], Aleksandra Pawlicka[1], Jörg Keller[3(✉)], Wojciech Mazurczyk[3,4], Rafał Kozik[1,2], and Michał Choraś[2,3]

[1] ITTI Sp. z o.o., Poznań, Poland
{apawlicka,michal.choras}@itti.com.pl
[2] Bydgoszcz University of Science and Technology, Bydgoszcz, Poland
marek.pawlicki@pbs.edu.pl, rkozik@utp.edu.pl
[3] FernUniversität in Hagen, Hagen, Germany
joerg.keller@fernuni-hagen.de
[4] Warsaw University of Technology, Warsaw, Poland
wojciech.mazurczyk@pw.edu.pl

Abstract. The increasing sophistication and frequency of cyberattacks pose significant risks to critical infrastructures, essential services, and digital ecosystems across the European Union. Addressing challenges such as polymorphic malware, encrypted malicious communications, and AI-generated attacks requires integrated, scalable, and trustworthy cybersecurity solutions. The PERUN (Protecting Sensitive Cyber Ecosystems from Upcoming Next Generation and AI-generated Malware Threats) project, funded within the Horizon Europe framework, develops advanced artificial intelligence-driven tools and methodologies for threat detection, vulnerability assessment, and incident response. The project focuses on enhancing the security and resilience of critical digital infrastructures while supporting European technological sovereignty and compliance with regulatory and ethical frameworks, by integrating innovative AI technologies and sector-specific protection strategies.

Keywords: Cybersecurity · Malware · Critical Infrastructure Protection · Research Project

1 Introduction

More than ever, Europe needs cybersecurity; with the constant development of new malicious software, cyberattacks have been on the rise [5,6]. Due to new attack vectors [2,17] and due to encryption [1,22], novel attacks are often undetectable by humans, simple rules, and signatures alike [10,18]. Without holistic and interdisciplinary cybersecurity solutions, the citizens' lives, well-being, money, and other assets are at risk [13].

© The Author(s), under exclusive license to Springer Nature Switzerland AG 2027
K. Renaud et al. (Eds.): EICC 2026, CCIS 3012, pp. 307–317, 2027.
https://doi.org/10.1007/978-3-032-28957-5_18

The PERUN project will strengthen the European cybersecurity by providing novel solutions powered by artificial intelligence (AI) and machine learning (ML), to analyse and counter emerging cyberthreats to software, firmware and hardware in an efficient and cost-effective way.

The PERUN project aims to enhance the security and resilience of digital infrastructures—national-level cybersecurity infrastructures, education and research networks, critical energy infrastructures, digital infrastructures of NGOs providing critical services and security operations centers (SOCs) across various sectors—by developing advanced cybersecurity solutions that leverage AI/ML technologies.

To this end, PERUN will produce a set of innovative methods and algorithms to analyse data and detect new types of malware, as well as a collection of high TRL tools and products using novel scientific approaches, all validated by representatives of the critical sectors.

The tangible results of PERUN will be applied in five real-life scenarios pertaining to critical sectors such as nonprofits (NGOs), energy, telecommunications, private SOCs and CERTs/CSIRTs.

The remainder of this paper is structured as follows. In Sects. 2, 3 and 4, we discuss the objectives, scope and methodology of the PERUN project, respectively. In Sects. 5 and 6, the concepts are presented, including models and use cases, respectively. Section 7 discusses the project's impact on Europe's cybersecurity while Sect. 8 gives conclusions and an outlook.

2 Objectives of the Project

Driven by the sophisticated novel cyberthreats, modern cybersecurity measures have evolved far beyond the simple signature-based and rule-based mechanisms [12]. The adoption of AI/ML methods in cybersecurity has brought new research directions, for example Explainable AI (xAI) has been identified as a crucial research frontier, enabling human analysts to interpret and validate the decisions made by ML-based detection systems [15,20]. Emerging research also demonstrates that coupling AI with secure architectural frameworks and trusted execution environments can substantially improve resilience against both software and hardware-level threats [21].

With growing legislation, ethical, privacy, and regulatory considerations have emerged as central concerns in the deployment of intelligent cybersecurity systems [24].

With this in mind, the main goals of the project are:

1. **Improved Malware Detection and Analysis:** To develop AI-based methods for detecting advanced threats like stegomalware, ransomware, and malware using encrypted communications. This includes creating custom malware analysis sandboxes as well as network intrusion detection systems (NIDS) with explainable AI frameworks.

2. **Enhanced Hardware and Software Security Engineering:** To utilise trusted hardware and software systems, such as RISC-V technology and Open FPGA Frameworks, to design resilient systems from the ground up.
3. **Automated Vulnerability Assessment and Risk Management:** To create systematic and automated approaches for risk assessment, including the development of new risk assessment methodologies like cyber-ranges and dynamic security assessments. This includes evaluating vulnerability combinations and critical paths in cybersecurity.
4. **AI-Driven Security Services:** To implement AI/ML methodologies such as predictive analytics, anomaly detection, and data fusion to provide predictive security services. This includes continuous security assessments and the integration of AI/ML solutions with existing systems and workflows.
5. **Privacy, Ethics, and Trustworthiness:** To ensure ethical and legal compliance of the PERUN solutions, and deliver enhanced regulation shaping through policy recommendations. Incorporate privacy considerations, ethical guidelines, and technical standards into the development and deployment of AI systems.

The PERUN project's solutions are well grounded in current technological capabilities and the strategic use of existing open-source tools and frameworks.

The project's emphasis on AI and ML for threat detection and anomaly analysis is also grounded in proven methodologies. Trustworthy AI-Powered Network Intrusion Detection Systems employ state-of-the-art AI techniques, including deep learning and reinforcement learning models that have been successfully applied in network security contexts. The integration of explainable AI frameworks ensures that the AI-driven solutions are transparent and trustworthy.

An especially novel threat class addressed by PERUN solutions is AI-generated malware, specifically malicious code produced with the help of LLMs and genAI systems. GenAI can synthesise functionally novel attack logic which evades current signature and behaviour-based detection. PERUN treats this category as a distinct threat class and aims to alleviate the threat with the PERUN-GENERATOR component. The objective is to use LLMs to generate diverse synthetic malware samples to enhance the training of detection systems, finding new ways to prevent such misuse of LLMs.

The feasibility of achieving these goals within the project timeline is enhanced by the involvement of experienced partners and collaboration with industry and academic institutions. The consortium includes national cybersecurity agencies like NASK and RoEduNet, leading energy providers like EDF Energy, and specialised organisations like the CyberPeace Institute. The places of origin of the consortium are showcased in Fig. 1. These partners bring extensive expertise, real-world insights, and access to operational environments for testing and validation.

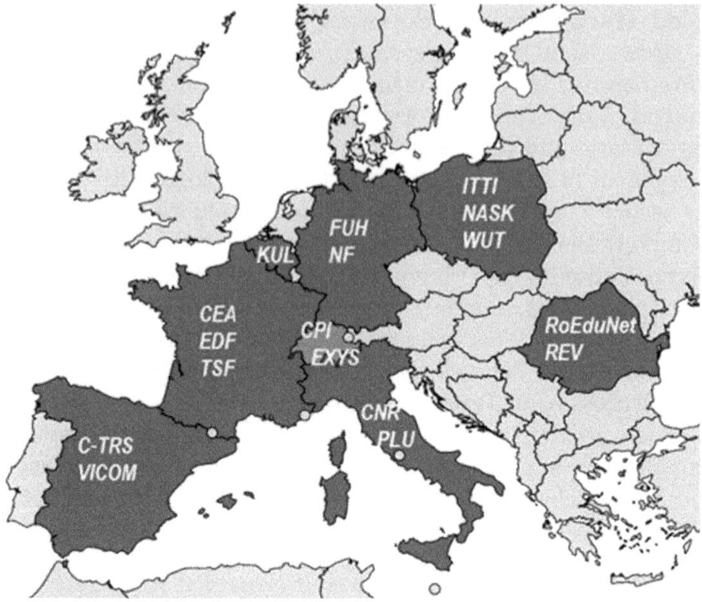

Fig. 1. Consortium Places of Origin

3 PERUN Scope

The scope of the PERUN project is grounded in a body of peer-reviewed research showing that systematic evaluation of hardware and software components, integrated secure engineering practices, and continuous security assessment frameworks are essential for building resilient digital ecosystems. Studies on firmware vulnerability assessment emphasise the need for holistic auditing and analysis techniques that span static, dynamic, and hybrid evaluation methods to uncover deep structural flaws in embedded systems and IoT (Internet of Things) devices [3]. The trends underline the need for resilient systems design, secure testing environments, automated vulnerability discovery, and continuous monitoring mechanisms, all of which inform the comprehensive scope of PERUN's activities [9]. Thus, the scope of PERUN encompasses the items found below.

1. **Improved hardware and software security engineering; resilient systems design**: PERUN analyses advanced malware behaviours, including encrypted and hidden communications, enhancing software security engineering by understanding sophisticated threats. PERUN provides information about firmware (e.g., Control Flow Graphs) and develops ML-based classifiers to detect malicious patterns, improving hardware security engineering through automated analysis and malware detection. PERUN enhances hardware security by developing hardware emulation and intelligent fuzzing techniques, enabling automatic detection of vulnerabilities in firmware and IoT devices.

2. **Improved access to testing of hardware and software in virtual, closed and secure environments**: PERUN develops new concepts for testing hardware and software using cyber-ranges, providing secure, simulated environments for risk assessment and vulnerability testing. PERUN implements hardware emulation and intelligent fuzzing, including IoT device firmware fuzzing, enabling automated vulnerability discovery in secure, virtual environments. PERUN deploys a custom malware analysis sandbox, offering a secure platform for dynamic malware analysis without putting operational systems at risk.
3. **Systematic and, where possible, automated study of vulnerabilities, software analysis, vulnerability discovery, and dynamic security assessment**: PERUN detects data-mining patterns in network traffic, automating the identification of malicious behaviours and contributing to systematic vulnerability analysis. PERUN gathers firmware and conducts static analysis, using machine learning for malware classification to automate the detection of malicious firmware components.
4. **Trustworthy certifiable hardware and software**: PERUN leverages open-source hardware architectures and contributes to developing hardware that is transparent, verifiable, and certifiable.
5. **AI-based security services e.g., predictive security, advanced anomaly and intrusion detection, system health checks**: PERUN creates trustworthy AI-powered Network Intrusion Detection Systems using deep learning and reinforcement learning, adapting to new threats and providing explainable AI outputs. PERUN implements an Incident Qualification Assistant using AI to automate information retrieval and summarization, supporting system health checks and faster incident response. PERUN applies ML-based real-time anomaly detection techniques to monitor peripheral connections, enhancing intrusion detection capabilities.

Current software development prioritises fast deployment over security, which results in vulnerabilities and insecure applications. Security engineering, both at the software and hardware levels, must be integrated in their development [19].

The EU should be able to rely on software and hardware that can be verified and audited as to their security [23]. In particular, the potential security implications of using open-source software and hardware, and security auditability in that context, should be further explored.

Software is subject to continuous updates, so the security posture cannot be assessed once and for all [16]. Hence, methods and tooling to perform continuous assessments of security are needed. In addition, security and privacy regulations also evolve, having to be factored in compliance approaches. The PERUN project integrates continuous security assessment methods and tools into its objectives to maintain an up-to-date security posture.

For example, the PERUN-NETWORK-AI component focuses on developing AI-powered Network Intrusion Detection Systems that adaptively respond to new threats using deep learning and reinforcement learning, providing ongoing monitoring and dynamic vulnerability assessment.

Whilst a great portion of the software and hardware used in the EU is developed outside the European Union [8], it must comply with the security requirements within the EU.

4 PERUN Methodology

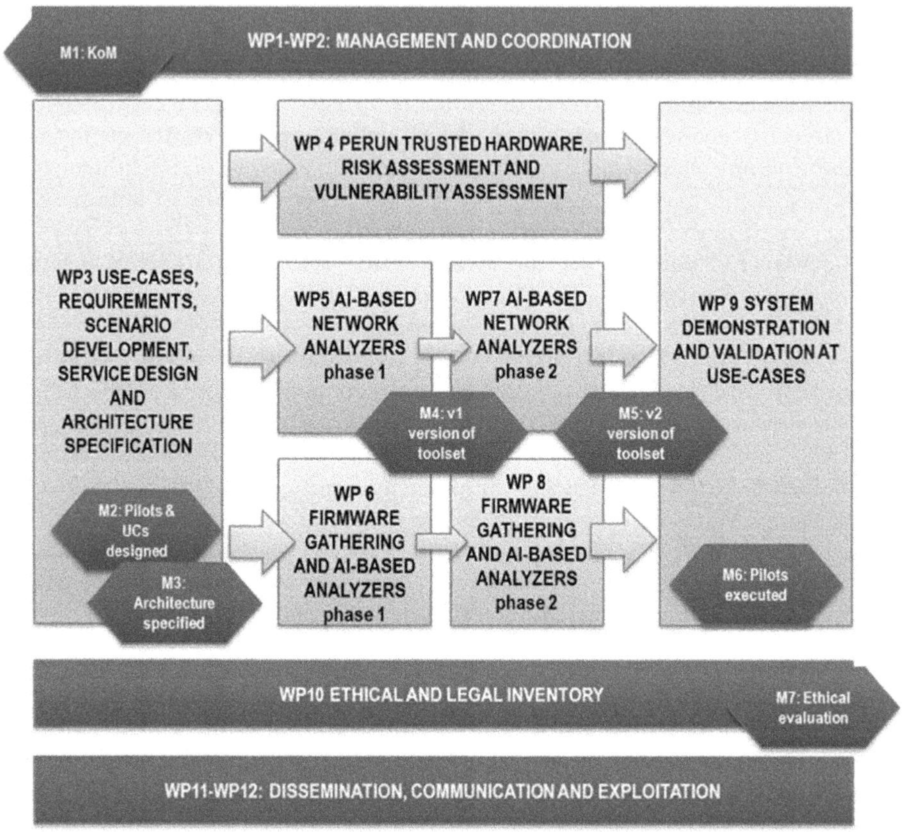

Fig. 2. High-level process flow diagram of PERUN

The PERUN project adopts a structured, phased approach aligned with its objectives and expected outcomes. This approach aligns with established research showing that embedding security practices into iterative development frameworks, such as Agile and DevSecOps, enhances flexibility, supports continuous risk assessment, and improves the early detection and mitigation of vulnerabilities across the lifecycle of complex systems [11]. The methodology encompasses:

- requirement analysis,
- design and development,
- integration and testing,
- pilot execution,
- and pilot evaluation

The methodology is aligned with Agile methodologies. The Activity diagram of PERUN is shown in Fig. 2.

To this aim, the following phases have been distinguished:

1. **Phase 1: Requirements Analysis and Use Case Definition** encompasses **Initial Planning, Gathering of Requirements, and Gap Analysis**, identifying specific needs and challenges within each use case. PERUN will conduct a thorough gap analysis to determine areas requiring technological advancements. PERUN will develop detailed scenarios for each use case, ensuring real-world applicability.
2. **Phase 2: Design and Development of AI-based Solutions.** PERUN will explore implementing trusted hardware (e.g., RISC-V technology) and open-source software, and evaluate the maturity and applicability of these technologies in security-critical environments. PERUN will employ AI and ML to identify malicious patterns and develop models for detecting malware communication and anomalies. PERUN will collect firmware samples from various architectures and perform static analysis to extract control flow graphs and relevant data. PERUN will apply ML classifiers to identify malicious patterns in firmware to enhance malware detection accuracy through advanced analysis.
3. **Phase 3: Integration and Testing.** PERUN will create intuitive interfaces for security analysts and implement visualisation techniques for improved situational awareness. PERUN will integrate the developed components into a cohesive platform, conducting rigorous testing in controlled environments to validate effectiveness.
4. **Phase 4: Pilot Execution and Evaluation**. PERUN will deploy solutions in real-world settings across the defined use cases, measuring performance against KPIs to assess success. PERUN will provide training for end-users to utilise the novel tools effectively.

At the time of writing this, PERUN is entering month six, and a structured requirements analysis has been conducted across all five use cases, involving direct engagement from the Pilot partners. Preliminary research results are also being published in [7,14] and [4].

5 Concepts, Models, and Assumptions

Within the duration of the PERUN project, we will take into account the following aspects:

1. **Agile Methodology:** The project adopts Agile methodologies to facilitate iterative development, frequent reassessment, and flexibility in responding to changes during the project lifecycle, by iteratively evaluating progress and incorporating stakeholder feedback and emerging insights to refine and optimise the development process continually. This approach is critical given the rapidly evolving nature of cybersecurity threats.
2. **AI and Machine Learning Models:** PERUN leverages advanced AI and ML techniques for anomaly detection, predictive analytics, and automated threat analysis. Models such as deep learning neural networks, reinforcement learning are utilised, including the latest advancements in Large Language Models.
3. **Interdisciplinary Collaboration:** PERUN integrates expertise from cybersecurity, AI/ML, hardware engineering, software development, and legal and ethical domains to ensure comprehensive solutions.

6 PERUN Use Cases

Each PERUN Use Case applies AI-based detection, prevention and response technologies to protect different critical sectors from modern cyberthreats. These include humanitarian non-profits, energy infrastructure, national research networks, and cybersecurity authorities.

1. **Use Case 1** targets NGOs providing essential services like humanitarian relief and healthcare.
2. **Use Case 2** protects operational technology (OT) and industrial systems in energy infrastructure.
3. **Use Case 3** improves cybersecurity for large research and education networks.
4. **Use Case 4** focuses on analysing malware samples offline using AI and static analysis.
5. **Use Case 5** focuses on enhancing Security Operations Centres.

The main challenges in the use cases involve:

- **Data Volume and Complexity** Modern networks and infrastructures generate vast amounts of security data, making analysis difficult and increasing the risk of overlooked threats.
- **Detection of Advanced and Unknown Threats** Polymorphic malware, encrypted traffic, and AI-generated malicious code challenge traditional detection methods.
- **Sector-Specific Operational Constraints** For example, NGOs face budget and expertise limitations, while critical infrastructure operators must maintain continuous service availability, limiting opportunities for security interventions.
- **High False Positive Rates and Alert Fatigue** Security systems often generate excessive alerts, reducing analyst efficiency.

- **Integration with Legacy and Distributed Systems** Many organisations rely on decentralised, hybrid, or outdated IT infrastructures.
- **Rapid Evolution of Cyber Threats** Threat actors continuously develop new attack methods, requiring cybersecurity systems to evolve dynamically.

7 How PERUN Will Impact EU Cybersecurity

The PERUN project is designed to strengthen the European Union's cybersecurity capabilities by developing advanced artificial intelligence (AI)-driven tools, secure testing environments, and standardised methodologies for vulnerability detection, threat analysis, and incident response. Through applied research, industrial collaboration, and real-world validation across multiple critical sectors, PERUN contributes to enhancing the EU's technological sovereignty, resilience of digital infrastructures, supply chain security, and cybersecurity culture.

PERUN develops European-based technologies capable of protecting critical infrastructure and reducing reliance on external proprietary solutions. In the short term, these technologies improve the cybersecurity posture of participating organisations and promote stronger collaboration among European cybersecurity stakeholders. Over the medium term, the adoption of PERUN tools across industries supports workforce development and strengthens Europe's position in the global cybersecurity market. Long-term outcomes include establishing the EU as a global leader in secure digital technologies and maintaining sustained technological independence.

PERUN tools allow organisations to proactively identify vulnerabilities, simulate cyberattack scenarios in cyber-range environments, and improve incident response efficiency. In addition, AI-based SOC assistance tools enhance threat prediction and incident management capabilities. These technologies promote safe adoption of innovative digital systems and contribute to increased public trust in emerging technologies.

Scientifically, PERUN advances EU research and innovation by delivering an integrated cybersecurity platform validated in operational environments at TRL7. The project demonstrates new methods for threat detection, vulnerability assessment, and AI-driven cybersecurity operations.

Where legally possible, PERUN will produce open datasets and make them available to the research community and the general public, including, for example, labelled network traffic captures.

8 Conclusions

In this paper, we have overviewed the Horizon Europe-funded project PERUN (Protecting Sensitive Cyber Ecosystems from Upcoming Next Generation and AI-generated Malware Threats). The project is active within the Civil Security for Society Cluster. It started in 2025 and will end in 2028.

PERUN establishes a comprehensive cybersecurity ecosystem combining advanced AI technologies, sector-specific protection strategies, certification frameworks, and collaborative knowledge-sharing mechanisms. By addressing current and emerging cyber threats, the project strengthens the EU's digital sovereignty, improves infrastructure resilience, enhances cybersecurity innovation, and supports the development of a secure and trustworthy digital society.

Acknowledgements. This research has been supported by the Horizon Europe PERUN Project, funded the European Union's Horizon Europe Research and Innovation programme under Grant Agreement No. 101225653.

References

1. Anderson, B., Paul, S., McGrew, D.: Deciphering malware's use of TLS (without decryption). J. Comput. Virol. Hacking Tech. **14**(3), 195–211 (2018)
2. Badar, L.T., Carminati, B., Ferrari, E.: A comprehensive survey on stegomalware detection in digital media, research challenges and future directions. Signal Process. **231**, 109888 (2025)
3. Bakhshi, T., Ghita, B., Kuzminykh, I.: A review of IoT firmware vulnerabilities and auditing techniques. Sensors **24**(2), 708 (2024)
4. Benedetti, G., et al.: Days of future past: towards robust detection of malware variants via LLM-based embedding generation. In: 2025 IEEE International Conference on Data Mining Workshops (ICDMW), pp. 1230–1237. IEEE (2025)
5. Bensaoud, A., Kalita, J., Bensaoud, M.: A survey of malware detection using deep learning. Mach. Learn. Appl. **16**, 100546 (2024). https://doi.org/10.1016/j.mlwa.2024.100546, https://www.sciencedirect.com/science/article/pii/S2666827024000227
6. Caviglione, L., et al.: Tight arms race: overview of current malware threats and trends in their detection. IEEE Access **9**, 5371–5396 (2020)
7. Caviglione, L., Comito, C., Benedetti, G., Falcone, A., Guarascio, M.: Deobfuscation of javascript code and identification of security weaknesses through large language models. Future Gener. Comput. Syst. 108318 (2025)
8. Di Marco, D., Thabit Gonzalez, S., Kotsev, A., Friis-Christensen, A., Minghini, M., King, M.: Open but not powerless: Towards a common understanding of eu digital sovereignty. Technical report, Ispra (Italy) (2025)
9. Jaigirdar, F.T., Jayatilaka, A., Babar, M.A.: Software vulnerability management in IoT systems: a systematic mapping study. Cybersecurity **9**(1), 96 (2026)
10. Khadka, K., Ullah, A.B.: Human factors in cybersecurity: an interdisciplinary review and framework proposal. Int. J. Inf. Secur. **24**(3), 1–13 (2025)
11. Lala, S.K., Kumar, A., et al.: Secure web development using OWASP guidelines. In: 2021 5th International Conference on Intelligent Computing and Control Systems (ICICCS), pp. 323–332. IEEE (2021)
12. Mohamed, N.: Artificial intelligence and machine learning in cybersecurity: a deep dive into state-of-the-art techniques and future paradigms. Knowl. Inf. Syst. 1–87 (2025)
13. Paraloganathan, V.: Cybersecurity management science-state of the art: A systematic literature review on cybersecurity management using bibliometric analysis and thematic literature review: Master thesis, stockholm university (2025)

14. Pawlicki, M., Choraś, R., Kozik, R., Choraś, M.: Survey on explainability-weaponising adversarial attack vectors against deep neural networks and artificial intelligence. In: Proceedings of the 18th International Conference on Agents and Artificial Intelligence - Volume 2: ICAART, pp. 1361–1370. INSTICC, SciTePress (2026). https://doi.org/10.5220/0014286400004052
15. Pawlicki, M., Pawlicka, A., Kozik, R., Choraś, M.: Advanced insights through systematic analysis: mapping future research directions and opportunities for xai in deep learning and artificial intelligence used in cybersecurity. Neurocomputing **590**, 127759 (2024). https://doi.org/10.1016/j.neucom.2024.127759, https://www.sciencedirect.com/science/article/pii/S0925231224005307
16. Prates, L., Pereira, R.: Devsecops practices and tools. Int. J. Inf. Secur. **24**(1), 11 (2025)
17. Puchalski, D., Caviglione, L., Kozik, R., Marzecki, A., Krawczyk, S., Choraś, M.: Stegomalware detection through structural analysis of media files. In: Proceedings of the 15th International Conference on Availability, Reliability and Security, pp. 1–6 (2020)
18. Rosenberg, I., Shabtai, A., Elovici, Y., Rokach, L.: Adversarial machine learning attacks and defense methods in the cyber security domain. ACM Comput. Surv. (CSUR) **54**(5), 1–36 (2021)
19. Saeed, H., Shafi, I., Ahmad, J., Khan, A.A., Khurshaid, T., Ashraf, I.: Review of techniques for integrating security in software development lifecycle. Comput. Mater. Continua **82**(1) (2025)
20. Sarker, I.H.: Deep learning: a comprehensive overview on techniques, taxonomy, applications and research directions. SN Comput. Sci. **2**(6), 1–20 (2021)
21. Shankar, V., Muralidhar, N.: Securing hardware with ai: intrusion detection, threat mitigation, and trust assurance. Int. J. Comput. Sci. Eng. **13**(4), 92–98 (2025)
22. Shen, M., et al.: Machine learning-powered encrypted network traffic analysis: a comprehensive survey. IEEE Commun. Surv. Tutorials **25**(1), 791–824 (2022)
23. Shukla, A., Katt, B., Nweke, L.O., Yeng, P.K., Weldehawaryat, G.K.: System security assurance: a systematic literature review. Comput. Sci. Rev. **45**, 100496 (2022)
24. Timmers, P.: Ethics of ai and cybersecurity when sovereignty is at stake. Mind. Mach. **29**(4), 635–645 (2019)

Privacy-Preserving Identity Management and Software Bill of Materials Vulnerability Detection: Practical Use Cases from the PRIVIDEMA Project

M. G. Belorgey[1(✉)], B. Cogliati[2], S. Demarty[2,3], L. Huguenin-Dumittan[1], O. Ozturk[4], S. Samiei[5], and O. Stan[6]

[1] TuneInsight, Lausanne, Switzerland
maria.georgievabs@gmail.com
[2] Thales DIS France SAS, Meudon, France
[3] Laboratoire de Mathématiques de Versailles, UVSQ, CNRS, Université Paris-Saclay, Versailles, France
[4] Sabanci University, Istanbul, Turkey
[5] Idakto, Montigny-le-Bretonneux, France
[6] Université Paris-Saclay, CEA LIST, 91120 Palaiseau, France

Abstract. **PRIVIDEMA** project (*Privacy-Preserving Identity Management for Digital Wallets and Secure Data Sharing and Processing for Cyber Threat Intelligence Data*) advances the state of the art in cryptographic and Privacy-Enhancing Technologies (PETs) to enable secure, interoperable, and trustworthy data exchange across sectors, with a focus on the domains of **Cyber Threat Intelligence** and **Digital Identity Management**. This paper presents two representative real-world use cases: **(1)** privacy-preserving digital identity management based on the European Digital Identity (EUDI) Wallet, and **(2)** privacy-preserving Cyber Threat Intelligence (CTI) sharing for Software Bill of Materials (SBOMs) and vulnerability datasets. Both use cases showcase how advanced PETs, including Fully Homomorphic Encryption (FHE), Federated Learning (FL), and Differential Privacy (DP), can be composed to protect sensitive data throughout its lifecycle while maintaining analytical and operational utility. Together, these use cases chart a practical course toward more scalable, standards-compliant, and privacy-preserving data ecosystems that align with Europe's vision for secure and trustworthy digital services.

Keywords: Cyber-Threat Intelligence · Digital Identity Management · PET · Homomorphic Encryption

1 Introduction and Context

The growing exchange of sensitive data, including personal identity attributes, biometric information, and cyber threat intelligence, across digital services,

© The Author(s), under exclusive license to Springer Nature Switzerland AG 2027
K. Renaud et al. (Eds.): EICC 2026, CCIS 3012, pp. 318–330, 2027.
https://doi.org/10.1007/978-3-032-28957-5_19

demands stronger guarantees for privacy, confidentiality, and regulatory compliance throughout the entire data lifecycle. Conventional protection mechanisms such as encryption in transit, encryption at rest, and anonymisation provide important safeguards, but they remain insufficient for modern distributed ecosystems in which data must also be processed, analysed, and shared among multiple stakeholders. In most existing architectures, this still requires decryption prior to computation, creating exposure points that are particularly problematic in high-assurance domains.

This challenge has become strategically important in the European context, where secure cross-border interoperability and trustworthy digital services are central policy objectives. Emerging initiatives in digital identity and collaborative cybersecurity increasingly depend on infrastructures capable of preserving confidentiality not only during storage and communication, but also during authentication, analytics, and cooperative decision-making. As a result, there is a growing need for practical Privacy-Enhancing Technologies (PETs) that support secure computation directly over protected data.

The PRIVIDEMA project, funded under Horizon Europe, addresses this need through the development of a modular and scalable privacy-preserving data sharing framework. Its architecture combines advanced PETs, including Fully Homomorphic Encryption (FHE), Federated Learning (FL), and Differential Privacy (DP), to ensure that sensitive data remains protected even during processing. By integrating these complementary techniques, the project aims to bridge the gap between state-of-the-art privacy-preserving cryptography and deployable solutions that satisfy real-world operational, performance, and interoperability constraints.

The motivation for this work lies in the absence of practical mechanisms that allow organisations and users to benefit from remote services without disclosing the sensitive data required to operate them. This gap is particularly visible in scenarios where authentication decisions or security analytics depend on highly confidential inputs that cannot be safely revealed to third-party infrastructures. Addressing this limitation is essential for enabling trustworthy digital services that remain aligned with Europe's principles of privacy by design, digital sovereignty, and secure collaboration.

To address these challenges, PRIVIDEMA develops and validates two representative real-world use cases:

- **Use Case 1:** Privacy-Preserving Digital Identity (EUDI Wallet), this use-case explores how secure Multi-Factor Authentication (MFA) can be enhanced through the integration of FHE-based biometric authentication, providing robust identity verification while preserving user privacy, even during remote or untrusted computations.
- **Use Case 2:** Software Bill of Materials Vulnerability Detection. A key way to assess system security is by using a Software Bill of Materials (SBOM). This SBOM assists in understanding what components are inside a software. One role of such an inventory is to help people identify all the possible vulnerabilities that could impact the said software. These vulnerabilities are generally

tracked against public databases such as MITRE's Common Vulnerabilities and Exposures database [12] (U.S.) or the European Union Vulnerability Database [9]. Most tools that assess for vulnerabilities, however, work offline and only help individual companies protect their own software. Using privacy-preserving technologies like homomorphic encryption allows stakeholders to match Software Bills of Materials with vulnerability databases.

These two promising real-world use cases provide a strong foundation for developing a safer digital Europe, while ensuring the protection of biometric identity privacy in full alignment with EU policy.

2 Use Case 1 – Privacy-Preserving Digital Identity (EUDI Wallet)

Europe is building a unified digital identity infrastructure through the EU Digital Identity (EUDI) Wallet, proposed by the European Commission under the eIDAS 2.0 regulation [2]. The EUDI Wallet enables interoperable digital identity across EU member states, involving three core parties: the *identity provider* (issues identity attributes), *relying parties* (services requiring identity verification), and the *holder* (the citizen). The Architecture and Reference Framework (ARF) [14] standardizes identity issuance and presentation protocols, relying on ISO 18013-5 and SD-JWT containers to cryptographically bind identity attributes to their legitimate holder through user binding information, typically a public key tied to user-controlled private credentials.

To achieve the required Level of Assurance (LoA), the ARF supports secure hardware components such as Secure Elements (SE), Trusted Execution Environments (TEE), or remote Hardware Security Modules (HSM). While biometric authentication is supported locally via SE or TEE, **biometric verification is currently absent from the EUDI Wallet ecosystem in remote cloud settings**. This gap stems from fundamental privacy tensions: delegating biometric verification to a remote server exposes sensitive personal data to an infrastructure the user does not control, risks activity tracking across sessions, and conflicts with the self-sovereign identity principle at the core of the EUDI vision. Existing standards do not provide mechanisms to perform remote biometric matching without disclosing biometric templates or behavioral patterns to the server, making cloud-based biometric unlock of HSM-protected keys an open and unaddressed problem.

In this project, we aim to fill this gap by integrating *Fully Homomorphic Encryption* (FHE) into the remote biometric authentication pipeline for the EUDI Wallet. FHE enables computation directly on encrypted data, allowing a remote authentication server to perform biometric matching without ever accessing plaintext biometric features, while the HSM retains exclusive control over decryption and key-release decisions. Concretely, we target the following objectives:

- Enable HSM key unlocking conditioned on remote biometric verification, supporting 1:1 *authentication* (query embedding matched against a single enrolled template), while keeping all biometric data encrypted throughout.
- Guarantee that a curious or compromised server learns nothing about the user's biometric characteristics, and cannot link authentication events across sessions.
- Demonstrate that the performance overhead introduced by FHE remains acceptable within the operational constraints of the EUDI Wallet use case.

The proposed solution aligns with both the zero-trust security model and the self-sovereign identity principle: the server performs its function, verifying identity, without acquiring the data that makes that verification possible in plaintext.

2.1 Face Recognition Framework

The biometric authentication pipeline is built on InsightFace [11], an open-source framework for deep face analysis, using ArcFace [8] as the core recognition model. ArcFace's additive angular margin loss produces highly discriminative 512-dimensional facial embeddings, making it well-suited for privacy-preserving matching under encryption. The pipeline is partitioned into two trust domains: an *on-device* domain responsible for all biometric capture and processing, and a *server-side* secure environment consisting of an authentication server that performs homomorphic similarity computation, and a remote HSM that holds the decryption key and performs threshold evaluation and key-release decisions. All biometric data is encrypted before leaving the device, and no plaintext biometric material is ever transmitted or processed outside it.

The pipeline proceeds through the following stages, illustrated in Fig. 1:

- **Capture and preprocessing.** The user's facial image is acquired via mobile camera and normalized (resizing, color correction) to ensure consistent input to the downstream recognition stages.
- **Detection and alignment** Locates the face and aligns it against five landmarks (eyes, nose, mouth corners), standardizing orientation prior to embedding extraction by RetinaFace [7].
- **Embedding generation.** ArcFace produces a compact 512-dimensional feature vector encoding the user's facial characteristics. This embedding is generated entirely *on the client device* and never transmitted in plaintext. The same procedure is applied both to the reference template, enrolled once and stored encrypted on the server, and to the fresh query image presented at each authentication attempt.
- **FHE encryption.** Both the query embedding and, during enrollment, the reference template embedding are encrypted on-device under the CKKS scheme, using a public key whose corresponding secret key is held exclusively by the HSM. Each encryption draws fresh randomness, ensuring that ciphertexts produced in distinct sessions are computationally indistinguishable. The resulting ciphertexts are then transmitted to the authentication server.

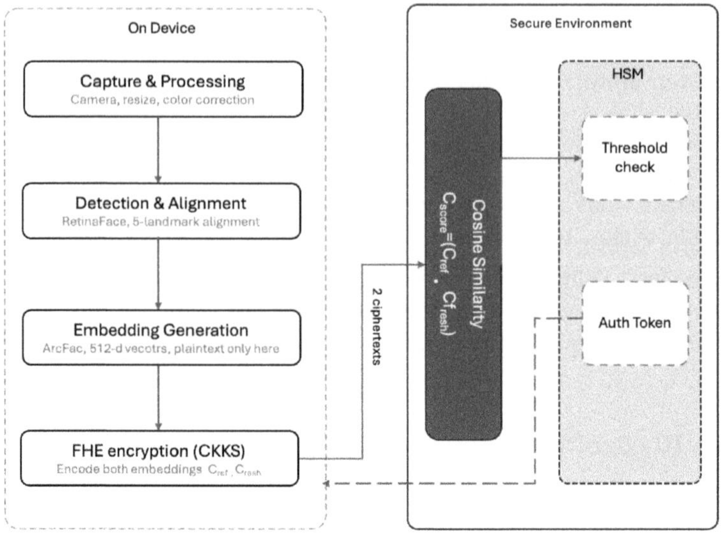

Fig. 1. Privacy-preserving biometric authentication pipeline for the EUDI Wallet.

– **Homomorphic matching.** The authentication server evaluates the cosine similarity between the encrypted query embedding and the encrypted stored template directly in ciphertext space, via an inner product over 512 CKKS-encoded coefficients. The resulting encrypted similarity score is forwarded to the HSM without decryption. The HSM decrypts the score within its protected boundary, compares it against the authentication threshold, and releases the protected key material only upon a successful match. No plaintext biometric data is accessible at any point outside the client device.

The face detection, alignment, and embedding extraction stages are executed entirely on the client device, where plaintext computation is both permissible and efficient. RetinaFace performs face localisation and five-landmark alignment on the raw image, after which ArcFace extracts the 512-dimensional embedding. Since these stages operate on plaintext data that never leaves the device, they incur no FHE overhead whatsoever. The sole stage requiring homomorphic evaluation is template matching: computing the cosine similarity between the encrypted query embedding and the encrypted template. This reduces to an encrypted inner product over 512 CKKS coefficients, implemented via the `prividema-fhe` library, a C++ implementation building on Lattigo [1]. The computation requires modest multiplicative depth and preserves matching accuracy within the approximation bounds of CKKS. This targeted use of FHE, applied only at the point where encrypted data must be compared on an untrusted server, avoids the prohibitive cost of evaluating CNN non-linearities homomorphically, which would otherwise constitute an intractable performance bottleneck.

Performance Considerations. Performance is a critical dimension of the proposed system, as the authentication sub-flow is directly embedded in user-facing wallet operations, including credential issuance and presentation under the VCI and VP workflows. An end-to-end latency budget must therefore be defined and validated against the operational requirements of the EUDI Wallet. This budget must be evaluated holistically across all phases of the pipeline: on-device biometric capture and preprocessing, FHE encryption, network transport to the authentication server, homomorphic matching, encrypted score transfer to the HSM, and HSM-side decryption and threshold evaluation. Meeting this target across the full pipeline, rather than optimizing individual stages in isolation, is a key research and engineering objective of this work.

2.2 Threat Model

We consider the remote authentication server as an *honest-but-curious* adversary: it executes the biometric matching protocol correctly but attempts to infer information about the user's biometric characteristics or activity patterns from the data it receives. Additionally, we consider a passive network attacker capable of intercepting all communications between the client device and the server. Compromised storage is also considered, where an attacker gains read access to the server-side encrypted template database.

Under these assumptions, our solution provides the following guarantees.

Template Confidentiality: the authentication server never accesses plaintext biometric embeddings. All data is encrypted under the CKKS scheme before transmission, ensuring that neither a curious server nor a storage attacker can recover facial features from the ciphertexts it holds.

Unlinkability: each authentication session produces a fresh ciphertext under independent randomness. Under the semantic security of the CKKS scheme, successive sessions are computationally indistinguishable to the server, preventing it from correlating authentication events or tracking user activity over time.

These properties directly fulfill the obligations from EU Regulation 2024/1183 (eIDAS 2.0) [2], specifically Article 5a(5), which prohibits wallet providers from obtaining any information about the use of electronic attestations, and Article 5a(14), which requires personal wallet data to remain logically separate from any other data held by the provider.

3 Use Case 2 – Privacy-Preserving Cyber Threat Intelligence Sharing

A Software Bill of Materials (SBOM) is an inventory of all components that have been used to build software. This includes all elements that are distributed as part of the final artifact, but also those that were used to develop and build it. One role of such an inventory is to help people identify all the possible vulnerabilities that would impact the said software. Such vulnerability detection checks whether the software of the SBOM includes known Common Vulnerabilities and Exposures (CVEs) reported in the MITRE database [12] or vulnerabilities published in the European Union Vulnerability Database [9]. However, a large proportion of the vulnerability detection tools are designed to be used offline as they assess the risk for one's product. The presented use case addresses this issue and shows a centralised yet privacy-preserving tool. The core of this Use Case is to build a privacy-preserving centralised SBOM analysis service that can also offer advanced features to evaluate the potential impact of new CVEs on its users.

3.1 SBOM Framework

The use case is split into several steps.

- Cyber Threat Intelligence Collection: The users collect and continuously update their SBOMs, the CVEs that affect them, along with relevant data about their interaction with the service provider.
- Data Ingestion and Preprocessing: The service provider continuously collects new CVEs.
- Secure Model Training: A machine learning model, shared among several users, is trained in a distributed manner across the participating users, leveraging FL techniques to preserve individual data privacy.
- Privacy-Preserving Threat Detection: The CVE database is used to analyze software components and potential and confirmed vulnerabilities.
- Early Warning and Threat Sharing: The detected vulnerabilities are securely shared with users promptly, enabling proactive remediation and threat mitigation.

To address those steps, PRIVIDEMA proposes four different scenarios tackling one subset of the project each. The first and second scenarios let the service provider send to the users detailed information about impacted elements of their SBOMs, either through one-time query or continuously by monitoring stored encrypted SBOMs. Third scenario helps the service provider assess an approximation of the percentage of its impacted users through Differential Privacy and Threshold homomorphic encryption. Finally, users collaboratively build an anomaly detection model to monitor the service provider's behaviour and detect any deviation from its supposedly normal demeanour, thanks to federated learning.

In this paper, we focus below on the first scenario and give details on its structure and the size and quantity of involved data.

3.2 Data Consideration

In this use case, two types of entities will be interacting:

- a centralised entity which provisions a service of SBOM analyser. It is thus named service provider.

- a flock of entities, called users, that all have a SBOM which is to be analysed by the service provider.

Before continuing, it is important to discuss the scale of manipulated data. Indeed, the service provider does use a large amount of CVEs stored in its internal database. According to [12], there are more than 300,000 known CVEs to this day. Also, the amount of discovered CVEs is rapidely growing, reaching 48,000 new ones in 2025. To ease the transfer of data between the service provider and its users, the main pieces of information of each CVE are encoded in a small bitstring of 72-bits. Compared to non-encoded CVEs (often stored in JSON that amount for a total of tens of thousands of bits), those smaller values facilitate the application of the OnionPIR protocol [13].

Finally, the way we designed this use case induces data duplication in the CVEs database, as each row corresponds to a version of a software and contains a set amount N of CVE that affect it. Thus each CVE may appear for two versions of the same software. Hence the database size is equal to the multiplication of the size of an encoded CVE, the variable N, and the number of rows. Taking $N = 1000$ (which allows each row to contain 1000 CVEs), we can generously upper bound the size of the database by 900 Mbytes, which corresponds to at most 100 000 rows of 9000 bytes each.

3.3 Scenario

The main scenario of this use case brings a solution to the base problem we mentioned earlier: finding a way to detect vulnerabilities in a manufacturer's software artifact by matching its SBOM with a database of CVEs. The flow, which is represented in Fig. 2, Fig. 3 and Fig. 4 is twofold. First, at the service provider level, some operations are run regularly:

- The service provider SP updates its database of software it monitors (cf. Fig. 2) .
- The service provider SP updates its CVEs database (cf. Fig. 3) .

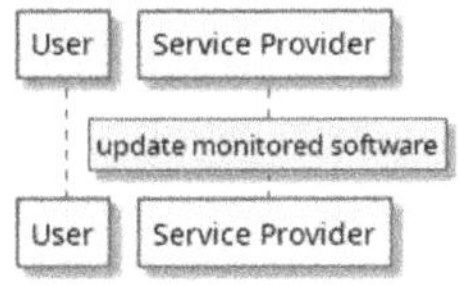

Fig. 2. Scenario 1: softwares collection.

Then, for a user to employ the service provided by SP, both SP and the user follows the protocol shown in Fig. 4 and detailed below:

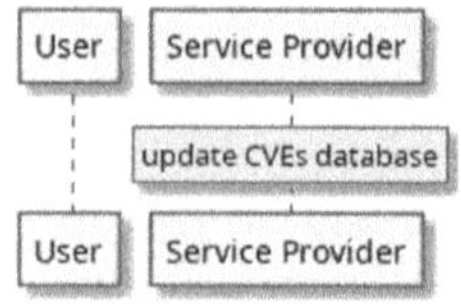

Fig. 3. Scenario 1: CVEs collection and preprocessing.

- The user asks the service provider for the encoding format needed for its SBOM.
- The user encodes its SBOM.
- The user encrypts its SBOM using its fully homomorphic encryption (FHE) public key.
- The user sends a query containing its SBOM and the public key to the service provider.
- The service provider runs the private information retrieval (PIR) protocol on its up-to-date database against the query in the encrypted domain.
- The service provider sends the answer back to the user.
- The user decrypts and decodes the answer.

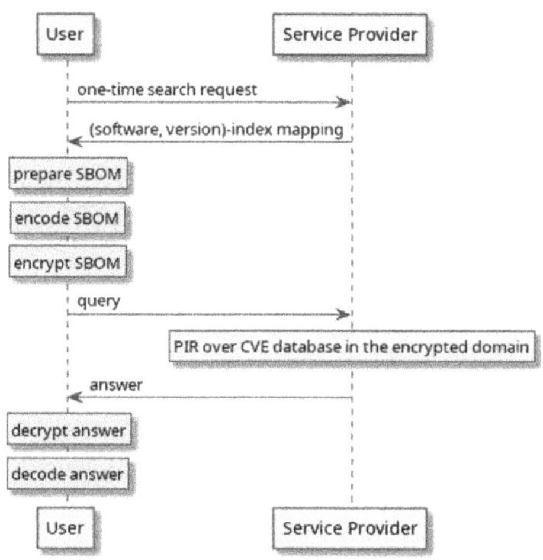

Fig. 4. Scenario 1: runtime operations.

To give a general idea of the PIR algorithm we use in this scenario and in the following ones, we consider that the service provider holds a database of CVEs as follows: each row index correspond to a pair (software, version of the software) and contains a fixed number of CVEs arbitrarily chosen (for instance the most critical ones). The PIR protocol evaluated in FHE essentially is a dot product between the database which is vectorised and the encrypted vectorised SBOM of the user. This is possible because of two points:

- in FHE, dot products are possible,
- the vectorised database of CVEs and the vectorised SBOM have the same format.

Indeed, the vectorised CVEs database basically is a vector where each element is a concatenation of encoded CVEs. When the user asks the service provider for the encoding format needed for its SBOM, it receives two pieces of information: the mapping (software, version)-index and the encoding table of CVEs. Thanks to the former, the user is able to produce a vector for its SBOM where each element is a zero except for the index corresponding to software that are in the SBOM. The vector elements at those indices are set to one. This is the vectorised SBOM that the user now encrypts and sends to the service provider. Finally, SP performs the PIR algorithm and sends the answer to the user, which uses the latter piece of information transmitted by the service provider to decode the result after decrypting it. It is possible to achieve better performance by compressing requests. One way of doing this is by representing data in a hypercube, as in the OnionPIR protocol [13].

Let us now describe more in details the homomorphic library build in PRIVIDEMA project which, together with Lattigo, will be used to implement both the biometric matching for use case 1 and the OnionPIR for use case 2.

4 Fully Homomorphic Encryption Library

The prividema-lib is the main unified FHE software library developed within the PRIVIDEMA project. It provides a common abstraction layer for multiple homomorphic encryption schemes, including BFV [10], CKKS [5] and TFHE/CGGI [6], through a shared GLWE/GGSW-based ciphertext representation. This unified design enables interoperability across schemes while preserving the strengths and optimizations specific to each of them.

A central feature of prividema-lib is its support for efficient scheme switching based on Chimera framework [4], allowing ciphertexts to transition from one scheme representation to another without requiring decryption. This capability enables hybrid encrypted computation workflows in which, for example, CKKS is used for approximate numerical processing, while TFHE is used for fast Boolean operations, comparisons, and control-flow logic.

The library is built on the base-2^k bivariate representation and relies on the spqlios-arithmetic [3] backend for high-performance FFT/NTT operations. This architecture significantly accelerates core primitives such as external products, key switching, automorphisms, relinearization, and bootstrapping-related transformations, offering substantial speedups compared with traditional full-RNS implementations.

At this stage of the project, prividema-lib almost implements the main cryptographic modules and arithmetic backends. The remaining work primarily focuses on scheme-switching interfaces and final optimization passes, hardware-specific acceleration, API stabilization, and integration with the higher-level privacy-preserving circuits required by the project use cases. Within the overall software stack, prividema-lib serves as the core cryptographic engine that bridges low-level arithmetic modules and high-level privacy-preserving applications, including biometric authentication, secure aggregation, and privacy-preserving cyber threat intelligence workflows.

4.1 Performance of OnionPIR with Prividema-Lib

We evaluated the OnionPIR workflow implemented with prividema-lib using the standard OnionPIR parameter set with $N = 4096$, which is representative of the SBOM vulnerability-detection use case. Preliminary benchmarking results show a significant performance improvement compared with the original implementation [13].

For a 500 GB encrypted database, the end-to-end PIR processing time is reduced to approximately 2600s, compared with 6500s in the earlier baseline, corresponding to a speedup of roughly $\times 2.5$. This gain is primarily attributed to the optimized arithmetic backend and the more efficient ciphertext transformation primitives available in prividema-lib.

In terms of throughput, we observe a peak bandwidth between 120 and 200 MB/s on standard cloud machines, while compute-optimized instances reach between 140 and 240 MB/s. These results significantly improve upon the previous 80–100 MB/s range, demonstrating the practical benefit of the new implementation for large-scale vulnerability database queries.

These first results confirm that prividema-lib enables OnionPIR-style private retrieval at scales compatible with industrial SBOM analysis services, while preserving the privacy guarantees required by the PRIVIDEMA threat-intelligence use case.

5 Conclusion and Future Work

This paper presented the PRIVIDEMA project's vision for practical privacy-preserving data processing, illustrated through two representative use cases: remote biometric authentication for the EUDI Wallet and privacy-preserving SBOM vulnerability detection. These two scenarios demonstrate how advanced Privacy-Enhancing Technologies (PETs), notably Fully Homomorphic Encryption, can be used to support secure, interoperable, and regulation-aligned digital services. A key enabler of this vision is prividema-lib, the project's unified FHE software framework, which provides efficient support for BFV, CKKS, TFHE, and scheme switching within a single cryptographic stack.

Future work will focus on four main directions. First, we will complete the scheme-switching interfaces and performance optimization layers, including hardware-aware acceleration and API stabilization. Second, we will extend the biometric authentication pipeline toward fully encrypted face detection and alignment, which remains the main computational bottleneck. Third, the SBOM analysis service will be enhanced with richer privacy-preserving analytics, including early-warning mechanisms for newly disclosed CVEs and privacy-preserving collaborative learning across stakeholders. Finally, large-scale validation on real deployment environments and cross-border interoperability scenarios will be carried out to assess scalability, usability, and compliance with the European digital identity and cybersecurity ecosystem.

Overall, PRIVIDEMA shows that advanced cryptographic techniques are moving from theoretical feasibility toward easily deployable, standards-compliant privacy-preserving infrastructures, contributing to Europe's broader objectives of digital sovereignty, privacy by design, and trustworthy data sharing ecosystems.

Disclosure of Interests. Funded by the European Union under Grant Agreement No. 101167964. Views and opinions expressed are however those of the author(s) only and do not necessarily reflect those of the European Union or the European Cybersecurity Competence Centre. Neither the European Union nor the European Cybersecurity Competence Centre can be held responsible for them.

Project. PRIVIDEMA.

Funding. Horizon Europe – Grant Agreement No. 101167964.

Website. https://prividema.eu.

References

1. Lattigo v6 (2024). https://github.com/tuneinsight/lattigo. EPFL-LDS, Tune Insight SA
2. Regulation (EU) 2024/1183 of the European Parliament and of the Council of 11 april 2024 amending regulation (EU) no 910/2014 as regards establishing the european digital identity framework. Official Journal of the European Union, L 1183, 30 Apr 2024, 2024. Entered into force (2024). https://eur-lex.europa.eu/eli/reg/2024/1183/oj
3. Belorgey, M.G., Carpov, S., Gama, N., Guasch, S., Jetchev, D.: Revisiting key decomposition techniques for FHE: simpler, faster and more generic. In: International Conference on the Theory and Application of Cryptology and Information Security, pp. 176–207. Springer, Cham (2024)
4. Boura, C., Gama, N., Georgieva, M., Jetchev, D.: Chimera: combining Ring-LWE-based fully homomorphic encryption schemes. J. Math. Cryptol. **14**(1), 316–338 (2020)
5. Cheon, J.H., Kim, A., Kim, M., Song, Y.: Homomorphic encryption for arithmetic of approximate numbers. In: ASIACRYPT, pp. 409–437
6. Chillotti, I., Gama, N., Georgieva, M., Izabachène, M.: TFHE: fast fully homomorphic encryption over the torus. J. Cryptol. **33**(2), 34–91 (2020)
7. Deng, J., Guo, J., Zhou, Y., Yu, J., Kotsia, I., Zafeiriou, S.: Retinaface: single-stage dense face localization in the wild (2019). https://arxiv.org/abs/1905.00641
8. Deng, J., Guo, J., Xue, N., Zafeiriou, S.: ArcFace: additive angular margin loss for deep face recognition. In: Proceedings of the IEEE/CVF Conference on Computer Vision and Pattern Recognition (CVPR) (2019)

9. European Union Agency for Cybersecurity. European Union Vulnerability Database (2025). Accessed 29 Oct 2025
10. Fan, J., Vercauteren, F.: Somewhat practical fully homomorphic encryption. Technical report 2012/144, IACR ePrint (2012)
11. InsightFace. InsightFace: State-of-the-art 2d and 3d face analysis project (2025). https://github.com/deepinsight/insightface
12. MITRE Corporation. Common Vulnerabilities and Exposures (CVE) (2025). https://cve.mitre.org/. Accessed 29 Oct 2025
13. Mughees, M.H., Chen, H., Ren, L.: OnionPIR: response efficient single-server PIR. In: Kim, Y., Kim, J., Vigna, G., Shi, E., (eds.) CCS 2021: 2021 ACM SIGSAC Conference on Computer and Communications Security, Virtual Event, Republic of Korea, November 15 - 19, 2021, pp. 2292–2306. ACM (2021)
14. European Digital IdentityWallet. Architecture and reference framework. Technical report, EUDI Wallet (2023)

Optimizing the Scalability of Cyber Range Deployments Using Architectural Patterns

Richard Stöckl[1(✉)], Tobias Pfaller[1], Lenhard Reuter[1], Florian Skopik[1], and Edgar Weippl[2]

[1] AIT Austrian Institute of Technology, Vienna, Austria
{richard.stoeckl,tobias.pfaller,lenhard.reuter,florian.skopik}@ait.ac.at
[2] University of Vienna, Vienna, Austria
edgar.weippl@univie.ac.at
https://ait.ac.at, https://univie.ac.at

Abstract. Cyber ranges are virtual IT/OT infrastructures that provide secure yet realistic environments for simulating cyber incidents, enabling the development of technical skills as well as the training of decision-making and communication processes under realistic conditions. Because cyber ranges must support diverse and frequently changing training scenarios, they impose high requirements on dynamism and scalability, which makes systematic evaluation essential. Benchmarks can be used to identify scalability limitations and to guide architectural improvements in such environments. This work, investigates architectural patterns for addressing scalability challenges and applies them to the AIT Cyber Range as a case study. A method based on *Design Science* was developed to classify architectural patterns according to their relevance for cyber ranges using defined *KOFs*. Based on this method, metrics were established and assessed by cyber range experts to rank the patterns according to their expected contribution to scalability. As a result, three architectural components – the *Proxy Jump Component*, the *Package Cache Component*, and the *Multi-Client Component* – were implemented in the AIT Cyber Range as adaptations of existing patterns from the literature. Benchmark-based evaluations were then conducted to quantify the scalability improvements introduced by each component and to validate the method-based predictions. The results show that the *Package Cache Component* delivers particularly significant scalability gains, while the *Proxy Jump Component* enables support for a larger number of participants, demonstrating both quantitative and qualitative benefits of the proposed approach.

Keywords: Cyber Range · Scalability · Patterns · Benchmarks

1 Introduction

Cyber ranges are virtual environments that enable exercises and training for students and professionals in infrastructures that are isolated from production

© The Author(s), under exclusive license to Springer Nature Switzerland AG 2027
K. Renaud et al. (Eds.): EICC 2026, CCIS 3012, pp. 331–351, 2027.
https://doi.org/10.1007/978-3-032-28957-5_20

systems. They primarily support the development of technical skills such as intrusion detection, digital forensics, log analysis, and the application of security best practices in enterprise environments. In addition, cyber ranges facilitate the training of incident response processes, organizational communication structures, and reporting channels in accordance with regulations such as the Network and Information Systems (NIS) Directive [6] and the Digital Operational Resilience Act (DORA) [8], while also improving communication between management and technical departments [44]. The increasing frequency of cyber attacks has further amplified the relevance of cyber ranges [36]. By isolating exercise environments from productive infrastructures, cyber ranges enable the execution of potentially hazardous scenarios without causing adverse effects on production systems [37].

To realistically mimic real-world environments, cyber ranges deploy scenarios consisting of numerous and heterogeneous services using specialized deployment automation tools. As the number of users and services grows, scalability challenges emerge and significantly increase the time required to provision exercise environments [33,39,41]. Deployment generally comprises infrastructure provisioning and software deployment, with the latter exhibiting substantially poorer scalability. Software deployment includes tasks such as account creation, key and certificate generation, firewall configuration, and service installation, which introduce considerable complexity and error potential. Because exercise environments are deployed and dismantled frequently and typically provision all services simultaneously, deployment duration becomes a critical factor, leading cyber ranges to rely heavily on Infrastructure as Code (IaC). Despite the existence of scalability improvement techniques, the lack of a systematic overview and comparison scheme makes identifying and evaluating the most effective patterns for cyber ranges a labor-intensive and exploratory process. Finally, cyber ranges are a complex and highly technology-driven topic that is therefore subject to change rapidly. The positioning of cyber ranges stands out particularly in EU projects, which necessitates further development and increased efficiency in their deployment.

This work makes several concrete contributions to the development of a methodology improving the scalability of cyber range deployments utilizing architectural patterns. First, it investigates how architectural patterns influence deployment scalability and applies pattern-based reasoning to derive insights that are transferable across platforms. Second, it proposes a structured, metric-driven process for ranking architectural patterns, based on the characteristics of cyber range environments. Third, it applies this process in practice by implementing selected high-impact patterns on the AIT Cyber Range and benchmarking their effects on deployment time. Finally, it provides quantitative evidence of scalability improvements and complements these results with a qualitative analysis of implementation challenges and side effects, offering practical guidance for applying architectural patterns in cyber range environments.

The remainder of the paper is organized as follows. Section 2 provides an overview of the lifecycle of cyber ranges and shows the current state of the AIT Cyber Range. Then, Sect. 3 elaborates in detail the process of selecting, implementing and evaluating the patterns. Section 4 implements the method on the AIT Cyber Range as an evaluation platform. Section 5 outlines the scalability improvements and any effects not related to the deployment time. Finally, Sect. 6 concludes the paper.

2 Background and Related Work

To understand the development and evaluation of the proposed methodology, it is essential to be familiar with both the lifecycle of cyber ranges and the current state of the AIT Cyber Range.

2.1 Lifecycle of Cyber Ranges

The establishment and execution of the method further necessitate clarification of the cyber range's lifecycle. The initial phase of the lifecycle of cyber ranges focuses on defining the scope of the exercise. This phase determines the types of tasks and attacks to be included and involves designing concepts that specify the required vulnerabilities and attack techniques. The outcome of this process is a timetable that defines the start and end times of individual attacks throughout the exercise.

After the planning phase, developers can plan, design, and integrate participant tasks in parallel. They select tasks from existing catalogs, derive them from newly identified vulnerabilities, or design them from scratch, while evaluating factors such as exploit complexity and defensive difficulty. If tasks do not yet exist, developers implement them in this phase and test them outside the exercise environment to accelerate development. Subsequent testing refines the anticipated scenarios, vulnerabilities, attack vectors, and overall exercise setup.

After refining the scenarios and attacks, developers integrate them into the cyber range deployment plan. In the AIT Cyber Range, this integration involves creating Terraform modules and Ansible playbooks within the IaC repository. This phase may include initial minimal deployments to validate newly integrated scenarios and configurations before full-scale execution.

Once the integration is complete, the cyber range undergoes a full deployment into a production environment. Participants then use this environment to perform the exercise by analyzing, defending against, and reporting predefined issues over a period of several days. After the exercise concludes, the infrastructure is dismantled to ensure clean conditions for future exercises and to reduce resource costs. Compared to long-lived production systems, cyber range deployments are short-lived, yet their deployment time remains substantial relative to their overall lifespan, making deployment efficiency a critical concern.

2.2 Current State of the AIT Cyber Range

Cyber ranges must support flexible use across a wide variety of scenarios in order to avoid redeveloping infrastructure code for each individual exercise. A cyber range should therefore dynamically adjust the number of deployed client machines to accommodate varying participant counts and to conserve resources when full capacity is not required. Empirical benchmarks demonstrate that the AIT Cyber Range exhibits significantly increased deployment times as the number of required client machines grows. Throughout this work, benchmarks are employed to assess the state of the AIT Cyber Range and to serve as a baseline for evaluating the impact of the proposed methods, with deployment time serving as the primary optimization objective. These benchmarks enable the identification of scalability bottlenecks and support approximating the sources of scalability limitations. The benchmark methodology evaluates the system along the following dimensions:

Number of clients Clients are the input size at which the AIT Cyber Range exhibits poor scalability characteristics. Figures show 0âĂŞ40 clients in steps of five on the x-axis.

Playbooks For playbooks, different diagrams are used. Each playbook has an effect on various parts, so they each behave differently in terms of the number of clients. Some playbooks in the AIT Cyber Range are optional. To make the significance as general as possible, only the playbooks are evaluated which are required for every scenario. They are described in Table 1.

Warm/Cold *Cold* runs execute immediately after infrastructure provisioning on clean hosts; *warm* runs repeat the same configuration on already configured hosts. This distinction helps identify whether scalability issues stem from the configuration workload itself or from the configuration management tools (e.g., repeated checks and idempotency overhead). Separate diagrams are provided to illustrate this dimension.

Table 1. AIT Cyber Range Playbooks

Playbook	Purpose
Bootstrap	Sets up proxies, disables firewalls temporarily and updates software. Affects all machines.
DNS	Installs and configures DNS services.
Client	installs required client software and configures resources such as browser-bookmarks, email-accounts and password-manager databases. Affects client machines.
Learners	Builds, installs and configures the platform for participation access. Affects the machine used for the noVNC platform for installation and all other machines for gathering facts in order to configure learners – the noVNC platform of the AIT Cyber Range.
Certificates	Generates and installs certificates on servers and distributes them to all machines. Affects all machines.
User Management	Creates the required user accounts. Affects all machines.
Post	Resets configuration only required during the deployment process. Affects all machines.

To quantify the severity of scalability issues, all playbooks were benchmarked to evaluate the current deployment behavior of the AIT Cyber Range. The following benchmarks establish an initial baseline that supports understanding the relevance of architectural patterns, estimating metric scores, and identifying irrelevant patterns at an early stage. Each diagram illustrates the deployment time of a playbook in cold and warm states as a function of the number of clients. Table 1 describes the purpose of each playbook. To reduce the influence of outliers, each measurement was repeated five times following the recommendations of Henning et al. [29]. In addition to outlier handling, the diagrams include linear and quadratic regressions to visualize deployment time trends.

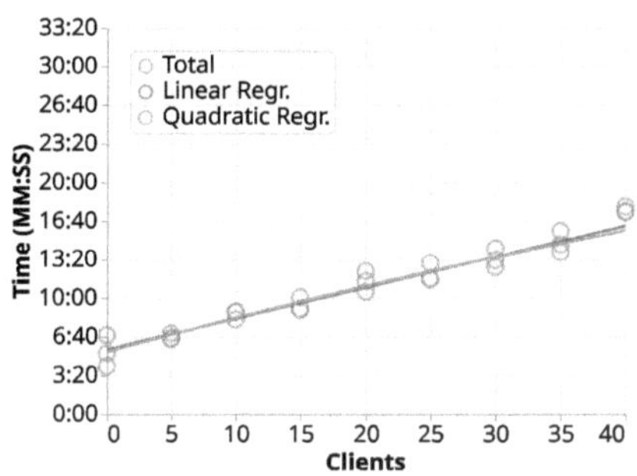

Fig. 1. Bootstrap Scaling Cold

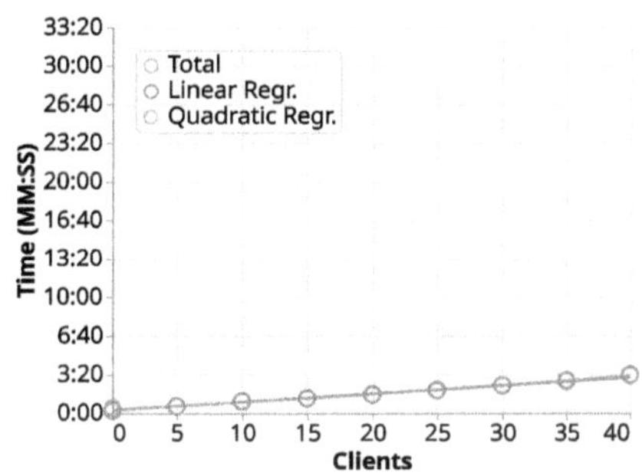

Fig. 3. User Management Scaling Cold

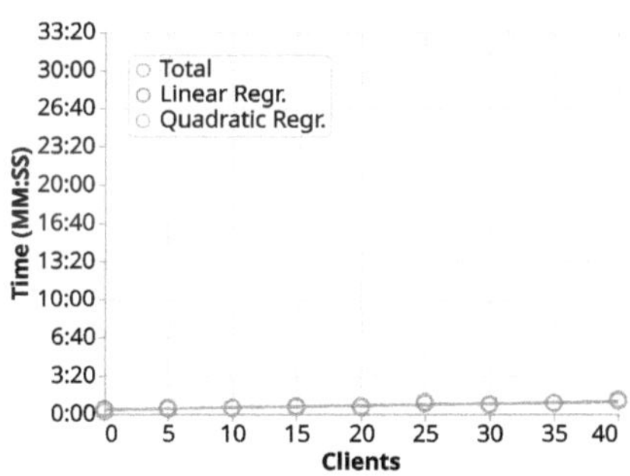

Fig. 2. Bootstrap Scaling Warm

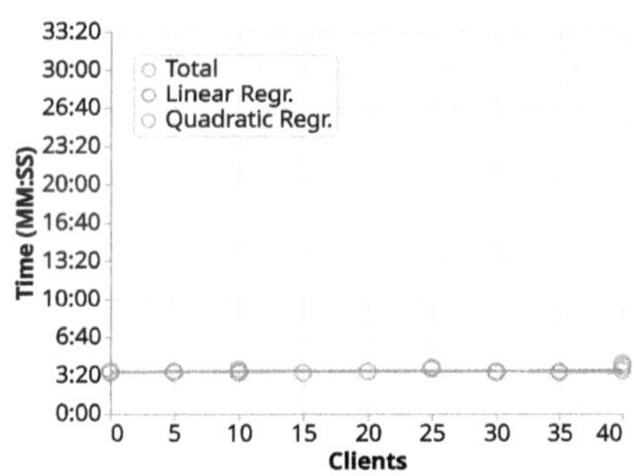

Fig. 4. Learners Scaling Cold

The AIT Cyber Range is utilized to test this with 0–40 clients in five-step increments. The most significant results from the cold benchmark diagrams are presented below, while the remaining diagrams, although evaluated, are omitted for brevity.

The benchmark results show that bootstrapping (Fig. 1) scales linearly but dominates overall deployment time, while the client deployment stage (Fig. 5) requires less absolute time yet exhibits quadratic scaling behavior. User management (Fig. 3) also scales linearly but with a comparatively large scaling factor, whereas certificate management (Fig. 6) scales linearly with a lower factor. The

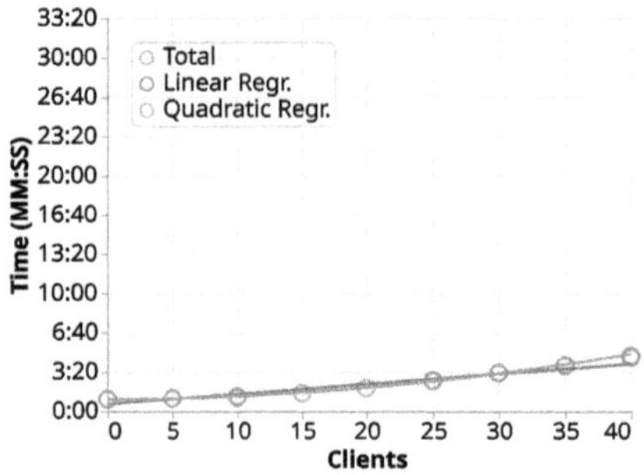

Fig. 5. Client Scaling Cold

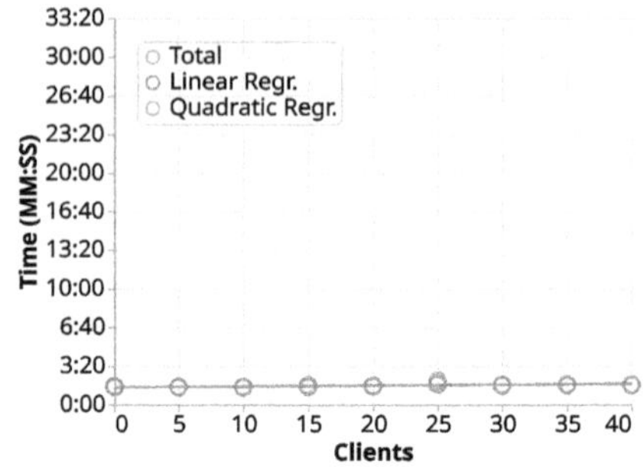

Fig. 6. Certificate Scaling Cold

DNS stage remains constant and under 30 s, rendering its impact negligible. Finally, the learners stage (Fig. 4) contributes a significant fixed deployment time and is largely independent of the number of clients.

The most significant outcomes of the benchmarks are the prolonged duration of the bootstrap playbook and the scalability issues of the client playbook. Therefore, these will receive the most attention in the subsequent sections. In general, it can be observed, that despite some diagrams such as Fig. 5 and Fig. 3 are showing increasing deployment times in dependence of the clients, their warm counterparts do as well. This indicates two possibilities:

1. The deployment workloads are not as heavy as expected. Instead, all the organizational work Ansible performs such as checking the current state, managing inventories or scheduling tasks is much more sophisticated and takes much more time than the actual work on the machines.
2. Ansible's idempotence checks do not work as expected which results in configuring the same things which are already done. The cause for this could be insufficient checks in the Ansible roles by being not careful enough, or by technical limitations which prevent such idempotence checks.

These observations indicate that measures affecting deployment scheduling rather than the deployment workload itself exert a greater influence on overall deployment time. Based on the current state of the AIT Cyber Range, several key takeaways emerge. First, pattern selection should prioritize organizational patterns over those that primarily target workload execution, as scheduling effects dominate scalability behavior. Second, the most severe scalability issues occur during bootstrapping and client deployment, as shown in Fig. 1 and Fig. 5, where execution time increases substantially with the number of deployed clients; this insight strongly influences both pattern selection and implementation planning. Finally, some deployment components considered in this work, such as the learners stage shown in Fig. 4, contribute a largely fixed overhead that is independent of the number of clients. This must be considered when interpreting the ranking and when discarding highly ranked patterns as impractical to evaluate.

2.3 Related Work

During the literature review on scalability issues in cyber ranges, no sources were identified that specifically analyze improvements to deployment scalability. A key challenge was aligning existing work with the definition of scalability adopted in this paper, as most studies focus on runtime workload scaling, whereas this work addresses the ability of resources to scale independently of runtime behavior. Nevertheless, the first three pages of results returned by Google Scholar contained several resources that, while not directly addressing deployment scalability, proved useful and are summarized in Table 2. Much of the identified literature concentrates on cloud deployment portability, such as migrations between on-premise environments and cloud platforms or across cloud providers, which lies outside the scope of this work. Despite these limitations, the book *Cloud Computing Patterns: Fundamentals to Design, Build, and Manage Cloud Applications* [21] offers a comprehensive and relevant collection of architectural patterns. As it systematically describes the problems addressed by each pattern, the corresponding solutions, and relevant implementation considerations, this work relies heavily on this source and shifts the focus of the literature review from pattern discovery to pattern selection.

Table 2. Related Work with Annotations

Literature	Annotation
Claudio A. Ardagna et al. [14]	Describes patterns of the view of a PaaS provider. Cyber ranges are no PaaS provider.
Sören Henning et al. [29]	No patterns, but how to benchmark scalability of cloud-native applications. States that ≤ 5 repetitions are required.
Alberto Avritzer et al. [15]	Interesting for performance when scaling compute resources in virtual environments, and Multi-Client Component.
Rui Han et al. [26]	Scale applications at run-time more efficiently. Also interesting for Multi-Client Component.
Rajesh Komar [34]	Comparison of *aaS. Reference for Multi-Cloud patterns.
Muhammad Ehsan Rana et al. [40]	No patterns, but scalability in cloud applications. Reasoning why architectural patterns are useful for cyber range scalability.
Christoph Fehling et al. [20]	Introduces cloud patterns but without scalability context.

3 Pattern Selection Methodology

The pattern selection process is important for identifying patterns in the literature that minimize the scalability issues that have been identified. This section enumerates the architectural patterns identified in the literature, the filtering process used to reduce the candidate set, the expert-based ranking of the remaining patterns, and the key considerations for selecting patterns from the final ranking. Conversely, prior to initiating the filtering procedure, there are several prerequisites that must be fulfilled.

3.1 Pattern Acquisition

Based on the results presented in Sect. 2.2, these benchmarks enable a filtering process that prioritizes architectural patterns by assigning higher scores to those with greater expected relevance. Next, the definition of suitable metrics establishes a ranking framework based on the potential usefulness of the patterns. This framework supports a targeted literature search and initiates the systematic filtering of candidate patterns. To guide the selection process and the metrics, the architectural requirements of cyber ranges are defined first. Both general requirements [32] and AIT Cyber RangeâĂŞspecific requirements underpin the subsequent analysis and pattern evaluation:

1. Adaptability is required to give cyber ranges the flexibility to change the exercise scenario. The more adaptable a cyber range is, the less maintenance effort is necessary for future scenarios.
2. The scenario has to be built in a way such that is configurable for participants. They have to solve tasks where components are misconfigured.
3. From a security perspective, cyber ranges have to ensure that only users have access to resources they actually should have to in order to prevent participants are impacting each other or even cheating.
4. Cyber ranges are customized and deployed per exercise. This makes it necessary to reduce the deployment time as much as possible.

3.2 Pattern Filtering

Given the large number of architectural patterns identified in the literature, this work applies a pre-selection step to reduce the effort required for the subsequent ranking process. The pre-selection estimates the relevance of each pattern to cyber range environments in order to narrow the candidate set. Although this relevance assessment does not rely on strict quantification, it provides a structured initial filter that supports the systematic reduction of patterns. The filtering process considers several influencing factors that cannot be expressed through precise metrics. The reasons why a pattern may not be applicable to cyber ranges are therefore captured using the following Kick-out Factors (KOFs):

(KOF 1) The pattern affects resources, such as long-time storage, that are not necessary for cyber ranges in general.

(KOF 2) The pattern is intended for distributed applications. Typically, cyber ranges are not required to manage state synchronization or maintain consistency across multiple components.

(KOF 3) The pattern is not applicable for deployment time but development or runtime instead.

(KOF 4) The pattern introduces safety or resilience features such as watchdogs. The challenge to solve such issues is part of the scenarios and must be solved by the participants.

(KOF 5) The pattern is designed to address the issue of managing large data streams. Cyber ranges do not handle massive data streams, aside from very specific scenarios.

3.3 Pattern Ranking

This work establishes a ranking to both improve the efficiency of the pattern selection process and to derive a structured outcome from the investigation. The ranking employs an estimation-based scoring system that defines several evaluation categories, which a pattern either fulfills or does not fulfill. If a pattern satisfies a category, it receives a score ranging from 1 to 5. The final pattern score equals the sum of all awarded points, with higher scores indicating greater relevance.

Risk This metric considers the security requirements of a cyber range. Potential risk vectors are cloud API access to cyber range components, single-point of failure or mechanisms that reduce the isolation of different companies or teams. When a pattern is expected to have no risk, it gains more points in this category.

Impact Whether a pattern affects many components. In this event, the likelihood of a more severe outcome increases. For instance, a pattern will receive a higher score if it encompasses clients, web servers, and application servers.

Dependencies When numerous components are dependent on a pattern, it may require a significant amount of effort to integrate it into existing platforms. As a result, the objective is accomplished in this instance when the dependencies are minimal. Things such as the implementation of a new protocol or the extensive reorganization of the architecture lead to a lower score.

Maintainability The less effort required to maintain a pattern after it has been implemented, the more straightforward it is to employ. This point might involve the pattern's actual structure. Declarative deployment methodologies that necessitate numerous variables and templates are perceived as less maintainable. This metric is adversely affected by the introduction of numerous new components or the risk of diverging configuration.

Flexibility The probability of a pattern being applicable in a broader range of concrete platforms increases as its abstraction increases. This metric is lowered by requirements for highly specialized protocols or highly specific configuration.

3.4 Pattern Selection

The final step of the process involves selecting patterns from the established ranking. Simply implementing the top-k ranked patterns is not always appropriate, as some patterns may conflict with one another, introduce redundancies, or require substantial changes to the existing technology stack. Therefore, a final evaluation step is necessary to assess compatibility and feasibility before selecting a coherent set of patterns for implementation.

4 Case Study: AIT Cyber Range

This case study applies the methodology from Sect. 3 to the AIT Cyber Range. Candidate architectural patterns are derived from the literature and subsequently assessed against the KOFs. The remaining candidates are then ranked by domain experts using the defined metrics, finishing with the final selection to decide on which patterns to implement.

4.1 Methodology Execution

Table 3 presents the patterns that were not filtered out by the KOFs and are therefore considered for further analysis. A brief description of each selected pattern is provided in Table 4. As part of this case study, four cyber range experts evaluated the patterns that passed the filtering stage by estimating a score for each metric. This expert assessment resulted in Table 5, which presents the mean scores of all expert evaluations per pattern and metric, as well as the corresponding total scores.

Before implementation, the candidate patterns must be selected, and several of the top-ranked patterns were excluded for specific reasons. The *Managed Configuration* pattern was omitted because its realization through tools such as Red Hat Ansible Automation Platform [7] or Ansible AWX [1] would primarily represent a technological upgrade rather than an architectural change, limiting comparability with other patterns. The *Loose Coupling* pattern was excluded due to the difficulty of objectively determining whether it has been implemented, as it functions more as a long-term design principle than a clearly measurable intervention. Finally, the *Provider Adapter* pattern was not considered because changing the cloud provider would produce benchmark results that are not directly comparable to the current state of the AIT Cyber Range.

4.2 Pattern Implementation

The patterns identified as having the highest potential for improving the scalability of the deployment time are presented in Sect. 4.1. Because this stage of the work requires concrete implementation decisions, the selected patterns are assigned new names to distinguish the literature-derived concepts from their specific implementations, namely Application Component Proxy as the Proxy Jump Component, Content Distribution Network as the Package Cache Component, and Elasticity Management Process as the Multi-Client Component.

Table 3. Pattern Relevance

Name	Rel./ KOF	Rejection Reason
Elasticity Management Process [22]	✓	
Content Distribution Network [21]	✓	
Loose Coupling [45]	✓	
Processing Component [30]	✓	
Managed Configuration [42]	✓	
Dedicated Component [21]	✓	
Multi-Component Image [12]	✓	
Shared Component [21]	✓	
Tenant-Isolated Component [21]	✓	
Provider Adapter [21]	✓	
User Interface Component [45]	✓	
Transaction-Based Processor [16]	✓	
Application Component Proxy [20]	✓	
Hybrid Processing [21]	1	Cyber ranges change the deployment platform frequently and components are hard to reuse.
Restricted Data Access Component [23]	2, 4	Not relevant for deployment time, nor tackling scalability.
Distributed Application [13]	2	Cyber ranges do not have necessarily applications to split.
Stateful Component [25]	2	State management has no focus in cyber ranges.
Stateless Component [23]	2	State management has no focus in cyber ranges.
Batch Processing Component [21]	5	Big data processing is not the focus of cyber ranges.
Data Access Component [24]	5	Big data processing is not the focus of cyber ranges.
Data Abstractor [14]	5	Big data processing is not the focus of cyberranges.
Idempotent Processor [30]	2	Pattern is about consistency, not scalability.
Timeout-Based Message Processor [43]	2	Message-oriented middleware are no core functionality of cyber ranges.
Message Mover [17]	2	Message-oriented middleware are no core functionality of cyber ranges.
Compliant Data Replication [30]	2	The scalability issues are not caused by accessing data.
Integration Provider [17]	2	The cyber range deployment is done in a single environment.
Elasticity Manager [21]	3	Workload is predictable and fully depends on scenario/participant registrations.
Elastic Load Balancer [21]	3	Workload is predictable and fully depends on scenario/participant registrations.
Elastic Queue [27]	3	Workload is predictable and fully depends on scenario/participant registrations.
Watchdog [18]	4	Participants are responsible for the infrastructure.
Feature Flag Management Process [21]	3	Does not effect deployment scalability.
Update Transition Process [31]	4	Downtimes are allowed during the deployment.
Standby Pooling Process [21]	3	Only effects provisioning speed during runtime, but not deployment time.
Resiliency Management Process [38]	4	Should be done by the participants.
Two-Tier Cloud Application [21]	5	Cyber ranges do not have large data streams.
Three-Tier Cloud Application [21]	5	Cyber ranges do not have large data streams.
Hybrid User Interface [28]	3	Not relevant for deployment time, nor tackling scalability.
Hybrid Data [21]	5	There is no 'data' scalability problem in the cyber range.
Hybrid Backup [21]	1	Cyber ranges are short living and do not have backups.
Hybrid Backend [21]	3	This is a runtime problem.
Hybrid Application Functions [21]	3	This is a runtime problem.
Hybrid Multimedia Web Application [46]	5	There are not much large multi-media files.
Hybrid Development Environment [35]	3	Only relevant during development, not deployment.

Table 4. Pattern Descriptions

Name	Description
Elasticity Management Process	Implement a component which automatically scales application instances according to the current workload in both directions. This helps to manage intense workloads when required while also remaining cost efficient.
Content Distribution Network	Cache frequently accessed resources within the private network. This reduces the internet usage, reducing cost and increasing the speed of streaming resources which are frequently required.
Loose Coupling	Reduce the dependencies between components as much as possible. This increases scalability and fault tolerance while reducing network load.
Processing Component	Distribute heavy workloads across multiple processing components. This enables the ability to scale out elastically on heavy workloads.
Managed Configuration	Provide a centralized configuration store for all components. This allows to control all components in a coordinated way without the need of re-configuring them while increasing consistency.
Dedicated Component	Instantiate components multiple times despite being multi-tenancy capable. This increases the isolation of the components to protect sensitive data.
Multi-Component Image	Instead of forging dedicated VM images for every different component, create a single image with all required software parts installed fitting all components. This allows each component to switch its task to another one which is required more urgently during runtime without the need of any reconfiguration.
Shared Component	Share components across multiple tenants. This reduces the deployment effort while saving cloud resources.
Tenant-Isolated Component	Share components across multiple tenants similar to SharedComponent. The difference is that the isolation prevents tenants from influencing each other.
Provider Adapter	Implement a generic interface for cloud providers to decouple software from specific vendor implementations. Prevents vendor lockin. User Interface Component Decouple the user interface components from the rest of the environment in a way such that it can be configured and customized completely independent from the rest of the application.
Transaction-Based Processor	Implement a mechanism to ensure consistency while processing messages such as deployment instructions.
Application Component Proxy	Provide a proxy which enables access to resources in an environment with restricted access. This allows to isolate the restricted environment from external access.

Proxy Jump Component. The AIT Cyber Range uses Ansible for software provisioning and relies on SSH for communication between machines, which enables the implementation of the *Proxy Jump Component* using native SSH proxy jump functionality. However, initial attempts to deploy this pattern led to instability in the deployment process, as connection attempts frequently failed

Table 5. Expert Evaluation of Patterns

Name	Risk	Imp.	Dep.	Main.	Flex.	Sum
Application Component Proxy	3.0	4.0	4.3	4.3	4.3	**20.0**
Content Distribution Network	4.2	3.6	3.6	4.2	4.4	**20.0**
Managed Configuration	4.5	3.8	3.0	3.4	3.8	**18.5**
Loose Coupling	3.8	3.0	3.4	3.6	3.6	**17.4**
Provider Adapter	3.5	3.8	3.0	2.8	3.2	**16.2**
Elasticity Management Process	3.2	3.7	2.8	3.2	2.4	**15.3**
Transaction-Based Processor	3.8	3.2	3.2	2.5	2.0	**14.8**
User Interface Component	4.0	3.2	3.0	2.8	1.5	**14.5**
Processing Component	4.0	2.6	2.8	1.6	2.6	**13.6**
Dedicated Component	3.2	2.0	3.2	2.6	1.6	**12.6**
Tenant-Isolated Component	2.2	2.2	2.4	1.8	1.6	**10.2**
Multi-Component Image	2.5	1.8	1.6	1.2	2.6	**9.7**
Shared Component	1.5	1.8	2.2	1.8	1.6	**8.9**

or timed out. To mitigate these issues, the proxy jump host required specific SSH configuration adjustments, including increasing `MaxSessions` to support extensive session multiplexing and raising `MaxStartups` to allow a large number of simultaneous unauthenticated connections during deployment.

Although the AIT Cyber Range already follows the *Proxy Jump Component* pattern, it remains one of the patterns with the highest potential impact and therefore warrants further evaluation. To isolate and assess its effect, benchmarks were conducted without using any proxy jump hosts, effectively removing the current implementation of the pattern. Under this setup, any observed improvement would contradict the expected benefits of the pattern and thus serve to validate its actual impact on scalability.

Package Cache Component. The *Package Cache Component* pattern introduces a central caching machine for Linux packages, and in the context of the AIT Cyber Range, this evaluation focuses on an `apt` cache due to the predominant use of Ubuntu. Several implementation options exist for apt caching, with Apt-Cacher-NG [2] initially selected because it is specifically designed for apt and offers advanced optimizations such as early removal of outdated packages and preloading of dependencies compared to generic HTTP proxies. However, this implementation exhibited unstable behavior, including package corruption, unresponsiveness, and crashes, caused by known issues such as segmentation faults, concurrency problems, and hash miscalculations [3]. As an alternative, Squid-deb-proxy [10], a preconfigured Squid-based proxy optimized for apt caching [9], was adopted and successfully integrated into the AIT Cyber Range using Ansible playbooks. To maximize its impact, the proxy is configured at the very beginning of the deployment process so that software updates and installations can already benefit from caching during deployment.

Multi-client Component. The initial step involved creating an environment that allows a single host to manage multiple client sessions by using Incus [5], a containerization platform forked from LXD. The client image was migrated to this component to enable on-demand client instantiation, and the noVNC platform now connects exclusively to the *Multi-Client Component* rather than to individual client machines. Since noVNC expects each client to be directly accessible via VNC, the *Multi-Client Component* employs systemd socket activation to listen on a single TCP port. When noVNC establishes a connection, a shell script creates a new client instance and transfers the connection, making the multi-session setup transparent to the noVNC platform, with client instantiation completing within a few seconds.

A key challenge in this implementation involved configuring the client images for participants after infrastructure provisioning, as dynamically generated client instances cannot be tracked by OpenStack or the dynamic Ansible inventory. To address this limitation, the implementation uses the `ansible.builtin.add_host` module to add all Incus-managed containers to the Ansible inventory. This approach enables post-provisioning configuration but may negatively affect deployment performance due to the required switch from the *free* to the *linear* Ansible execution strategy.

4.3 Pattern Benchmarks

This section presents the benchmark results for each architectural pattern applied to the AIT Cyber Range. Unless explicitly stated otherwise, the results show no significant deviations from the baseline and are therefore omitted for brevity.

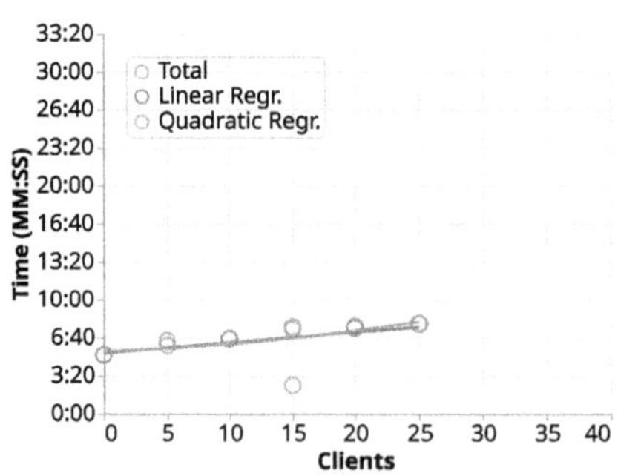

Fig. 7. Bootstrap Scaling Cold (without Proxy Jump Component)

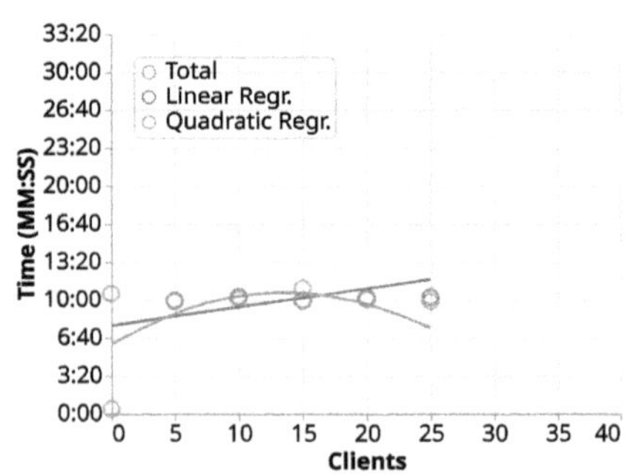

Fig. 8. Learners Scaling Cold (without Proxy Jump Component)

Proxy Jump Component. As observed in the benchmark diagrams, the number of provisioned clients is limited to 25 when the pattern is not implemented due to the stability issues discussed in Sect. 5.1. Figure 7 shows that at 25 clients the average total deployment time increases from approximately 7:50 to approximately 12:20 compared to Fig. 1, corresponding to an increase of more than 50%, while the deployment time for 0 clients remains unchanged, indicating that only the scaling factor deteriorates. Similarly, Fig. 8 illustrates that at 25 clients the total deployment time rises from a constant value of approximately 3:40 in Fig. 4 to approximately 10:00 on average, representing an increase of roughly 170% that is independent of the number of clients.

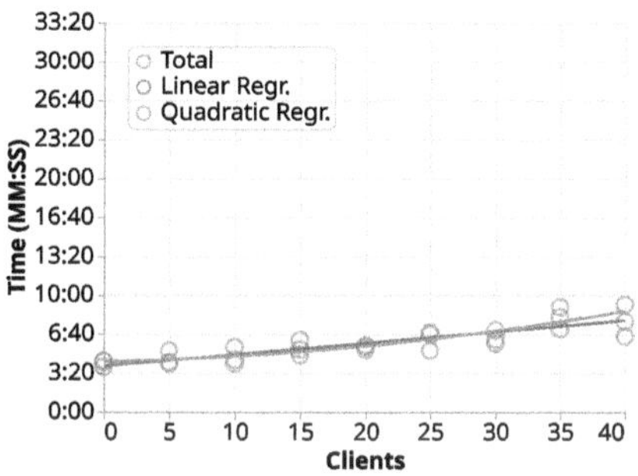

Fig. 9. Bootstrap Scaling Cold (Package Cache Component)

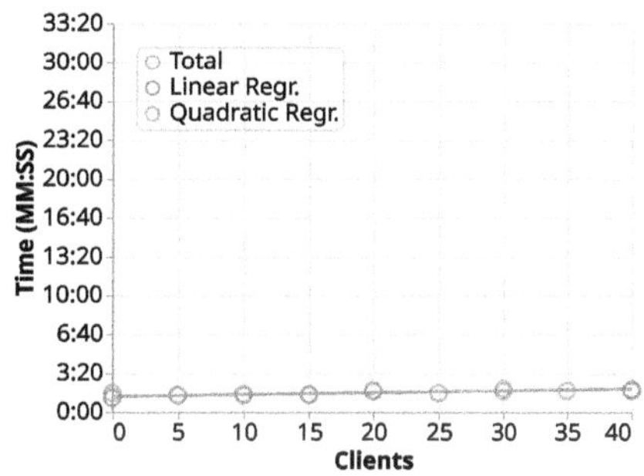

Fig. 10. Bootstrap Scaling Warm (Package Cache Component)

Package Cache Component. Figure 9 shows that the pattern significantly reduces deployment time, with average duration decreasing by about 20% for 0 clients and roughly 50% for 40 clients, albeit with increased variability. In contrast, Fig. 10 indicates higher deployment times, with increases of approximately 300% for 0 clients and 60% for 40 clients.

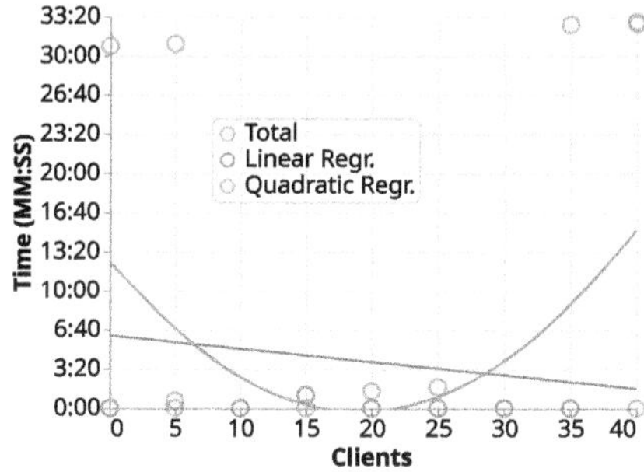

Fig. 11. User Management Scaling Cold (Multi-Client Component)

Multi-client Component. All benchmark results exhibit relatively high variability due to the instability of the *Multi-Client Component* implementation. During deployment, frequent hangs and numerous network timeouts occurred, which prevented the execution of reliable and reproducible benchmarks. Although Fig. 11 represents the most extreme outlier, similar instability affected all other benchmark results.

5 Evaluation and Discussion

In addition to the quantitative results presented in Sect. 4.1, this section discusses expectations, side effects of the applied patterns, and potential limitations of the methodology.

5.1 Applicability in the AIT Cyber Range

Each pattern introduces changes in deployment time that in some cases align with expectations and in others do not. Beyond their impact on deployment duration, the patterns also produce distinct effects on the AIT Cyber Range when implemented.

Proxy Jump Component. The *Proxy Jump Component* introduces an intermediate jump host that tunnels SSH connections between nodes, thereby hiding sensitive systems from public networks. Because all stages of software provisioning rely on SSH, the pattern is expected to affect the entire deployment process, particularly stages with many small tasks that could benefit from SSH session multiplexing. However, implementing the *Proxy Jump Component* results in unchanged or increased deployment times, with degradation observed only during the cold bootstrap and cold learners stages, while the overall complexity class remains unchanged. This behavior can be attributed to increased network latency caused by routing traffic through an additional hop and the extra computational overhead required to process proxied network traffic.

Despite its negative impact on deployment time, the *Proxy Jump Component* removes a critical limitation in the AIT Cyber Range by increasing the maximum number of deployable clients. Without the pattern, deployments are limited to approximately 25 clients due to OpenStack floating IP constraints, whereas the jump host reduces the need for direct external connectivity. Additionally, the pattern lowers the number of required public or floating IP addresses, which can reduce operational costs, provided that access controls are carefully configured to mitigate man-in-the-middle risks [11].

Package Cache Component. The *Package Cache Component* introduces a dedicated node that caches software packages for other nodes in order to reduce the time required for software installation and updates. The pattern primarily targets stages such as bootstrapping, where extensive package installation and updating occur, while stages unrelated to software management, such as certificate handling, DNS configuration, or user management, are not expected to benefit. Implementing the *Package Cache Component* improves the scalability of cold bootstrapping as expected, despite slightly increased relative variability, whereas warm bootstrapping scales marginally worse. This behavior can be explained by the additional proxy step introduced by the cache, as warm bootstrapping typically performs no package installations but still requires synchronizing package metadata, where the proxy overhead outweighs the caching benefits.

In the current implementation, the *Package Cache Component* is deployed as part of the exercise infrastructure and is therefore recreated for each scenario. While this approach ensures reproducibility of the benchmarks, it limits the potential reduction in deployment time. A long-running, independently deployed cache instance would allow packages to remain cached across exercises, further improving scalability and reducing deployment time for subsequent scenarios that rely on similar software stacks.

Multi-client Component. The *Multi-Client Component* is expected to reduce deployment time for stages primarily responsible for client configuration, while simultaneously increasing the processing time required to configure the component itself due to the need to set up the containerization platform. The deployment time should decrease since fewer machines require configuration and client-specific stages execute only once per container host, even when serving multiple clients. In practice, however, the scalability behavior of the *Multi-Client Component* proved to be highly unpredictable. Although deployment time was reduced in isolated cases, frequent failures occurred because Ansible intermittently failed to reach containers within the containerization platform, for reasons that could not be consistently reproduced. This instability prevents reliable automated deployments and makes the pattern unsuitable in its current state for the AIT Cyber Range.

Despite these limitations, the *Multi-Client Component* provides significant advantages through dynamic client provisioning, where clients are instantiated on demand from a preconfigured container, enabling scalability to an unpredictable number of participants while requiring provisioning for only a single base instance. The implementation utilizes Btrfs [4] copy-on-write features to reduce storage usage and accelerate instantiation, although containerization introduces additional CPU and memory overhead. Because multiple clients share a single host, ensuring proper isolation is essential, and for multi-organization scenarios, separate container hosts are recommended to prevent cross-participant interference.

5.2 Applicability in General

A potential approach to improving scalability is the simultaneous implementation of multiple architectural patterns, with selecting the top-k ranked patterns appearing as a straightforward strategy. However, this is not always feasible, as patterns may be mutually exclusive or conflict in their design principles such as *Shared Component* and *Tenant-Isolated Component*, and technical constraints, such as Ansible's limitations with nested connections, can further restrict compatibility. Moreover, since patterns are evaluated in isolation, similar patterns may receive comparable scores, leading to redundancy in the top-ranked set. Therefore, ranking alone is insufficient, and additional evaluation is required to ensure compatibility and feasibility when combining patterns.

Although this work incorporates implementation effort through *dependencies* in the metrics, it cannot fully capture the actual effort required to introduce patterns into cyber ranges. Additional factors such as inherent pattern complexity, non-conflicting but effort-increasing pattern interactions, and deployment stability issues significantly influence implementation effort. Moreover, deployment-oriented patterns provide greater flexibility than programming-language patterns, which increases design freedom but also planning complexity. Unlike static code-based implementations, deployment patterns operate in dynamic environments with changing system states, further increasing implementation effort.

While this work discusses and justifies the chosen implementations, they represent only one possible realization of each architectural pattern, as patterns allow flexibility in how they are implemented and may therefore exhibit different scalability characteristics depending on the chosen approach. To validate or refute such discrepancies, future work should implement identical patterns using alternative methodologies and systematically compare their benchmark results. Furthermore, because the Design Science methodology requires evaluation on a specific platform, pattern behavior may vary across environments due to differences in architectural design, scenario complexity, and technology choices such as component software stacks and cloud platforms.

The benchmark diagrams are organized by playbook, separating stages such as client deployment, user management, and the noVNC platform. However, since many playbooks share common tasks (e.g., data transfer and package installation), a task-level analysis would provide a more precise view of pattern effects than the current host-based perspective.

Benchmark reliability is further limited by Ansible's profiling output, which is not machine-readable and requires error-prone parsing, as well as by unstable deployments that often require reruns due to network issues. These factors increase evaluation effort and may bias results, highlighting the need for a dedicated benchmarking framework with structured output and more robust execution control.

6 Conclusion

This work developed a method for improving the scalability of cyber range deployments through the systematic application of architectural patterns. The method combines a structured literature review with a metric-based pattern selection process refined using the current state of the AIT Cyber Range and expert rankings. The three highest-ranked patterns—the *Proxy Jump Component*, the *Package Cache Component*, and the *Multi-Client Component*—were implemented and evaluated on the AIT Cyber Range, demonstrating deployment time reductions of up to 50%. The evaluation also revealed limiting factors such as network traffic imbalance and caching overhead, as well as additional effects including the removal of client limits and reduced compute resource costs.

Future work will build on these results by identifying and evaluating additional architectural patterns using the metrics, implementing a broader set of patterns to further validate and refine the proposed metrics, and examining long-term maintenance aspects of pattern-based cyber range architectures. A major future collaboration opportunity is to evaluate the introduced methodology on different cyber ranges in order to identify the mentioned possibilities in other deployment technologies and to evaluate the validity of the methodology independently of the used technology stack.

Acknowledgments. This work is based on the thesis *Analysis of Scalability in Declarative Deployment Environments: in the Case of the AIT Cyber Range*, which was written at the Vienna University of Technology. This work is conducted under the EU-funded project CyberUnity, Project Number 101128024, which aims to federate multiple cyber ranges across Europe.

Disclosure of Interests. There are no competing interests to declare.

References

1. Ansible AWX community documentation. https://ansible.readthedocs.io/projects/awx/en/latest/
2. Apt-cacher-ng. https://github.com/ashang/apt-cacher-ng
3. Apt-cacher-ng Bugs. https://bugs.debian.org/cgi-bin/pkgreport.cgi?pkg=apt-cacher-ng
4. BTRFS documentation. https://btrfs.docs.kernel.org
5. Linux Containers - Incus. https://linuxcontainers.org/incus/
6. NIS Directive — enisa.europa.eu. https://www.enisa.europa.eu/topics/cybersecurity-policy/nis-directive-new
7. Red Hat Ansible Automation Platform — redhat.com. https://www.redhat.com/en/technologies/management/ansible
8. Regulation - 2022/2554 - EN - DORA - EUR-Lex — eur-lex.europa.eu. https://eur-lex.europa.eu/eli/reg/2022/2554/oj?uri=CELEX:32022R2554
9. squid: Optimising Web Delivery — squid-cache.org. https://www.squid-cache.org/
10. Squid-deb-proxy. https://github.com/mvo5/squid-deb-proxy
11. ssh(1) General Commands Manual (2024). https://man.openbsd.org/ssh
12. Abbott, M.L., Fisher, M.T.: The Art of Scalability: Scalable Web Architecture, Processes, and Organizations for the Modern Enterprise (2009). 978-0-13-703139-9
13. Andrew, T.S., Van Steen, M.: Distributed Systems-Principles and Paradigms, 2nd edn (2007)
14. Ardagna, C.A., Damiani, E., Frati, F., Rebeccani, D., Ughetti, M.: Scalability patterns for platform-as-a-service. In: 2012 IEEE Fifth International Conference on Cloud Computing, pp. 718–725 (2012). https://doi.org/10.1109/CLOUD.2012.41

15. Avritzer, A., et al.: Scalability assessment of microservice architecture deployment configurations: a domain-based approach leveraging operational profiles and load tests. J. Syst. Softw. **165**, 110564 (2020). https://doi.org/10.1016/j.jss.2020.110564
16. Bernstein, P.A., Newcomer, E.: Principles of Transaction Processing (2009). 978-0-08-094841-6
17. Chappell, D.A.: Enterprise Service Bus: Theory in Practice (2004). 978-1-4493-9109-6
18. Douglass, B.P.: Real-time Design Patterns: Robust Scalable Architecture for Real-time Systems (2003). 978-0-201-69956-2
19. Fehling, C., Ewald, T., Leymann, F., Pauly, M., Rütschlin, J., Schumm, D.: Capturing cloud computing knowledge and experience in patterns. In: 2012 IEEE Fifth International Conference on Cloud Computing, pp. 726–733 (2012). https://doi.org/10.1109/CLOUD.2012.124
20. Fehling, C., Leymann, F., Retter, R., Schumm, D., Schupeck, W.: An architectural pattern language of cloud-based applications. In: Proceedings of the 18th Conference on Pattern Languages of Programs, PLoP 2011, pp. 1–11 (2011). https://doi.org/10.1145/2578903.2579140
21. Fehling, C., Leymann, F., Retter, R., Schupeck, W., Arbitter, P.: Cloud Computing Patterns: Fundamentals to Design, Build, and Manage Cloud Applications (2014). https://doi.org/10.1007/978-3-7091-1568-8. 978-3-7091-1567-1 978-3-7091-1568-8
22. Fehling, C., Leymann, F., Rütschlin, J., Schumm, D.: Pattern-based development and management of cloud applications. Future Internet **4**(1), 110–141 (2012). https://doi.org/10.3390/fi4010110
23. Fielding, R.T., Taylor, R.N.: Principled design of the modern web architecture. ACM Trans. Internet Technol. **2**(2), 115–150 (2002). https://doi.org/10.1145/514183.514185
24. Fowler, M.: Data access routines. IEEE Softw. **20**(6), 96–98 (2003). https://doi.org/10.1109/MS.2003.1241375
25. Fowler, M.: Patterns of Enterprise Application Architecture (2012). 978-0-13-306521-3
26. Han, R., Guo, L., Ghanem, M.M., Guo, Y.: Lightweight Resource Scaling for Cloud Applications, pp. 644–651 (2012). https://doi.org/10.1109/CCGrid.2012.52
27. Hapner, M., Burridge, R., Sharma, R., Fialli, J., Stout, K.: Java message service. Sun Microsystems Inc., Santa Clara, CA, vol. 9 (2002)
28. Hatzelhoffer, L., Lobeck, M.: Smart City in Practice: Innovation Lab Between Vision and Reality (2012)
29. Henning, S., Hasselbring, W.: A configurable method for benchmarking scalability of cloud-native applications. Empir. Softw. Eng. **27**(6), 143 (2022). https://doi.org/10.1007/s10664-022-10162-1
30. Hohpe, G., Woolf, B.: Enterprise Integration Patterns: Designing, Building, and Deploying Messaging Solutions (2004). 978-0-321-20068-6
31. Humble, J., Farley, D.: Continuous Delivery: Reliable Software Releases through Build, Test, and Deployment Automation (2010). 978-0-321-67022-9
32. Kampourakis, V., Gkioulos, V., Katsikas, S.: A step-by-step definition of a reference architecture for cyber ranges. J. Inf. Secur. Appl. **88**, 103917 (2025). https://doi.org/10.1016/j.jisa.2024.103917
33. Karakus, M., Durresi, A.: A survey: control plane scalability issues and approaches in software-defined networking (SDN). Comput. Netw. **112**, 279–293 (2017)

34. Komar, R., Patil, A.: Emerging trends in cloud computing: a comprehensive analysis of deployment models and service models for scalability, flexibility, and security enhancements. J. Intell. Syst. Appl. Data Sci. **1**(1) (2023). https://jisads.com/index.php/1/article/view/10
35. Krafzig, D., Banke, K., Slama, D.: Enterprise SOA: Service-oriented Architecture Best Practices (2005). 978-0-13-146575-6
36. Kuchar, K., Blazek, P., Fujdiak, R.: From playground to battleground: cyber range training for industrial cybersecurity education. In: ACM International Conference Proceeding Series, pp. 209–214 (2023)
37. Leitner, M., et al.: AIT cyber range: flexible cyber security environment for exercises, training and research. In: EICC 2020, pp. 1–6 (2020). https://doi.org/10.1145/3424954.3424959
38. Leymann, F., Roller, D.: Production Workflow - Concepts and Techniques (2000). http://www2.informatik.uni-stuttgart.de/cgi-bin/NCSTRL/NCSTRL_view.pl?id=BOOK-2000-01&engl=0
39. Patibandla, R.S.M.L., Kurra, S.S., Mundukur, N.B.: A study on scalability of services and privacy issues in cloud computing. In: Ramanujam, R., Ramaswamy, S. (eds.) Distributed Computing and Internet Technology, pp. 212–230 (2012)
40. Rana, M.E., Farooq, U., Rahman, W.N.W.A.: Scalability enhancement for cloud-based applications using software oriented methods. IJEAT **8**(6), 4208–4213 (2019). https://doi.org/10.35940/ijeat.F8869.088619
41. Stephens, B., Cox, A.L., Rixner, S., Ng, T.E.: A scalability study of enterprise network architectures. In: ANCS 2011, pp. 111–121 (2011). https://doi.org/10.1109/ANCS.2011.28
42. Varia, J.: Cloud architectures. White Paper of Amazon, jineshvaria. s3. amazonaws. com/public/cloudarchitectures-varia. pdf **16** (2008)
43. Varia, J.: Architecting for the cloud: best practices (2010)
44. Vykopal, J., Vizvary, M., Oslejsek, R., Celeda, P., Tovarnak, D.: Lessons learned from complex hands-on defence exercises in a cyber range. In: FIE 2017, pp. 1–8 (2017). https://doi.org/10.1109/FIE.2017.8190713
45. Weerawarana, S., Curbera, F., Leymann, F., Storey, T., Ferguson, D.F.: Web Services Platform Architecture: SOAP, WSDL, WS-Policy, WS-Addressing, WS-BPEL, WS-Reliable Messaging and More (2005). 978-0-13-148874-8
46. Wessels, D.: Web Caching (2001). 978-1-56592-536-6

Generalization of Tree-Based Machine Learning Models for Obfuscated Zero-Day Malware Detection

Selene Falchetti[1](✉), Gabriele Lenzini[1], and Christian Hennig[2]

[1] University of Luxembourg, Esch-sur-Alzette, Luxembourg
{selene.falchetti,gabriele.lenzini}@uni.lu
[2] University of Bologna, Bologna, Italy
christian.hennig@unibo.it

Abstract. We test whether tree-based machine learning models—Gradient Boosting, Random Forest, and CatBoost– when trained on a small portion of a dataset of memory dump of one malware subtype, can detect unseen malware variants on the full datasets, the CIC-MalMem-2022 and BCCC-Mal-NetMem-2025, datasets that contain the memory dumps produced by several code-obfuscated malware types. We apply both Gaussian noise and SHAP-guided perturbations to the most influential features, to assess the robustness of our analysis, while ensuring SHAP-based explainability and dimensionality-reduction visualizations with t-SNE and UMAP. The results show that tree-based models generalize effectively to unseen malware families, achieving up to 100% accuracy and recall while remaining computationally efficient. CatBoost exhibits the strongest robustness under targeted perturbations. Our work suggests that machine learning tree-based models show a practical and robust capability of capturing key malware features even when trained on the traces left on memory by a single malware variant, giving evidence that they could detect obfuscated malware's zero-day attacks.

Keywords: Obfuscated malware detection · zero-day malware · machine learning · explainable AI · visualization techniques

1 Introduction

With the rapid evolution of technology, malware attacks are becoming increasingly sophisticated: they exfiltrate data, spy on users' behaviour online, and disrupt system performances. Malware attacks are also better at avoiding detection, with obfuscation (*e.g.,* packing, polymorphism), hiding strategies (*e.g.,* rootkit-based concealment, DLL injection), and evasive tactics (*e.g.,* sandbox detection, anti-debugging), thus posing zero-day risks. It is therefore essential to develop solutions to detect malware that are efficient and adaptive.

Current malware detection solutions rely on either static or behavioural (dynamic) analysis, or a combination of both. In static analysis, a malware's

© The Author(s), under exclusive license to Springer Nature Switzerland AG 2027
K. Renaud et al. (Eds.): EICC 2026, CCIS 3012, pp. 352–368, 2027.
https://doi.org/10.1007/978-3-032-28957-5_21

code is inspected for malicious characteristic signatures or suspicious structures. A static analysis is fast and scalable but can miss recognizing code-obfuscated, hidden, or concealed malware [31].

Dynamic analysis looks at how malware interact with a system. Interaction events such as memory usage, API calls, or process activity offer in this case rich, high-dimensional data about complex behavior over time that can reveal, after analysis, malicious actions from malware that have escaped a static analysis.

However, defining rules to detect malicious behavioural patterns is challenging, or even insufficient, with unseen or certain obfuscated malware variants: in fact, obfuscation often disrupts code-level patterns and weakens methods that depend on static or family-specific signatures [31]. Here, Machine Learning (ML) offers a natural solution: by learning patterns directly from the data, ML models can capture subtle indicators beyond already known characterizing signatures. Even if code obfuscation can distort dynamic evidence with "specific interference" in a way that behavior-based classifiers fail to detect [28], ML can succeed in capturing peculiar traces of those interferences [5].

The major problem is another. Training ML takes time and requires large datasets. Better would be to use ML models that have proven capabilities of learning malware features from minimal data sets, as they can be used to train models faster. This requirement motivates our study: we investigate whether models trained on limited data from a single malware variant have or have not the ability to detect other variants and unseen types. If a ML model performs well in this setting, it is less likely to rely only on narrow variant-specific regularities and more likely to capture execution-level properties that transfer across families. This is plausible because malware must eventually execute its core logic — allocating memory, invoking system calls, or manipulating processes — leaving functional artefacts in memory that persist even under code transformations.

ML models that have such a faster "generalization" ability could be useful in zero-day detection [12], because from the analysis of *e.g.*, memory dumps only of a subset of one malware variant they could extract enough information to detect samples from many other families and variants. Indeed, such possibility has been shown already for one particular ML model [22].

Contribution. This paper studies extensively the "generalization" ability of several ML models when trained on behavioral data, in particular memory dumps of code-obfuscated malware. The focus on obfuscated malware seems unnecessary, but we claim that studying obfuscated malware leads to stricter and more meaningful findings about generalization in the context of behavioural analysis.

In fact, obfuscated malware is the most realistic and challenging scenario for memory-based detection: these are the samples that are most likely to evade static defenses and reach execution, making them the most relevant test case for evaluating whether memory-based generalization holds under adversarial conditions. Obfuscated malware provides a stronger and more meaningful test of whether an ML model is truly capable of generalization.

We systematically evaluate whether tree-based ML models (Gradient Boosting, Random Forest, CatBoost) trained on a single malware subtype generalize to unseen variants across two memory-analysis datasets. Our experiments show that the models achieve near-perfect or perfect accuracy and recall, while offering good interpretability via SHAP, strong robustness under perturbations, and latent structure via t-SNE and UMAP.

2 Related Work

Several ML models have been proven to be effective in malware detection, including Random Forest (RF), Extreme Gradient Boosting (XGBoost), K-Nearest Neighbors (KNN), Decision Trees (DT), Naïve Bayes (NB), and AdaBoost. RF achieved 97.68% accuracy on the UNSWNB15 dataset [6], while Decision Trees, SVMs, and CNNs reached 96–99% on Canadian Institute for Cybersecurity datasets [1]. On Kaggle datasets, RF, Stochastic Gradient Descent, Extra Trees, and Gaussian NB reached 100% on accuracy, precision, recall, and F1 [2]. These results establish ML as a strong baseline for malware detection across different data types.

In particular, when observing memory-based behavioral features as such in the CIC-MalMem-2022 dataset (which includes obfuscated and hidden memory dumps), ML detection accuracies are reported to be between 99.00% and 100.00% [9,15,29,32,34], with Gradient Boosting (GBM) and RF achieving 100.00% [18,22]. Similar results are achieved with the BCCC-Mal-NetMem-2025 dataset (containing evasive samples memory dumps), where the models show 96–98% accuracy [13].

ML models can therefore capture complex statistical patterns from training data, enabling them to detect similar attacks and adapt to evolving behaviors [12], making them promising to become a line of defence against zero-day malware attacks. For example, autoencoders achieve 8999% accuracy on NSL-KDD and 7598% on CICIDS2017 [17]. Several classifiers trained on the CIC-AWS-2018 dataset, which includes fourteen intrusion types, were tested on eight zero-day attacks from a real-world network. The decision tree achieved the best performance, with a true-positive rate up to 96% and 5% false positives, though results vary by attack type [33].

This work is also motivated to find out how ML can be employed in detecting zero-day malware, although we did not test directly our results on zero-day malware; rather we test how specific ML models (*i.e.*, tree-based) can efficiently capture general characteristics shared by different malware variants, and therefore potentially also by unseen ones. In this line of research, the work most related to ours is Madamidola et al. [22]. It combines memory-analysis features with interpretable ML. They trained a RF for each of the 15 malware subtypes in CIC-MalMem-2022 and evaluated on the full dataset; 11/15 models exceeded 99.00% accuracy with the Transponder subtype reaching the highest accuracy (99.84%) and Zeus the lowest (97.99%), showing RF can detect unseen malware.

These generalization challenges are amplified in adversarial settings, where even high-performing malware and intrusion detection models can degrade under carefully crafted perturbations [23]. For example, Aryal et al. [4] showed that feature attribution can enhance evasion by identifying influential regions in Windows PE malware files and selectively injecting perturbations. By combining SHAP-based feature refinement with robustness testing under Gaussian noise, FGSM, and PGD, convolutional models are robust to perturbations, while tree-based models generalize better across malware families.

Beyond accuracy and robustness, another critical dimension is the detection speed. According to Madamidola et al., for the CIC-MalMem-2022 dataset, inference times reported in the literature range from 738μs per instance for the Dilated CNN [25], 384 μs for RobustCBL, and 255μs for CompactCBL [29], to 8μs for Linear Regression [9], 5.7μs for Random Forest [22], and 0.4μs for XGBoost [3]. These results show that tree- and ensemble-based models can provide sufficiently fast inference for real-time systems.

We built on Madamidola et al. and extended it to a wider range of ML models, namely Gradient Boosting, Random Forest, and CatBoost. Following the need for neutral benchmark studies [7], we focused on a systematic comparison of established models rather than proposing new algorithms. We incorporate hyperparameter tuning, a second dataset, Gaussian noise and SHAP-guided perturbations to assess robustness under random and targeted manipulations, aiming to provide a rigorous benchmark for tree-based ML in memory-based zero-day malware detection.

To our knowledge, no existing framework combines subtype-specific training, memory-based representations, and both random and SHAP-guided perturbations to jointly assess robustness, interpretability, and generalization in obfuscated, hidden, and evasive malware detection.

3 Methodology

Our analysis is defined on the following variables: i) malware dataset and pre-processing steps; ii) ML models used; iii) evaluation criteria (metrics to assess quality and performance); iv) model explainability using SHAP; v) perturbation; vi) visualization techniques.

3.1 Malware Dataset and Pre-processing

We used two datasets: CIC-MalMem-2022 by Carrier et al. [9] and BCCC-Mal-NetMem-2025 by Lashkari et al. [13]. Both consist of memory dumps—snapshots captured during program execution, providing realistic system views for malware behavior analysis.

CIC-MalMem-2022 contains hidden and obfuscated malware with 57 sandbox-extracted statistical features (e.g., callbacks, handlers per process, EPROCESS pool anomalies), while BCCC-Mal-NetMem-2025 targets evasive malware with 258 memory-extracted features (many overlapping with CIC, such as handles, services, callbacks). Benign and malware distribution is shown in Fig. 1.

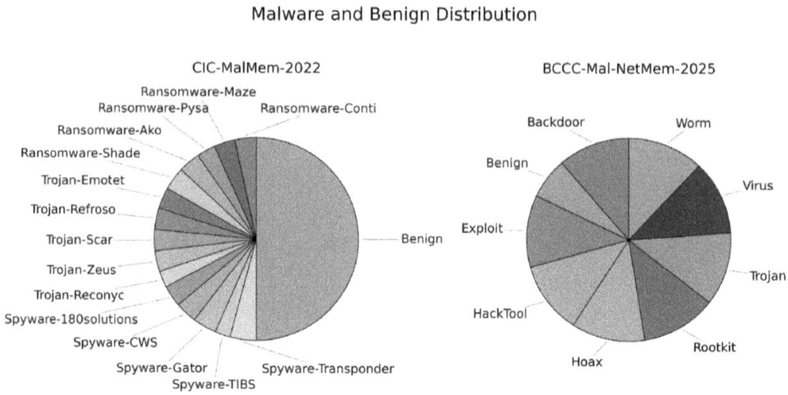

Fig. 1. Distribution of Malware and Benign memory dumps across both datasets.

Before training, we applied several pre-processing steps, including removing invariant features, normalizing and scaling numerical features, and converting the target variable to numerical format. We then created subtype-specific datasets. As a preliminary analysis for GBM models, we replicated the strategy of Madamidola et al. [22]: for each malware subtype, select all malware samples from the entire dataset and add an equal number of randomly selected benign memory dumps to form balanced training sets, splitting each into 80% training and 20% preliminary testing. A final evaluation was performed on the entire dataset excluding observations used during training.

However, since this approach discards a portion of the benign samples during training—potentially losing useful information—we developed our own strategy. We split the entire dataset into two equal halves, reserving one for final evaluation. From the other half, for each malware subtype, we selected all malware samples plus all available benign samples to form subsets, shuffled and divided into 80% training and 20% preliminary testing.

3.2 Machine Learning Models

We employed three supervised ensemble models: Gradient Boosting Machine (GBM), Random Forest (RF), and CatBoost. GBM sequentially fits decision trees, where each tree corrects the errors of the previous ones by minimizing a specified loss function [10,16,26]; for our binary classification task, we used binary cross-entropy (log-loss) and selected decision trees as base learners due to their ability to capture complex, non-linear relationships. RF constructs multiple decision trees using bootstrap sampling and random feature selection, reducing variance and improving generalization [8,16]. CatBoost is a gradient boosting algorithm designed to efficiently handle categorical features while reducing overfitting [27]; it employs ordered boosting to prevent target leakage, uses symmetric tree structures to improve regularization and computational efficiency, and optimizes binary cross-entropy loss for binary classification.

Implementation. To reduce overfitting and manage class imbalance, in each model we tuned key hyperparameters, used 10-fold cross-validation, and applied weights inversely proportional to class frequency during training. Classification thresholds were set a priori: 0.3 for CIC-MalMem-2022 and 0.8 for BCCC-Mal-NetMem-2025, chosen based on the class imbalance structure of each dataset — a lower threshold increases sensitivity when malware is the minority class, while a higher threshold does so when benign samples are underrepresented. For perturbation experiments, all models were retrained with default parameters, no class weighting, and a standard threshold of 0.5. Additional preliminary configurations include a non-standard GBM threshold of 0.3 on BCCC-Mal-NetMem-2025, tuned and untuned CatBoost variants with class weighting 1:20 and threshold 0.5, and GBM under the preliminary dataset construction strategy. GBM and RF were implemented in R (`gbm3`, `ranger`, `caret`), CatBoost and perturbed models in Python (`CatBoost`, `scikit-learn`).

3.3 Evaluation Criteria and Performance

We assessed model performance using confusion matrices, accuracy, precision, recall, F1 score, and prediction speed. The confusion matrix provides the counts of true positives, true negatives, false positives, and false negatives, forming the basis for all the derived metrics. Accuracy reflects general correctness, while precision and recall capture false-positive and false-negative behavior, respectively. The F1 score combines precision and recall into a single measure, which is especially useful for unbalanced datasets [16].

For our use case of zero-day malware detection, recall is the primary metric of interest because missing an infection (false negative) can have severe security consequences. While the F1 score balances precision and recall, a high F1 alone is insufficient if recall is low, as this could indicate that many threats are not being detected. Accuracy is also reported to allow direct comparison with existing literature, where it remains the most commonly used metric.

We computed these metrics on both datasets, evaluating the models on the subset of the known malware subtypes and on the full dataset, including the unseen samples.

Performance. We measured the inference speed of the GBM models (evaluated on the CIC-MalMem-2022 dataset) and of the CatBoost models (evaluated on the BCCC-Mal-NetMem-2025) as the average prediction time per instance. We divided the total prediction time, recorded using R's `system.time()` and Python's `time.time()`, by the number of test samples, providing an estimate of the practical latency of the models in real-world malware detection scenarios.

The R-based experiments were performed locally on a macOS machine with an Apple M4 CPU (10 cores) and 16 GBRAM, running R version 2025.09.1+401 in RStudio. The Python experiments were executed on Google Colab, accessed via Safari on macOS. The computations ran on Colab's virtual machine, equipped with an Intel(R) Xeon(R) CPU @ 2.20GHz, 12.7 GBRAM, and Python 3.12.12.

3.4 Model Explainability Using SHAP

Using ML algorithms alone is often insufficient; understanding how models make decisions is essential to verify performance, enhance credibility, and analyze feature discrimination between malware and benign instances.

We used SHAPley Additive exPlanations (SHAP), based on Shapley values from cooperative game theory [30], offering a theoretically sound method for feature attribution [20]. SHAP estimates each feature's marginal contribution across all possible combinations, ensuring local accuracy, missingness, and consistency. For tree-based models, TreeSHAP efficiently computes exact SHAP values by traversing tree structures [19].

We applied `shap.TreeExplainer` to GBM, RF, and CatBoost models on both datasets to identify top discriminative features (then used for the perturbation), and applied `shap.Explainer` for CatBoost on CIC-MalMem-2022 for deeper analysis.

3.5 Perturbation

To evaluate the generalizability of our models, we applied both Gaussian random noise and SHAP-guided perturbation methods [20] to the entire test set.

Gaussian Random Noise. For each numeric feature x_i, the perturbed value x_i' is computed as:

$$x_i' = x_i + \epsilon, \quad \epsilon \sim \mathcal{N}\big(0, \sigma \cdot \mathrm{std}(x_i)\big)$$

where $\mathrm{std}(x_i)$ is the standard deviation of feature x_i and σ is a scaling factor controlling the noise magnitude (0.10, 1, 5, ...).

SHAP-Guided Perturbation. For each model, we identified the top 10 features with the highest SHAP values and perturbed only the malware samples. The perturbed value for feature $x_{i,j}$ of malware sample i is:

$$x_{i,j}' = x_{i,j} + \mathrm{sign}\Big(- \mathrm{SHAP}_{i,j}\Big) \cdot |\epsilon|, \quad \epsilon \sim \mathcal{N}\big(0, \sigma \cdot \mathrm{std}(x_j)\big)$$

where $\mathrm{SHAP}_{i,j}$ is the local SHAP value of feature j for sample i, $\mathrm{std}(x_j)$ is the standard deviation of feature j, and σ is the same scaling factor used for Gaussian noise.

This methodology tests the robustness of the model against both random perturbations and targeted, explanation-informed attacks. The perturbed datasets were then used to evaluate accuracy, recall, precision, F1-score, robustness, and total misclassification rates across both datasets and all models (CatBoost, GBM, and RF).

3.6 Visualization Techniques

Visualization techniques such as t-SNE and UMAP, though less common in malware dataset analysis, reveal hidden patterns, highlight relationships, and enhance model interpretability. These nonlinear dimensionality reduction methods project high-dimensional data into 2D/3D spaces, capturing nonlinear structures. t-SNE models pairwise similarities to preserve local structure [21], while UMAP–rooted in Riemannian geometry–maintains both local and global structure with superior scalability and efficiency [24].

Hamid and Muthukumarasamy applied t-SNE to KDD99, improving SVM-based intrusion detection [14]. Ghani et al. used t-SNE and UMAP on 5G network traffic to uncover patterns boosting detection accuracy [11].

We applied both methods to each dataset using R's `Rtsne` and `umap`, visualizing with `ggplot2` (2D) and `plotly/rgl` (3D). We analyzed either 1,000 balanced samples or the full dataset, using strategies such as benign vs. malware, family-level vs. benign, malware-type vs. benign, malware-only families, and a 15-subtype view.

4 Results

4.1 Models' Performance

All the models showed near-perfect accuracies across the respective test set, containing the same malware types they were trained on, with results tightly concentrated above 98.5% and many reaching 100%, confirming that the classifiers learn the training variants very effectively.

When evaluated on the full CIC-MalMem-2022 dataset, all treebased models maintain very high accuracy and recall across most malware variants (Fig. 2). GBM exhibits only modest drops on a few challenging types (e.g., Refroso, Scar), while CatBoost and RF remain the most stable models, with values tightly clustered near 100% for both accuracy and recall over all malware types. On BCCC-Mal-NetMem-2025, the overall pattern is more heterogeneous (Fig. 2). GBM shows clear weaknesses on specific families such as Backdoor, Exploit and Rootkit, whereas it performs strongly on others like HackTool and Worm, producing a markedly uneven profile. RF and especially CatBoost present the most uniform and elevated values across families, confirming their stronger ability to generalize to unseen malware types in this more challenging dataset.

4.2 SHAP Analysis

Figure 3 shows beeswarm plots for the CatBoost model trained on Transponder malware and evaluated on the entire CIC-MalMem-2022 dataset, including unseen variants. Each dot represents an instance, positioned by its SHAP value on the x-axis and colored by feature value, with vertical stacking indicating density. Blue dots (low values) to the right of the vertical line indicate feature contributions toward malware detection. SHAP analyses yielded consistent

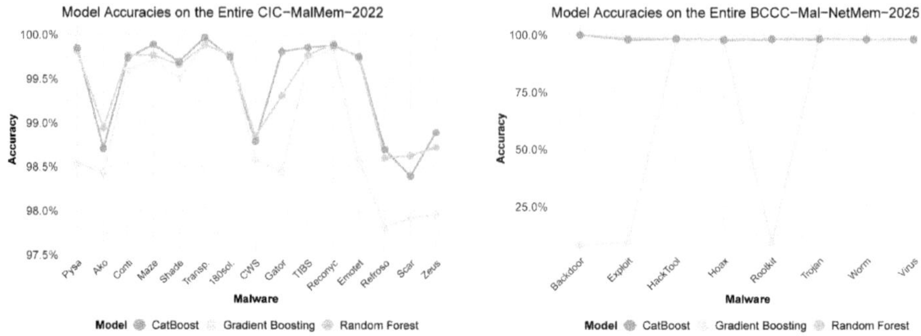

Fig. 2. Accuracies of the CatBoost, Gradient Boosting and Random Forest models on the CIC-MalMem-2022 (left) and BCCC-Mal-NetMem-2025 (right) datasets, detecting malware unseen during training.

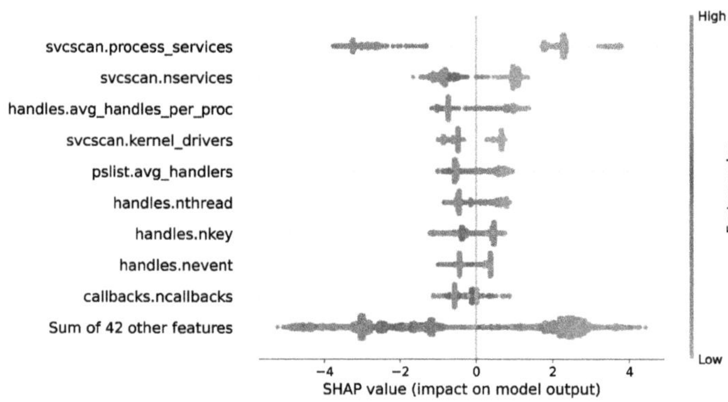

Fig. 3. Beeswarm plot for the CatBoost model trained on Transponder malware and tested on the full CIC-MalMem-2022 dataset.

results across all trained models on both test sets and the full dataset. The most discriminative features for CIC-MalMem-2022 include:

```
svcscan.nservices: total number of running services
svcscan.shared\_process\_nservices: services running in shared processes
svcscan.kernel\_drivers: number of kernel drivers
svcscan.process\_nservices: number of services running in separate processes
handles.nsemaphore: number of semaphore handles
dlllist.avg\_dlls\_per\_proc: average number of DLLs loaded per process
```

Lower values generally indicate higher malware probability, consistent with strategies like minimal environments and service bypassing. Similarly, for the BCCC-Mal-NetMem-2025 dataset, SHAP analyses highlighted system-level artifacts including privilege usage, process activity, active user sessions, memory modules not listed in loader structures, and file-system organization such as Master File Table entries. These features reflect anomalies and behaviors indicative of malicious activity.

4.3 Model Speed

The detection speed was evaluated using GBM on CIC-MalMem-2022 and CatBoost on BCCC-Mal-NetMem-2025. For GBM, speeds ranged from 0.83 μs per instance for Refroso to 7.01 μs for Maze, with the most accurate models achieving around 2 μs, demonstrating a strong balance between accuracy and efficiency. CatBoost on BCCC-Mal-NetMem-2025 showed uniform detection times of 8.65 μs across all malware types, slower than GBM but still within a practical range for real-time applications.

These results confirm that tree-based ensemble models can provide sufficiently fast inference for real-time detection. However, as noted in Sect. 3.3, inference times were measured on different hardware – an Apple M4 CPU for GBM and a virtualized Intel Xeon on Google Colab for CatBoost – limiting direct comparability.

4.4 Perturbation

Figure 4 summarizes how model accuracy degrades as perturbation strength increases for different malware families and datasets.

Under Gaussian noise accuracy remains relatively stable for small values of σ and only gradually declines at higher noise levels, indicating that all models are reasonably robust to random perturbations. In contrast, SHAP-guided perturbations cause a steeper, more systematic drop in accuracy as σ increases, since they specifically target the most influential features for each sample. Under Gaussian noise, CIC-MalMem-2022 degraded slightly more, but the trend is reversed under SHAP-guided perturbations.

Across both settings, CatBoost consistently preserves higher accuracy than the other models, highlighting its stronger resilience to both random and explanation-driven feature disruptions.

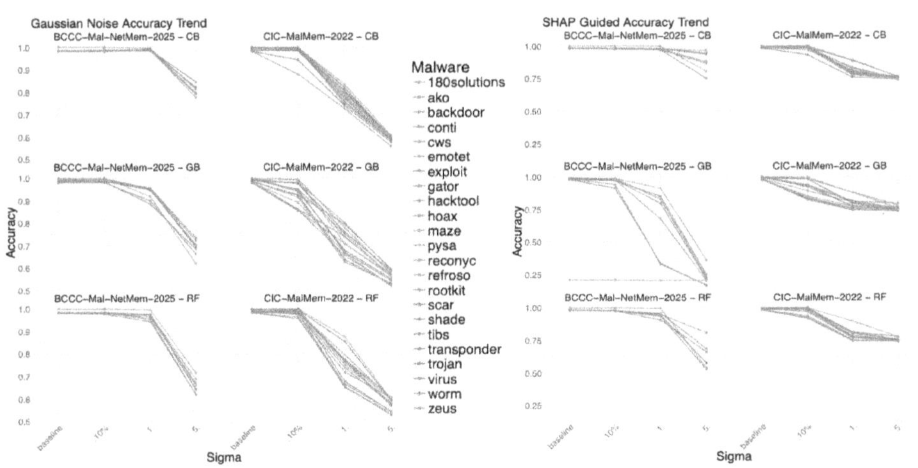

Fig. 4. Accuracy of CatBoost, Gradient Boosting and Random Forest under increasing levels of Gaussian noise (left) and SHAP-guided feature perturbations (right).

Fig. 5. t-SNE (left) and UMAP (right) visualizations separating benign and malware family samples in a 2D projection of a 3D result.

4.5 Visualization Techniques

Figure 5 shows the 2D projection from 3D results of malware families (ransomware, spyware, trojans) alongside benign samples from the CIC-MalMem-2022 dataset using t-SNE and UMAP. Both techniques clearly distinguish benign from malicious instances; however, malware families show no clear boundaries between categories.

Across all visualizations, in CIC-MalMem-2022, malware samples appear scattered with significant overlap, obscuring structure and complicating subtype differentiation. In contrast, for BCCC-Mal-NetMem-2025, visualizations reveal that backdoor instances are more distant from other malware variants and occasionally close to or overlapping with benign samples, suggesting greater behavioral similarity to benign activity.

5 Discussion

We interpret the results across both datasets, examining how the choice of training malware variant influences generalization, distinguishing between accuracy and recall as complementary indicators of zero-day detection performance, and assessing robustness to methodological choices.

CIC-MalMem-2022. All three models show strong detection capabilities for previously unseen obfuscated and hidden malware. GBM achieves accuracy between 0.9781 (Refroso) and 0.9989 (Transponder), RF matches or slightly improves this (0.9860 Refroso to 0.9990 Reconyc), and CatBoost performs best (0.9840 Scar to 0.9997 Transponder).

Recall values remain high across models, particularly ranged from 0.9566 Refroso to 0.9981 Transponder in GBM, from 0.9723 Refroso to 0.9984 Reconyc for RF, and from 0.9839 Scar to 0.9997 Transponder indicating rare false negatives–critical for zero-day detection.

BCCC-Mal-NetMem-2025. GBM shows low accuracy for Backdoor, Rootkit and Exploit (0.08210.0966), reflecting poor recall, though it performs relatively well for Hacktool (0.9739) and Trojan (0.9826). CatBoost demonstrates stable performance with accuracy 0.9768 (Hoax) to 1.0000 (Backdoor) and recall 0.9874 (Hoax) to 1.0000 (Backdoor). RF similarly reports good results, with accuracy 0.9816 (Hoax, Worm) to 1.0000 (Backdoor) and recall 0.9802 (Hoax, Worm) to 1.0000 (Backdoor).

Impact of Training Malware Variants. The choice of training malware variant has a notable impact on model performance, particularly in BCCC-Mal-NetMem-2025. Backdoor malware presents the most challenging case: t-SNE and UMAP visualizations reveal that Backdoor samples partially overlap with benign instances, suggesting that their runtime behavior is closer to legitimate processes. This behavioral ambiguity likely explains why GBM struggles to generalize. In contrast, CatBoost and RF achieve perfect performance even on this difficult variant, indicating greater capacity to learn fine-grained decision boundaries under class overlap – and when exposed to other malware variants that are more behaviorally distinct from benign samples, this advantage translates into strong generalization across the full dataset. Exploit and Rootkit show similarly challenging behavior, though no clear structural patterns emerge from the visualizations to fully explain their difficulty. For the remaining variants, the consistently high accuracy and recall and the overlapping observed among malware samples in the visualizations suggest that their behavioral signatures are sufficiently distinct from benign activity to be reliably captured by all models.

Accuracy, Recall, and Robustness to Methodological Choices. Accuracy provides useful context, but recall is the critical metric for zero-day detection. Across both datasets, CatBoost and RF achieve the highest and most stable recall, making them the most reliable models for detecting unseen malware. GBM is competitive when training data are abundant but shows instability when trained on variants with limited or ambiguous feature separation.

In the preliminary analysis (see Appendix 7) we tested an alternative dataset construction strategy, and different configurations including untuned CatBoost, CatBoost with different weighting and threshold, and GBM with a non-standard threshold. We obtained similar performance, confirming that these findings, even if not directly comparable, do not change substantially based on the methodology used, with no remarkable differences in accuracy or recall. Overall, the results confirm that tree-based models – particularly CatBoost and RF – best capture the complex behavioral patterns required for zero-day malware detection.

5.1 Comparison with Previous Work

Table 1 reports detection accuracies and F1-scores from previous studies alongside our results. Despite adopting the more challenging strategy of training on individual malware subtypes rather than all variants, our models achieve state-of-the-art performance on both datasets. On CIC-MalMem-2022, our results slightly exceed those of Madamidola et al. [22] (0.9997 vs. 0.9984), the closest comparable work in terms of training strategy, though direct comparison remains limited by differences in features, optimization, and dataset sampling. For BCCC-Mal-NetMem-2025, comparison with Lashkari et al. [13] is limited, as they trained on both network data and memory dumps, making their setup substantially different from ours, and reported F1 only.

Table 1. Detection accuracies or F1-score reported in literature

Dataset	Authors	Accuracy / F1	Methods
CIC-MalMem-2022	Zakaria et al. [32]	0.9998	RF/XGBM
		0.9996	SVM
	Carrier et al. [9]	0.97	RF/DT
		0.99	Ensemble
	Shafin et al. [29]	0.9996	RobustCBL
		0.9992	CompactCBL
	Ozturk et al. [34]	0.9999	XGBM
	Hasan et al. [15]	0.9999	RF/MLP/XGBM
	Hossain et al. [18]	1.0000	GBM
		0.9999	BG/VT
		0.9996	RF
	Madamidola et al. [22]	1.0000	RF
		0.9984	RF (Transponder)
	Our study	0.9990	RF (Reconyc)
		0.9997	CatBoost (Transponder)
BCCC-Mal-NetMem-2025	Lashkari et al. [13]	F1: 0.98	RF (multi-level)
	Our study	F1/Acc.: 1.0000	CatBoost/RF (Backdoor)

5.2 Complementary Analyses

GBM averaged ~2 μs per instance (0.83–7.01 μs) and CatBoost 8.65 μs; though measured on different hardware, both confirm suitability for real-time detection. SHAP analysis revealed consistent discriminative features across all models: in CIC-MalMem-2022, low values for service counts, shared-process usage, and kernel drivers correlated with malicious behavior, supporting transparency and aligning with Madamidola et al. [22]. Under perturbations, all models maintain strong performance up to $\sigma = 1$ for both Gaussian and SHAP-guided strategies, with CIC-MalMem-2022 particularly stable; at $\sigma = 5$, accuracy declines as expected under extreme manipulation, with CatBoost showing the strongest resilience overall. Finally, t-SNE and UMAP confirm clear separation between benign and malicious samples; overlaps across malware subtypes reflect shared behavioral patterns, further supporting generalization to unseen families.

5.3 Limitations and Future Research

Despite encouraging results, several limitations exist. Inevitably, our analysis depends on the datasets, which may not fully capture real-world complexity. Also, the near-perfect or perfect performance that we observed likely reflects dataset-specific characteristics, such as controlled sandbox environments and static memory dumps, rather than true model superiority, raising concerns about generalization to noisier settings. The class imbalance in BCCC-Mal-NetMem-2025, where benign samples are underrepresented relative to malware – contrasting with real-world conditions – may artificially inflate detection performance and limit generalization validity, necessitating evaluation under more realistic class distributions. Moreover, inference speed was measured across different hardware environments, limiting direct comparability. Future work should incorporate dynamic telemetry (e.g., API calls and system logs), more diverse and recent malware, and temporal models, as well as benchmarking tree-based ensembles against deep learning approaches (e.g., LSTMs or Autoencoders) to help determine whether current feature engineering imposes a performance ceiling.

6 Conclusion

We assessed the capability of several tree-based machine learning models—trained on two datasets of memory dumps of only one subtype of obfuscated malware—to detect unseen malware variants on the whole data sets (*i.e.*, across CIC-MalMem-2022 and BCCC-Mal-NetMem-2025).

All models achieved very high performance, with accuracies of 0.9997 and 1.0000 on the data sets, respectively. The same results have been measured for the recall. Thus there is evidence that models effectively capture patterns that can be found in new malware variants. Our results are interpretable through SHAP analysis and visualization techniques. Low values of features such as the number of running services, services in shared processes, and kernel drivers strongly increased the likelihood of malware classification, highlighting behavior-based detection mechanisms. Low inference times suggest that the models may be capable of detecting new malware in real-time; perturbation tests (up to $\sigma = 1$ Gaussian and SHAP-guided noise) prove robustness.

Our research—because of the extensive training, including the presence of perturbed data, and the stress under which we tested the models—confirms, more extensively than previous research, that tree-based machine learning models can be a useful tool to detect new malware variants. For instance, one possible way to further assess this capability would be to collect newer malware samples and evaluate them using the same model without additional training, in order to verify whether the observed performance remains consistent. However, this predictive hypothesis still requires further experimental validation.

Acknowledgments. We thank S. Z. El Mestari and H. Hammouchi for valuable discussions. Research funded by Luxembourg Government's PSYBER/CyberResearch Hub.

Disclosure of Interests. The authors have no competing interests to declare that are relevant to the content of this article.

7 Appendix

See Fig. 6.

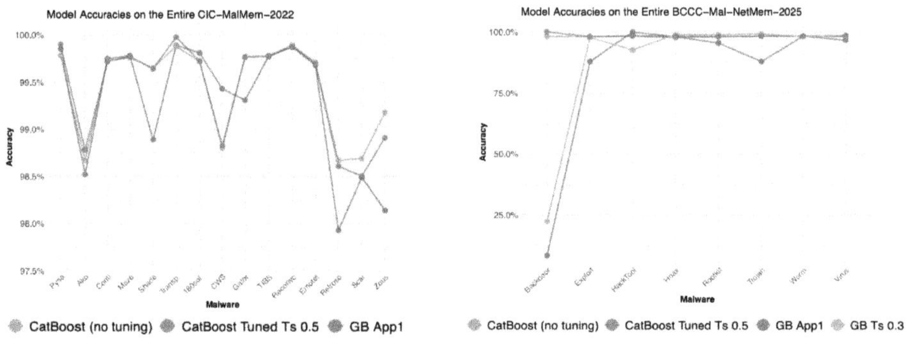

Fig. 6. Accuracy of preliminary analysis of CatBoost and Gradient Boosting models on the full CIC-MalMem-2022 (left) and BCCC-Mal-NetMem-2025 (right) datasets, detecting unseen malware.

References

1. Akhtar, M., Feng, T.: Malware analysis and detection using machine learning algorithms. Symmetry **14**(11), 2304 (2022). https://doi.org/10.3390/sym14112304
2. Akhtar, M., Feng, T.: Evaluation of machine learning algorithms for malware detection. Sensors **23**(2), 946 (2023). https://doi.org/10.3390/s23020946
3. Alani, M., Mashatan, A., Miri, A.: Xmal: a lightweight memory-based explainable obfuscated-malware detector. Comput. Secur. **133**(2), 103409 (2023). https://doi.org/10.1016/j.cose.2023.103409
4. Aryal, K., Gupta, M., Abdelsalam, M., Saleh, M.: Explainability guided adversarial evasion attacks on malware detectors. In: Proceedings of the 33rd International Conference on Computer Communications and Networks (ICCCN) (2024). https://doi.org/10.1109/ICCCN61486.2024.10637577
5. Aurangzeb, S., Aleem, M.: Evaluation and classification of obfuscated android malware through deep learning using ensemble voting mechanism. Sci. Rep. **13**, 3093 (2023). https://doi.org/10.1038/s41598-023-30028-w
6. Azeem, M., Khan, D., Iftikhar, S., Bawazeer, S., Alzahrani, M.: Analyzing and comparing the effectiveness of malware detection: a study of machine learning approaches. Heliyon **10**(20), e23574 (2023). https://doi.org/10.1016/j.heliyon.2023.e23574

7. Boulesteix, A.L., Lauer, S., Eugster, M.: A plea for neutral comparison studies in computational sciences. PLoS ONE **8**(4), e61562 (2013). https://doi.org/10.1371/journal.pone.0061562
8. Breiman, L.: Random forests. Mach. Learn. **45**(1), 5–32 (2001). https://doi.org/10.1023/A:1018054314350
9. Carrier, T., Victor, P., Tekeoglu, A., Lashkari, A.: Detecting obfuscated malware using memory feature engineering. In: Proc. of the 8th Int. Conf. on Information Systems Security and Privacy (ICISSP). SCITEPRESS (2022). https://doi.org/10.5220/0010908200003120
10. Friedman, J.H.: Greedy function approximation: a gradient boosting machine. Ann. Stat. **29**(5), 1189–1232 (2001). https://doi.org/10.1214/aos/1013203451
11. Ghani, H., Salekzamankhani, S., Virdee, B.: Critical analysis of 5G networks' traffic intrusion using PCA, t-SNE, and UMAP visualization and classifying attacks. In: Proceedings of Data Analytics and Management (2024). https://doi.org/10.1007/978-981-99-6544-1_32
12. Guo, Y.: A survey of machine learning-based zero-day attack detection: challenges and future directions. Comput. Commun. **198**, 175–185 (2023). https://doi.org/10.1016/j.comcom.2022.11.001
13. Habibi Lashkari, A., Shafi, M., Li, Y., et al.: Unveiling evasive malware behavior: toward generating a multi-sources benchmark dataset and evasive malware behavior profiling using network traffic and memory analysis. J. Supercomputing **81**(6), 782–802 (2025). https://doi.org/10.1007/s11227-025-07267-x
14. Hamid, Y., Muthukumarasamy, S.: A t-SNE based nonlinear dimension reduction for network intrusion detection. Int. J. Inf. Technol. **12**(1), 1–10 (2019)
15. Hasan, S., Dhakal, A.: Obfuscated malware detection: investigating real-world scenarios through memory analysis. In: Proceedings of of IEEE International Conference on Telecommunications and Photonics (ICTP). IEEE (2023)
16. Hastie, T., Tibshirani, R., Friedman, J.: The Elements of Statistical Learning: Data Mining, Inference, and Prediction, 2nd edn. Springer, New York (2009)
17. Hindy, H., Atkinson, R., Tachtatzis, C., Colin, J.N., Bayne, E., Bellekens, X.: Utilising deep learning techniques for effective zero-day attack detection. Electronics **9**(10), 1684 (2020). https://doi.org/10.3390/electronics9101684
18. Hossain, M., Islam, M.: Enhanced detection of obfuscated malware in memory dumps: a machine learning approach for advanced cybersecurity. Cybersecurity **7**, 16 (2024). https://doi.org/10.1186/s42400-024-00205-z
19. Lundberg, S., Erion, G., Lee, S.I.: Consistent individualized feature attribution for tree ensembles. In: Proceedings of the 32nd International Conference on Neural Information Processing Systems (NeurIPS'18) (2019). https://doi.org/10.48550/arXiv.1802.03888
20. Lundberg, S., Lee, S.I.: A unified approach to interpreting model predictions. In: Proceedings of the 31st International Conference on Neural Information Processing Systems (NeurIPS'17). Curran Associates Inc. (2017). https://doi.org/10.48550/arXiv.1705.07874
21. van der Maaten, L., Hinton, G.: Visualizing data using t-SNE. J. Mach. Learn. Res. **9**, 2579–2605 (2008)
22. Madamidola, O., Ngobigha, F., Ez-zizi, A.: Detecting new obfuscated malware variants: a lightweight and interpretable machine learning approach. Intell. Syst. Appl. **25**(6), 200472 (2025). https://doi.org/10.1016/j.iswa.2024.200472
23. Martins, N., Cruz, J., Cruz, T., Henriques Abreu, P.: Adversarial machine learning applied to intrusion and malware scenarios: A systematic review. IEEE Access **8**, 35403–35419 (2020). https://doi.org/10.1109/ACCESS.2020.2974752

24. McInnes, L., Healy, J., Melville, J.: Umap: Uniform manifold approximation and projection for dimension reduction (2020)
25. Mezina, A., Burget, R.: Obfuscated malware detection using dilated convolutional network. In: Proceedings of the 14th International Congress on Ultra Modern Telecommunications and Control Systems and Workflows (ICUMT), vol. 2022, p. 14 (2022). https://doi.org/10.1109/ICUMT57764.2022.9943443
26. Natekin, A., Knoll, A.: Gradient boosting machines. Tutorial. Front. Neurorobotics **7**, 21 (2013). https://doi.org/10.3389/fnbot.2013.00021
27. Prokhorenkova, L., Gusev, G., Vorobev, A., Dorogush, A., Gulin, A.: CatBoost: unbiased boosting with categorical features. In: Proceedings of 32nd International Conference on Neural Information Processing Systems (NeurIPS 2018). Curran Associates Inc. (2019)
28. Qiang, W., Yang, L., Jin, H.: Efficient and robust malware detection based on control flow traces using deep neural networks. Comput. Secur. **122**, 102871 (2022). https://doi.org/10.1016/j.cose.2022.102871
29. Shafin, S., Karmakar, G., Mareels, I.: Obfuscated memory malware detection in resource-constrained IoT devices for smart city applications. Sensors **23**(11), 5348 (2023). https://doi.org/10.3390/s23115348
30. Shapley, L.: A value for n-person games. In: Kuhn, H., Tucker, A. (eds.) Contributions to the Theory of Games (AM-28). Princeton University Press (1953). https://doi.org/10.1515/9781400881970-018
31. Sihwail, R., Omar, K., Zainol Ariffin, K.: A Survey on Malware Analysis Techniques: Static, Dynamic, Hybrid and Memory Analysis. Int. J. Adv. Sci. Eng. Inf. Technol. **8**(4-2), 1662–1674 (2018). https://doi.org/10.18517/ijaseit.8.4-2.6827
32. Zakaria, M., Mohamed, M., Hussein, S., Salama, G.: Obfuscated file-less malware detection using integrating memory forensics data with machine learning techniques. Int. J. Comput. Digit. Syst. (2025). https://doi.org/10.12785/ijcds/XXXXXX
33. Zhou, Q., Pezaros, D.: Evaluation of Machine Learning Classifiers for Zero-Day Intrusion Detection: An Analysis on the CIC-AWS-2 (2019). https://doi.org/10.48550/arXiv.1905.03685
34. Öztürk, A., Hızal, S.: Detection and analysis of malicious software using machine learning models. J. Comput. Inf. Sci. **7**(2), 264–276 (2024). https://doi.org/10.35377/saucis/1489237

Implementation and Benchmarking of Privacy Techniques in the Project PrivacyUmbrella

Lena Wiese[1,4(✉)], Jürgen Riegel[2], Torsten Panholzer[2], Sai Chintha[3], Hayate Sato[3], Kai Wo Yau[1], Despina Tawadros[1,4], Pronaya Prosun Das[4], Mohan Xu[4], Axel Zieschank[5], and Holger Storf[5]

[1] Goethe-Universität Frankfurt, Institute of Computer Science, 60325 Frankfurt, Germany

[2] IMBEI-Medical Informatics, University Medical Center Mainz, 55131 Mainz, Germany

[3] MCS Data Labs GmbH, 10627 Berlin, Germany

[4] Fraunhofer Institute for Toxicology and Experimental Medicine, Nikolai-Fuchs-Str.1, 30625 Hannover, Germany

lwiese@cs.uni-frankfurt.de

[5] Institute of Medical Informatics (IMI), Goethe University Frankfurt, University Medicine, Frankfurt am Main, Germany

https://www.dbda.cs.uni-frankfurt.de/, https://www.unimedizin-mainz.de/imbei/informatik/, https://mcs-datalabs.com, https://www.item.fraunhofer.de, https://www.imi-frankfurt.de

Abstract. This paper reports on the outcomes of the project PrivacyUmbrella that explores decentralized privacy-preserving data processing for digital health applications. A software demonstrator enables data uploading, anonymization strategy configuration, and flexible attribute-level anonymization, supporting a well-established registry system. Moreover, tumbling Window k-anonymization on mobile devices is introduced to ensure privacy-aware data sharing without compromising analytical utility, strengthened by federated learning and local differential privacy to protect sensitive data. Synthetic Data Generation (SDG) using GANs creates secure synthetic data enhanced by Rényi Differential Privacy. Last but not least a Bloom filter-based record linkage method allows sensor data transmission without identifiers. Overall, the project results integrate diverse approaches to enhance healthcare technology with privacy guarantees.

Keywords: Anonymization · Federated Learning · Differential Privacy · Privacy-preserving Record Linkage

1 Introduction

The funded project "PrivacyUmbrella" addressed challenges in the field of data protection versus data usability, specifically in the context of medical and personal data. The focus was on the development of a comprehensive anonymization

© The Author(s), under exclusive license to Springer Nature Switzerland AG 2027
K. Renaud et al. (Eds.): EICC 2026, CCIS 3012, pp. 369–384, 2027.
https://doi.org/10.1007/978-3-032-28957-5_22

system that combines various established methods, such as k-anonymity, differential privacy, and privacy-preserving record linkage, while integrating standardized privacy metrics. In addition, the project analyzed the applicability of federated learning and synthetic data generation. In particular, the project laid several technical foundations to enable sensitive medical data to be used for further downstream analysis without compromising the privacy of individual patients.

Contributions. To achieve the project goals, several modules were developed.

- First, an optimization model was developed, which, using the heuristics "Simulated Annealing" automates the selection of appropriate generalization levels to achieve k-anonymity.
- Furthermore, the project explored methods to evaluate anonymized datasets in terms of their susceptibility to potential re-identification attacks using machine learning techniques and to improve resilience against such threats.
- The project also addressed specific challenges in handling nested and graph-based medical data formats. In this context, solutions were developed based on HL7 FHIR resources and the open-source registry platform OSSE, demonstrating how complex structured datasets can be effectively anonymized while retaining usability.
- Another key goal of the project was to examine the applicability of distributed anonymization techniques directly on mobile devices and wearables. Federated learning was utilized to train machine learning models in a decentralized manner, thereby ensuring that sensitive data are processed locally on user devices.
- We have also refined the concept of distributed, decentralised creation of Bloom filters as pseudonyms based on identifying data (IDAT) such as names and date of birth. This enables data records to be linked in a privacy-preserving manner from decentralised locations without transferring the respective IDAT. In addition, this record linkage can be performed on a probability basis and in an error-tolerant manner with regard to the IDAT entries. By normalising the IDAT and analysing various Bloom filter combination strategies, we were also able to significantly improve the robustness and error tolerance of the privacy-preserving record linkage.

The developments were prototypically implemented and further optimized through practical testing. In the upcoming sections, results are presented that show how the above concepts successfully provides the targeted data protection without significantly restricting data usability for analytical purposes. The research outcomes lay the groundwork for further improvements and the application of the developed systems in a variety of medical and other data-intensive scenarios.

Outline. This article explores innovative methods for privacy-preserving data processing, particularly in digital health applications, through five key areas. In Sect. 2, we discuss an optimization model for achieving k-anonymity, focusing on secure and flexible anonymization configurations to safeguard sensitive data. Section 3 examines the implementation of distributed anonymization methods tailored for mobile devices and wearables, highlighting techniques like Tumbling Window k-anonymization and federated learning that prioritize privacy without sacrificing utility. Section 4 delves into Synthetic Dataset Generation using Generative Adversarial Networks (GANs), enhanced with Rényi Differential Privacy, to address privacy risks while maintaining statistical data fidelity. In Sect. 5, we investigate privacy-preserving record-linkage approaches, emphasizing Bloom filters for secure data integration and transmission. Finally, Sect. 6 provides an overview of findings and outlines future directions, including optimizing resource efficiency, scaling advanced machine learning models, and leveraging Large Language Models (LLMs) to design customized anonymization strategies within ethical and legal frameworks. Overall, this article showcases the technological feasibility and promising potential of distributed and privacy-conscious systems for real-world applications.

2 Optimization Model for k-Anonymity

Data anonymization, involving the alteration or masking of sensitive information in datasets, is a crucial strategy to protect individual privacy and prevent data misuse or privacy breaches. The overall goal of data anonymization is to make data records untraceable back to the original subjects. To this end, the field has developed various privacy-preserving models, such as k-anonymity [27], l-diversity [14], t-closeness [12], δ-differential privacy [6]. Despite advancements in data privacy models, k-anonymity's simplicity and effectiveness keep it a frequent choice for organizations and researchers dealing with sensitive data [8]. In a k-anonymized dataset, each record is indistinguishable from at least $k-1$ other records with respect to certain identifying attributes; see also Sect. 3.2. This is achieved by modifying or generalizing quasi-identifiers (like age, zip code, etc.) in the dataset.

2.1 Simulated Annealing

We implemented a method that applies the simulated annealing algorithm within the search space of generalization hierarchies, incorporating a limit on suppression, to identify the optimal solution that enhances classification accuracy[1]. We employed the AUCROC score, a metric for classification accuracy, as the optimization objective [28]. Comparing the scores before and after dataset anonymization directly reflects the impact of various anonymization processes on model accuracy and provides an assessment of the quality of the anonymized datasets. Determining the optimal generalization level for features, while considering k-anonymity and suppression constraints, poses a complex NP-hard challenge that intensifies with feature count and generalization options. The algorithm begins by preprocessing the dataset, including class rebalancing, removing

[1] https://github.com/desstaw/DataPrivacy_SimulatedAnnealing.

direct identifiers, and establishing a Domain Generalization Hierarchy (DGHA) for quasi-identifiers with constraints like k-anonymity and suppression limits.

Simulated annealing serves as a practical solution to efficiently explore solution spaces, aiming to achieve near-optimal configurations balancing privacy and classification accuracy. Simulated annealing can overcome local optima and explore a wide range of anonymization strategies, making it particularly suitable for the highly nonlinear optimization landscape in the tension between data privacy and data usability. Temperature management involves an exponential cooling strategy where the algorithm gradually decreases the temperature across iterations to refine solutions. Iterative exploration systematically alters generalization levels for quasi-identifier attributes through randomized selection, evaluating solutions via an objective function combining classification accuracy and privacy compliance. The Metropolis criterion probabilistically accepts or rejects changes, balancing exploration and exploitation to avoid premature convergence to local optima while navigating complex solution spaces. This approach optimizes classification utility and privacy constraints through a dynamic and iterative search process, leveraging simulated annealing for effective anonymized data analysis. In particular, Simulated Annealing seeks an optimal balance between maximum privacy and minimal information loss.

2.2 Anonymization Webservice via REST API

A particular challenge in registry-based research is to ensure that datasets can be shared or analyzed without compromising individuals' privacy. Techniques such as k-anonymity or other anonymization models are often used to reduce the risk of re-identification. However, the implementation of these techniques requires the calculation of complex optimization problems: We established a generally applicable anonymization system that guarantees k-anonymity based on simulated annealing described above[2]. It is designed for use with the registry software OSSE (Open Source Registry System for Rare Diseases) [26]. The system offers an API interface that allows external applications to upload data and their schemas, request anonymization strategies, and receive them. It includes basic access control and allows for the processing of flexible anonymization configurations at the attribute level. Its performance is evaluated against the already established ARX tool [20] using two example datasets. The UCI Adult and UCI Obesity Estimation datasets are utilized, with metrics for classification quality (accuracy, balanced accuracy, and F1-score) and data usability (Discernibility Metric). In this context, simulated annealing presents itself as an attractive method to tackle the mentioned challenges: By gradually reducing the temperature parameter, simulated annealing converges to solutions that offer an acceptable compromise between analytical validity and protection against re-identification. This aligns with OSSE's core principle of enabling secure yet useful data exchanges. To ensure compatibility with OSSE, a Java-based web application, the architecture of this project follows a similar service-oriented design.

[2] https://github.com/tina5980g/MA_OSSE/tree/main.

Specifically, the anonymization system is implemented as an independent web server that provides its functionality through a RESTful API. This approach enables OSSE and similar systems to submit anonymization requests over HTTP, thereby ensuring platform interoperability and easy integration. The system is designed to receive datasets in CSV format, accompanied by metadata in JSON format that describes the data structure. This metadata includes definitions of quasi-identifiers, levels of generalization, and classification objectives.

2.3 Distributed Search Across OSSE Registries

One of the main focuses of the project was its application in research into rare diseases, i.e., diseases that are often life-threatening or chronically debilitating for those affected and have a low prevalence (less than 1 in 2,000). The high number of diseases that fall under this definition makes research in this area an important but challenging task [18]. The low prevalence in particular necessitates national and international cooperation in the collection and exchange of research data. To simplify the exchange between registries, a search broker was implemented as part of this project. As a central point of contact, it provides information about the data elements used in the connected distributed OSSE registries, in accordance with the specifications in the respective metadata repositories. Based on this metadata, registered users can submit search queries to the distributed registries. These are received and stored by the search broker. The connected OSSE registries can query the search broker for existing search queries. Those responsible for the registries can then decide whether to respond to the request and which data to provide. In addition, the registry operators can define the anonymization strategy. This approach is intended to ensure both data sovereignty and data protection. Communication between the individual components takes place via a REST API, which ensures secure communication between the services and at the same time provides easy integration for external services.

3 Distributed Anonymization for Mobile Devices and Wearables

Conventional privacy-preserving approaches typically rely on centralized architectures in which sensitive data are collected, stored, and processed in a central repository [21,27]. Although effective in many scenarios [11], such architectures introduce a single point of failure and increase the risk of data breaches, unauthorized access, and re-identification attacks [17]. Moreover, transferring large volumes of sensitive data to centralized servers creates additional risks during transmission and raises concerns regarding data ownership and user control.

To address these concerns, decentralized processing becomes essential. By performing anonymization directly on user devices (e.g., smartphones or wearables), sensitive information remains under user control and does not need to be transmitted in raw form. However, decentralized anonymization introduces two primary challenges: (1) maintaining strong privacy guarantees without excessively degrading data utility, and (2) implementing anonymization techniques in resource-constrained environments characterized by limited computational power, memory, and battery capacity. To address these challenges, *Tumbling Window k-anonymization* for time-series data was developed in this project.

3.1 Mathematical Formulation of Privacy–Utility Trade-Off

We model the biometric data stream as $X = \{x_t\}_{t=1}^{n}$ and define local anonymization as a transformation $\tilde{X} = \mathcal{A}_{k,T}(X)$, parameterized by anonymity level $k \in \mathbb{Z}^+$ and tumbling window size T (expressed in seconds). Table 1 summarizes the privacy and utility components and the resulting optimization objective.

Table 1. Privacy–Utility Trade-off Formulation

Component	Formulation	Interpretation
Privacy model	$\mathcal{P}(k)$	Larger $k \Rightarrow$ stronger k-anonymity guarantee
Utility model	$\mathcal{U}(k, T) = \mathrm{corr}(X, \mathcal{A}_{k,T}(X))$	Higher value $\Rightarrow$ better signal fidelity
Optimization	$\max_{T} \mathcal{U}(k, T)$ for given k	Find T that maximises signal fidelity at a chosen privacy level k

The parameter k creates the central trade-off: larger values strengthen the k-anonymity guarantee but reduce signal fidelity. Increasing T provides more samples per window, enabling k-anonymity at lower generalization levels and thereby recovering utility without sacrificing privacy. The evaluation in this work focuses on the interaction between k and T; although federated learning with local differential privacy (LDP) can be applied as an independent additional privacy layer, its effect on data utility is outside the scope of the present evaluation.

3.2 Local *K*-anonymous Anonymization of Time-Series Data on Mobile Devices

K-anonymity is a well-established privacy model commonly applied to demographic data (e.g., age, gender, ZIP code) to prevent individual re-identification [19]. A dataset satisfies k-anonymity if each record is indistinguishable from at least $k-1$ other records with respect to quasi-identifiers [27]. This property is typically achieved through generalization or suppression.

Since the PrivacyUmbrella system collects not only demographic attributes but also continuous biometric signals and streams them to a central server, conventional k-anonymization techniques are not directly applicable. Therefore, a real-time anonymization method suitable for time-series data and resource-constrained mobile environments was required.

The proposed tumbling window k-anonymization algorithm processes streaming data in fixed-duration windows and applies hierarchical generalization to ensure k-anonymity within each window. Given a window $W_t = \{x_i\}$, each value x_i is mapped to a generalized representation using a predefined hierarchy:

$$g_\ell(x_i),$$

where ℓ denotes the level of generalization. Lower levels correspond to finer ranges, while higher levels represent coarser intervals. For each value, the algorithm selects the minimum generalization level ℓ_i such that at least k values in the window share the same generalized representation:

$$|\{j : g_{\ell_i}(x_j) = g_{\ell_i}(x_i)\}| \geq k$$

This level-by-level process ensures that each value is indistinguishable from at least $k-1$ others within the window. The generalized outputs are represented as value ranges. To obtain a numerical signal, a post-processing step is applied:

$$\tilde{x}_i = \begin{cases} \text{midpoint}(g_{\ell_i}(x_i)) & \text{if a range is assigned,} \\ \text{mean}(W_t) & \text{if the value is suppressed.} \end{cases}$$

In practice, users can adjust k and T to balance privacy and utility. Figure 1 illustrates the effect of different parameter configurations on ECG signals. Higher k values (e.g., $k = 20$) visibly distort the signal, whereas increasing the window size (e.g., $T = 30$ s) preserves the overall waveform more effectively.

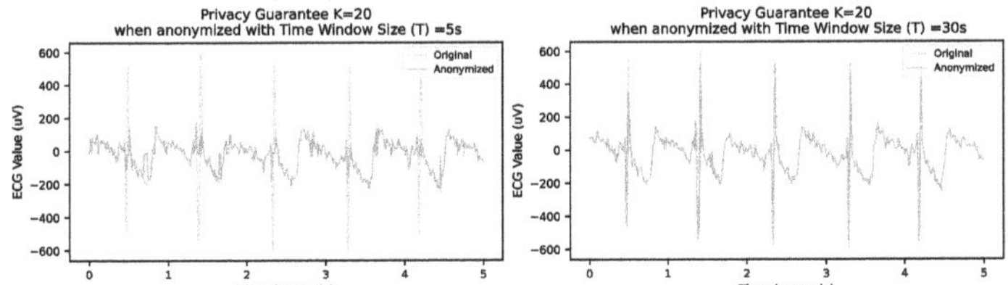

Fig. 1. Comparison of raw and locally anonymized ECG signals using Tumbling Window k-anonymization.

3.3 Federated Learning with Local Differential Privacy

The privacy-preserving architecture was further strengthened by integrating federated learning (FL) with local differential privacy (LDP) [16]. Federated learning is a decentralized machine learning paradigm in which a global model is distributed to participating devices and trained locally using on-device data. After local training, only model updates are transmitted to a central server for aggregation into a new global model [9,15]. This approach enables collaborative model training without transferring raw data.

From a privacy perspective, federated learning reduces data exposure by transmitting only model updates rather than raw data, thereby limiting what a central server can observe about individual users. When combined with LDP, the leakage risk is further reduced by injecting calibrated noise into model updates, controlled by a privacy budget $\varepsilon \in \mathbb{R}^+$: smaller values provide stronger protection at the cost of increased noise.

However, federated learning remains vulnerable to attacks such as model inversion and membership inference, which may reveal sensitive information from model updates [7,24]. To mitigate these risks, local differential privacy is applied prior to transmitting model updates. The overview of the system architecture is available in Fig. 2.

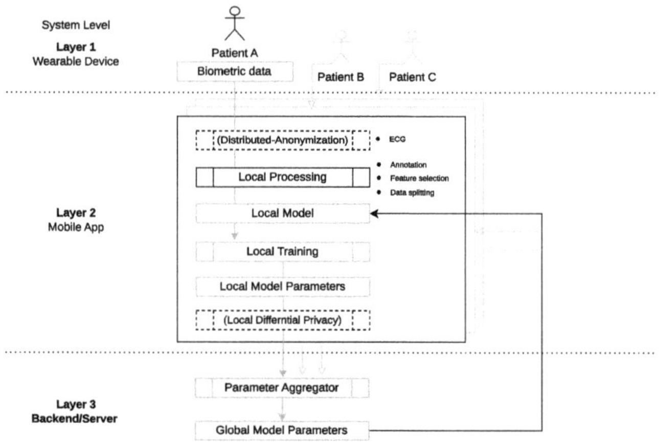

Fig. 2. Decentralized training architecture integrating Federated Learning and Local Differential Privacy.

In the implemented system, a decision tree classifier was used as the local model [10]. Biometric signals (including heart rate, SpO_2, PPG, anonymized ECG, and three-axis accelerometer data) served as input features for a binary classification task that detects whether the user performed physical exercise during a recording session. While more complex models such as deep neural networks could potentially improve predictive performance, a computationally lightweight model was intentionally selected to reduce processing overhead on mobile devices.

LDP was applied to feature importance values and leaf node predictions of the decision tree using the Laplace mechanism [2]. The magnitude of injected noise is controlled by the privacy budget ε. This mechanism [5] ensures that individual contributions cannot be inferred from transmitted model updates while preserving collaborative learning capabilities. The validation of privacy efficacy and the effect of FL with LDP on data utility are, however, outside the scope of the present work and remain a direction for future evaluation.

3.4 Evaluation of Privacy–Utility Trade-Offs

Following the formulation above, $\mathcal{U}(k, T)$ is evaluated empirically by measuring Pearson correlation[3] between original and k-anonymized ECG signals across different combinations of k and T (see Fig. 3).

Consistent with prior research [25], increasing k reduces signal fidelity, whereas larger window sizes (T) mitigate this effect. Higher correlation values are observed for larger T, even at higher k, indicating that window size plays a key role in preserving utility. Selecting larger window durations (e.g., $T = 30$ s) therefore enables stronger privacy settings while maintaining acceptable analytical value.

3.5 Evaluation of Device Performance

To assess resource efficiency, CPU usage, energy consumption, and memory usage were evaluated on an Android reference device (TCL T767H) using Apptim [1]. The results (see Table 2) indicate that resource consumption increases with longer tumbling windows.

Under the recommended privacy configuration ($k = 20$, $T = 30$ s), local anonymization increased battery consumption by approximately 73% compared to data collection without anonymization. Although this reveals potential for further optimization, the overall resource demand remains within a range that supports continuous operation in practical mobile health scenarios.

These results demonstrate that privacy-preserving processing directly on mobile devices is feasible, albeit with measurable resource overhead that must be considered in system design.

3.6 Federated Learning for Heterogeneous Devices

Another of our approaches examines a heterogeneous model aggregation method [29] based on partial training, where clients only train submodels of a global

[3] The simulated annealing optimisation in Sect. 2 uses AUCROC as its utility measure, which requires a downstream classification task. Since no classification task was performed on the anonymized time-series data, Pearson correlation is adopted here as a task-agnostic measure of signal fidelity, directly quantifying how well the anonymized signal preserves the structure of the original.

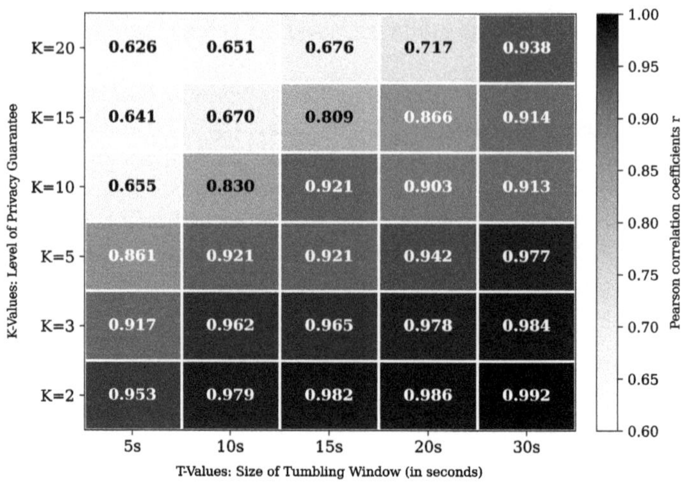

Fig. 3. Pearson correlation between original and anonymized ECG signals across different K and T Configurations.

model to reduce computational and communication costs. We propose the adaptive partial training algorithm (cFedPT), which considers both current and historical activations when selecting neurons, thereby avoiding overtraining or undertraining. Experiments on image and audio classification tasks confirm the effectiveness of cFedPT in federated learning environments with heterogeneous clients.

4 Synthetic Dataset Generation

Table 2. Resource Overhead of On-Device Time-Series Anonymization Across Privacy Configurations

Activity (Configuration)	Avg CPU (%)	Avg Energy (Apptim Score)	Avg Memory (MB)
No Anonymization (Reference Activity)	138.40 ± 5.93 (Reference)	416.72 ± 42.84 (Reference)	471.23 ± 46.93 (Reference)
Anonymization activated ($K = 20$, $T = 5s$)	142.14 ± 7.84 (+3.74%)	487.73 ± 82.09 (+71.01%)	527.01 ± 57.60 (+55.78%)
Anonymization activated ($K = 20$, $T = 30s$)	156.82 ± 11.82 (+18.42%)	489.84 ± 62.80 (+73.12%)	616.74 ± 84.37 (+145.51%)

Note: Percentage changes are as reported by Apptim and reflect the tool's internal scoring methodology relative to the reference activity; they are not computed directly from the raw score values shown in the table.

Synthetic patient data were provided in the standard HL7 FHIR format. It is a modern, hierarchically organized, web-based standard for exchanging healthcare data using interoperable formats like JSON and XML [4]. We use it to test data processing as well as the use of encryption and anonymization techniques. The use of synthetic data allowed for unrestricted utilization under realistic conditions without affecting the rights of patients. Ethical or data protection approvals are not required for this usage. The employed HL7 FHIR standard enables harmonization of data formats, thereby forming a solid basis for data integration. Additionally, methods for converting FHIR data into a tabular CSV format were tested and applied. For the implementation of selected anonymization techniques, the CSV format proved to be more suitable than the hierarchically organized FHIR format. Within the FHIR data structures, personally identifiable data fields were systematically identified, which represent potential points of attack for re-identification risks. A differentiation was made between direct and indirect identifiers to enable targeted application of anonymization measures. The datasets used were also compared with the core dataset of the Medical Informatics Initiative as a reference model. This ensured the consistency and quality of the data, which is a crucial prerequisite for valid data analysis and processing processes. At the same time, this supports a high level of interoperability between different systems and applications.

We implemented a specific method to compute synthetic datasets as follows [3]. The methodology developed in our work focuses on integrating Rényi-Differential Privacy (RDP) with a Generative Adversarial Network (GAN) similar to [13] but additionally enhanced by a Variational Autoencoder (VAE), resulting in our RDP-VAEGAN architecture. This setup aims to generate high-quality synthetic data that preserves the statistical properties of the original data while ensuring robust privacy guarantees. The architecture combines the strengths of GANs and VAEs for synthetic data generation, where the GAN learns the distribution of input data and generates new instances that mimic this distribution, while the VAE transforms input data into a continuous latent space, facilitating the management of mixed data types (continuous and discrete) commonly found in medical datasets. The GAN consists of two components:

- **Generator (G):** Takes random noise z and outputs synthetic data $\hat{x}$, aiming to generate data indistinguishable from real data.
- **Discriminator (D):** Receives either real data x or synthetic data from the generator and attempts to differentiate between the two, outputting a probability that indicates whether the input is real or synthetic.

The VAE is integrated to transform data into a continuous latent space, simplifying the GAN-based generation process, especially for handling discrete and mixed data types. It has two components:

- **Encoder:** Converts real data into a latent representation.
- **Decoder:** Reconstructs data from the latent space for use in the generator's training process.

The main goal of this system is to maintain privacy during the training of the GAN and VAE, achieved by applying Rényi-Differential Privacy. Our work demonstrates that training the discriminator with differential privacy and using the VAE to preprocess input for the generator ensures the entire system prevents the extraction of sensitive information from the synthetic data.

5 Privacy-Preserving Record-Linkage

Privacy-preserving record linkage is used to connect data from different sources without compromising individuals' privacy; the following subsections describe the techniques we applied to protect sensitive information during the process.

5.1 Bloom Filters as Pseudonyms

Until now, identifying patient data (IDAT), such as name and date of birth, had to be transmitted to a central body (trustee) and stored there in order to generate stable pseudonyms – including in the event of subsequent changes, such as a change of name. The centrally assigned pseudonyms are then returned to the bodies that collect medical data (MDAT), so that these can later be clearly merged (record linkage). However, the central storage and transmission of IDAT is problematic for data protection and security reasons, as these are potential targets for cyberattacks. The situation is further complicated when patient-managed devices are involved (e.g. blood glucose monitors). Furthermore, changes to IDAT must be reliably reported to prevent the inadvertent creation of multiple pseudonyms for the same person.

Instead of central storage of IDAT, we have developed a decentralized concept of pseudonymisation in which Bloom filters [23] are used to generate pseudonyms from IDAT at local data storage locations. This enables privacy-friendly record linkage without the transfer of IDAT (Privacy-Preserving Record-Linkage). The relevant parameters for generating Bloom filters were examined: when the Bloom filter is created with the same parameters at each location, the pseudonyms are unique across all sites. However, they are also deterministic, which means that IDAT could potentially be reconstructed through dictionary attacks. Therefore, the parameters for creation must only be known among partner locations. We have developed a concept for securely transmitting the necessary parameters for generation to the decentralized sites.

5.2 Bloom Filters in Privacy Preserving Record Linkage

Furthermore, we have improved the search for an already existing Bloom filter [22] for privacy-preserving record linkage, ensuring comparability even in cases of minor deviations or typos in the IDAT. To achieve this, we examined normalization of the IDAT prior to the creation of the Bloom filters, which involved the removal of titles and hyphens, conversion to ASCII characters, and standardization of capitalization. We also investigated whether the first and last names might have been mistakenly swapped during input. Similarity was quantified using the Sørensen-Dice coefficient. Additionally, we analyzed how Bloom filters behave when a common filter is generated from all relevant IDAT components, compared to an approach using separate Bloom filters for each IDAT component, which are then merged. By using short q-grams (2–3), a sufficiently large Bloom filter and the Sørensen-Dice coefficient as a measure of similarity, it was possible to achieve results relevant to practical applications. The aim was to assess the robustness of the methods to individual typos and positional transpositions, as well as their suitability for record linkage.

5.3 Bloom Filter Demonstrator

As a demonstrator, we developed a compact source code in Python that generates a Bloom filter as a pseudonym from the input IDAT[4]. The identified relevant parameters for Bloom filter generation were implemented, enabling the federated and decentralized production of Bloom filters (after normalization). A graphical user interface was created for this purpose. The program is linked to a database, where subsets of the input IDAT can be created. With these two sets, similarity searches can be performed. To test the demonstrator, particularly the decentralized pseudonymization, a customizable data generator was developed, which can create artificial personal datasets (ID, last name, first name, date of birth, gender, street, house number, postal code, city) in the desired quantity (in CSV output format). A total of 1000 personal datasets were generated as examples, both with and without Bloom filter pseudonyms. Through this technical groundwork, the subsequent project aims to facilitate knowledge transfer and applicability to sensor devices (wearables, smartwatches) and edge devices (smartphones).

6 Conclusion

The works of the project demonstrate the feasibility of decentralized privacy-preserving data processing for digital health applications. The software demonstrator, which implements our optimization model for anonymization variants, offers an API interface to enable any external applications to (1) upload data and their schemas, (2) configure anonymization strategies, and (3) process flexible anonymization configurations at the attribute level. It is designed for use with OSSE (Open Source System for Electronic Data Capture), but it can also be integrated into other systems via the API and further developed in subsequent projects.

By applying Tumbling Window k-anonymization to ECG signals and performing signal aggregation at the wearable level, the system enables privacy-aware data sharing without sacrificing analytical utility. The integration of federated learning combined with local differential privacy further strengthens the

[4] https://github.com/deFlip99/PPRL-using-Bloomfilters.

privacy guarantees by ensuring that local data remain on user devices and that transmitted model updates do not leak sensitive information. Importantly, the evaluation of device performance shows that local anonymization and privacy-preserving learning are achievable within the computational and energy constraints of modern mobile devices. Although resource consumption increases with stronger privacy configurations, the overhead remains within acceptable operational limits. This demonstrates the practical viability of distributed anonymization approaches for real-world mobile health systems. While the federated learning and LDP components were implemented as proof-of-concept solutions, the results indicate that the proposed architecture provides a promising foundation for privacy-sensitive digital health monitoring.

Synthetic Data Generation (SDG) is another approach we applied to address privacy concerns by creating synthetic data that retains statistical properties of real data while preventing sensitive information sharing. At its core, SDG uses Generative Adversarial Networks (GANs) to generate high-fidelity synthetic data that closely resembles real data while protecting privacy by hiding the actual data distribution. However, GANs alone may have vulnerabilities that could compromise confidentiality during data generation. To overcome these limitations, we incorporated Rényi Differential Privacy into GAN architectures, resulting in the novel RDP-enhanced Variational Autoencoder GAN (RDP-VAEGAN).

Additionally, another software product focuses on expanding the privacy-preserving record linkage technique using Bloom filters. By integrating this into a mobile application, sensor data can be locally linked using a Bloom filter and transmitted to medical institutions without direct identity data.

In follow-up projects, this foundation will be expanded through a practice-oriented approach that leverages real data and emphasizes the commercialization of the research outcomes. Future work will focus on further optimizing resource efficiency, validating the system under real-world deployment conditions, and extending the framework to support more advanced machine learning models within federated environments. The findings from the previous project have shown that purely algorithmic solutions for anonymization processes are effective but often lack the flexibility required to meet the diverse demands of real-world application scenarios. This is where the further development comes into play: using a Large Language Model (LLM) to assess the sensitivity of data in relation to the intended use or release within the framework of legal and ethical boundaries and recommend an optimal anonymization strategy.

Acknowledgments. This work was funded by the German Federal Ministry of Education and Research and NextGenerationEU by the European Union under grant number 16KISA001K (PrivacyUmbrella).

Disclosure of Interests. The authors have no competing interests.

References

1. Apptim: Apptim (version 2.5.0): Mobile application performance testing software (2026). https://www.apptim.com/, Accessed 12 Feb 2026
2. Chamikara, M.A.P., Liu, D., Camtepe, S., Nepal, S., Grobler, M., Bertok, P., Khalil, I.: Local differential privacy for federated learning (2022). http://arxiv.org/abs/2202.06053, arXiv preprint arXiv:2202.06053
3. Das, P.P., Tawadros, D., Wiese, L.: Privacy-Preserving Medical Data Generation Using Adversarial Learning. In: Athanasopoulos, E., Mennink, B. (eds.) Information Security. ISC 2023. Lecture Notes in Computer Science, vol. 14411. Springer, Cham (2023). https://doi.org/10.1007/978-3-031-49187-0_2
4. Duda, S.N., et al.: HL7 FHIR-based tools and initiatives to support clinical research: a scoping review. J. Am. Med. Inform. Assoc. **29**(9), 1642–1653 (2022)
5. Dwork, C.: Differential Privacy. In: Bugliesi, M., Preneel, B., Sassone, V., Wegener, I. (eds.) ICALP 2006. LNCS, vol. 4052, pp. 1–12. Springer, Heidelberg (2006). https://doi.org/10.1007/11787006_1
6. Dwork, C., McSherry, F., Nissim, K., Smith, A.: Calibrating Noise to Sensitivity in Private Data Analysis. In: Halevi, S., Rabin, T. (eds.) TCC 2006. LNCS, vol. 3876, pp. 265–284. Springer, Heidelberg (2006). https://doi.org/10.1007/11681878_14
7. Fredrikson, M., Jha, S., Ristenpart, T.: Model inversion attacks that exploit confidence information and basic countermeasures. In: Proceedings of the 22nd ACM SIGSAC Conference on Computer and Communications Security, pp. 1322–1333. ACM (2015). https://doi.org/10.1145/2810103.2813677
8. Gkoulalas-Divanis, A., Loukides, G., Sun, J.: Publishing data from electronic health records while preserving privacy: a survey of algorithms. J. Biomed. Inform. **50**, 4–19 (2014)
9. Jafarigol, E., Trafalis, T., Razzaghi, T., Zamankhani, M.: Exploring machine learning models for federated learning: A review of approaches, performance, and limitations (2023). http://arxiv.org/abs/2311.10832, arXiv preprint arXiv:2311.10832
10. Kwatra, S., Torra, V.: A k-anonymised federated learning framework with decision trees. In: Data Privacy Management, Cryptocurrencies and Blockchain Technology, Lecture Notes in Computer Science, vol. 13140, pp. 106–120. Springer, Cham (2022). https://doi.org/10.1007/978-3-030-93944-1_7
11. LeFevre, K., DeWitt, D.J., Ramakrishnan, R.: Mondrian multidimensional k-anonymity. In: 22nd International Conference on Data Engineering (ICDE'06), pp. 25–25. IEEE (2006). https://doi.org/10.1109/ICDE.2006.101
12. Li, N., Li, T., Venkatasubramanian, S.: t-closeness: Privacy beyond k-anonymity and l-diversity. In: Proceedings of the 23rd IEEE International Conference on Data Engineering (ICDE), pp. 106–115. IEEE (2007)
13. Ma, C., Li, J., Ding, M., Liu, B., Wei, K., Weng, J., Poor, H.V.: RDP-GAN: A rényi differential privacy based generative adversarial network. IEEE Trans. Dependable Secure Comput. **20**(6), 4838–4852 (2023)
14. Machanavajjhala, A., Kifer, D., Gehrke, J., Venkitasubramaniam, M.: l-diversity: Privacy beyond k-anonymity. ACM Trans. Knowl. Discovery Data (tkdd) **1**(1), 3–es (2007)
15. McMahan, H.B., Moore, E., Ramage, D., Arcas, B.A.y.: Federated learning of deep networks using model averaging (2016). http://arxiv.org/abs/1602.05629, arXiv preprint arXiv:1602.05629
16. McMahan, H.B., Ramage, D., Talwar, K., Zhang, L.: Learning differentially private recurrent language models (2018). http://arxiv.org/abs/1710.06963, arXiv preprint arXiv:1710.06963

17. Narayanan, A., Shmatikov, V.: How to break anonymity of the netflix prize dataset (2007). http://arxiv.org/abs/cs/0610105, arXiv preprint arXiv:cs/0610105
18. Nguengang Wakap, S., et al.: Estimating cumulative point prevalence of rare diseases: analysis of the orphanet database. Eur. J. Hum. Genet. **28**(2), 165–173 (2020)
19. Olatunji, I.E., Rauch, J., Katzensteiner, M., Khosla, M.: A review of anonymization for healthcare data (2021). http://arxiv.org/abs/2104.06523, arXiv preprint arXiv:2104.06523
20. Prasser, F., Eicher, J., Spengler, H., Bild, R., Kuhn, K.A.: Flexible data anonymization using arx—current status and challenges ahead. Software: Pract. Experience **50**(7), 1277–1304 (2020). https://doi.org/10.1002/spe.2812
21. Samarati, P.: Protecting respondents' identities in microdata release. IEEE Trans. Knowl. Data Eng. **13**(6), 1010–1027 (2001). https://doi.org/10.1109/69.971193
22. Schnell, R.: Randomized response and balanced bloom filters for privacy preserving record linkage. In: 2016 IEEE International Conference on Data Mining Workshops (ICDMW). IEEE, Barcelona, Spain (2017). presented at ICDMW 2016, 12–15 Dec 2016
23. Schnell, R., Bachteler, T., Reiher, J.: Privacy-preserving record linkage using bloom filters. BMC Med. Inf. Decis. Making 41 (2009). https://doi.org/10.1186/1472-6947-9-41
24. Shokri, R., Stronati, M., Song, C., Shmatikov, V.: Membership inference attacks against machine learning models (2017). http://arxiv.org/abs/1610.05820, arXiv preprint arXiv:1610.05820
25. Slijepčević, D., Henzl, M., Klausner, L.D., Dam, T., Kieseberg, P., Zeppelzauer, M.: k-anonymity in practice: How generalisation and suppression affect machine learning classifiers. Comput. Secur. **111**, 102488 (2021). https://doi.org/10.1016/j.cose.2021.102488
26. Storf, H., Schaaf, J., Kadioglu, D., Schreiweis, B., Ückert, F., Dugas, M.: Osse - open source registry system for rare diseases. Bundesgesundheitsblatt - Gesundheitsforschung - Gesundheitsschutz **60**(5), 523–531 (2017). https://doi.org/10.1007/s00103-017-2536-3
27. Sweeney, L.: k-anonymity: a model for protecting privacy. Internat. J. Uncertain. Fuzziness Knowl.-Based Syst. **10**(5), 557–570 (2002). https://doi.org/10.1142/S0218488502001648
28. Tawadros, D., Yang, W., Wiese, L., Meyer, V.: Optimizing classification accuracy with simulated annealing in k-anonymity. In: International Database Engineered Applications Symposium, pp. 71–84. Springer, Cham (2025)
29. Xu, M., Wiese, L.: Constrained adaptive partial training for federated learning on heterogeneous clients. In: International Symposium on Algorithmic Aspects of Cloud Computing, pp. 43–57. Springer, Cham (2025)

Author Index

© The Editor(s) (if applicable) and The Author(s), under exclusive license to Springer Nature Switzerland AG 2026

K. Renaud et al. (Eds.): EICC 2026, CCIS 3012, pp. 385–386, 2026.
https://doi.org/10.1007/978-3-032-28957-5

GPSR Compliance
The European Union's (EU) General Product Safety Regulation (GPSR) is a set of rules that requires consumer products to be safe and our obligations to ensure this.

If you have any concerns about our products, you can contact us on

ProductSafety@springernature.com

In case Publisher is established outside the EU, the EU authorized representative is:

Springer Nature Customer Service Center GmbH
Europaplatz 3
69115 Heidelberg, Germany

www.ingramcontent.com/pod-product-compliance
Ingram Content Group UK Ltd.
Pitfield, Milton Keynes, MK11 3LW, UK
UKHW021642180926
475650UK00002B/85
* 9 7 8 3 0 3 2 2 8 9 5 6 8 *